Talking About Life

Conversations on Astrobiology

With over 450 planets now known to exist beyond the Solar System, spacecraft heading for Mars, and the ongoing search for extraterrestrial intelligence, this timely book explores current ideas about the search for life in the Universe.

It contains candid interviews with dozens of astronomers, geologists, biologists, and writers about the origin and range of terrestrial life and likely sites for life beyond Earth. The interviewees discuss what we've learnt from the missions to Mars and Titan, talk about the search for Earth clones, describe the surprising diversity of life on Earth, speculate about post-biological evolution, and explore what contact with intelligent aliens will mean to us.

Covering topics from astronomy and planetary science to geology and biology, this book will fascinate anyone who has ever wondered, "Are we alone?"

CHRIS IMPEY is a University Distinguished Professor and Deputy Head of the Department of Astronomy at the University of Arizona. His research interests are observational cosmology, gravitational lensing, and the evolution and structure of galaxies. He has 160 research papers to his name, has coauthored two introductory textbooks, and wrote the well-received astrobiology book *The Living Cosmos* (Random House, 2007). His most recent popular science book, *How It Ends* (Norton, 2010), covers endings, from personal to cosmic.

Talking
About Life

Conversations on
Astrobiology

CHRIS IMPEY

University of Arizona

CAMBRIDGE
UNIVERSITY PRESS

CAMBRIDGE UNIVERSITY PRESS
Cambridge, New York, Melbourne, Madrid, Cape Town, Singapore,
São Paulo, Delhi, Dubai, Tokyo, Mexico City

Cambridge University Press
The Edinburgh Building, Cambridge CB2 8RU, UK

Published in the United States of America by Cambridge University Press, New York

www.cambridge.org
Information on this title: www.cambridge.org/9780521514927

First published 2010

Printed in the United Kingdom at the University Press, Cambridge

A catalog record for this publication is available from the British Library

Library of Congress Cataloging in Publication data
Talking about life : conversations on astrobiology / [edited by] Chris Impey.
    p.   cm.
Includes bibliographical references and index.
ISBN 978-0-521-51492-7 (hardback)
1. Astrobiology.   2. Life–Origin.   3. Astronomers–Interviews.
4. Geologists–Interviews.   5. Biologists–Interviews.   I. Impey, Chris.   II. Title.
QH326T35 2010
576.8´39–dc22   2010016396

ISBN 978-0-521-51492-7 Hardback

Contents

Preface

Scientists tell stories with the power to inspire. Geologists can tell a story of the titanic forces that have shaped the planet. Chemists can tell the stories of the almost infinite complexity that result from the combinations of fewer than a hundred ingredients. Physicists tell a story where the microworld of forces and interactions is based on a pleasing symmetry and unity. Biologists have their own story of unity, where the diversity of life is based on a single genetic code. The story astronomers tell is perhaps the grandest of all, since it plays out over 13.7 billion years in a universe that utterly dwarfs us.

One vital part of this story has yet to be told: the role of life in the universe. We have a high degree of confidence that the laws of physics and chemistry are universal. The visible universe contains about 10^{22} stars, a number so large that it's hard to grasp intuitively, and theory and observation indicate that many of them will have orbiting planets. Life on Earth is tenacious and pervasive; it's found in a bewildering array of environments, yet Earth is the only place we know of with life. That may be about to change.

Astrobiology is the scientific study of biology in its broadest context. It is a young field. Sixty years ago, we had little idea of how life on Earth started and no idea of the unity of life at a molecular level. Thirty years ago, Earth seemed to be the only plausible site for biology in the Solar System. And just fifteen years ago, we knew of no planets orbiting stars other than the Sun. Currently, a fleet of large telescopes and a small armada of spacecraft are starting to identify and scrutinize plausible sites for life in the Solar System and beyond.

There are excellent textbooks, popular books, and scholarly monographs on astrobiology. Such writing is authoritative, but a third-person narrative can be impersonal and sometimes sterile. *Talking About Life* is different. It's a selection of interviews where scientists tell the story of the search for life in their own words. Taken as a whole, it's a snapshot of the state of the search for life in the universe a decade into the twenty-first century. It has the immediacy of the first-person voice. It describes how science works not in theory but in practice. It conveys the excitement of asking deep questions about nature and the challenge of doing research, which defines the boundary between what we do and don't know.

The first part of the book sets the scene. Timothy Ferris gives his view as a writer and chronicler of the field for the past thirty years. Iris Fry reveals how modern ideas of the origin of life came about. Steven Dick recounts the history of astrobiology and Ann Druyan explores the excitement of communicating science. Pinky Nelson gives us the view of one of Earth's lucky life-forms who has travelled beyond Earth, and in Neil deGrasse Tyson we hear from the foremost popularizer of astronomy. Then Steven Benner and William Bains remind us that the toolkit of biology in other settings might be quite different than it is on Earth.

In the second part, insights into the history of life on Earth are presented. Roger Buick and Joe Kirschvink talk about the oldest rocks and tell how the changes in the biosphere are intimately coupled to the geological record. Lynn Rothschild and John Baross remind us that the typical organisms on Earth are extremophiles: microbes that can handle physical extremes fatal to us. Next, Andrew Knoll gives the broad sweep of evolution over the past 3 billion years, and Simon Conway Morris discusses the tension between the ideas of contingency and convergence. Even if microbial life proves to be abundant in the universe, some people will likely be disappointed; part of the lure of astrobiology comes from a craving for companionship. As Roger Hanlon and Lori Marino remind us, even as we search for intelligent life out in space, we share a planet with highly functioning creatures; they are the "aliens among us."

The third part turns to the search for life in our backyard, the Solar System. Chris McKay recounts what we can learn from Solar System analogs on Earth, and David Grinspoon talks about what makes a planet habitable. After an overview by Jonathan Lunine, Carolyn Porco takes us with her on the adventure of the Cassini mission to Saturn and its moons. Laurie Leshin and Guy Consolmagno talk about the ingredients for life that are found in primordial material from the formation epoch, and Peter Smith gives us an insider's view of the recent Phoenix mission to Mars.

The biggest revolution in astrobiology is the routine discovery of exoplanets after decades of fruitless searching. Alan Boss presents the theorist's view; he has been struggling to understand the often unusual systems found so far. Geoff Marcy and Debra Fischer give us a sense of the thrill of that chase. Young researchers Sara Seager and David Charbonneau then describe how the discovery phase is giving way to characterization, and how the detection of terrestrial planets is finally within reach. Vicki Meadows closes the fourth part by describing how astronomers hope to find life on distant planets by detecting spectral biomarkers.

The final part of the book gathers perspectives on astrobiology that are more "out of the box," starting with Jill Tarter and Seth Shostak giving an update on the search for extraterrestrial intelligence. Then Ray Kurzweil and Nick Bostrom consider the implications of the fact that humans or species like us elsewhere might pass through a transition to a post-biological stage. Next, Paul Davies and Martin Rees present the cosmic context for life, including the strange alignment of fundamental physical quantities around values necessary for biology. The last word goes to writers: Ben Bova considers how science fiction has anticipated astrobiology and Jennifer Michael Hecht has a provocative take on the implications of contact with intelligence from afar. Our imaginations may have trouble keeping up with what the universe has conjured up.

I am a novice in many of the subject areas of this book, so I am grateful to my colleagues for their patience and tutelage. The work and words of many of them are contained here, but an equal number who I interviewed are not; their omission is due to the constraints of space and is no reflection on the quality or importance of their research.

The interviews are true to the original digital recordings and so represent a historical record, but a large amount of "invisible" work was required to get them ready for publication. Each interview was transcribed and then edited three or four times to improve clarity and flow. Every attempt was made not to alter the scientific content in this process, but I take responsibility for any mistakes or distortions that remain. Erin Carlson kept track of all the digital files and kept me organized as I juggled this project with many others. I am particularly grateful to Laura Robb, who did the bulk of the transcription and early editing; her talent and attention to detail resulted in material that was already in good shape when I began my work. Katherine Larson did much of the early transcription and editing and served a larger role as the inspiration for this project; I'm indebted to her. Thanks go to my agent Anna Ghosh for encouragement and to Vince Higgs for patiently shepherding this project to completion.

Part I INTRODUCTION

1

Timothy Ferris

Timothy Ferris has experienced the best of two worlds. His scientific writing earned him the American Institute of Physics Prize, the American Association for the Advancement of Science writing prize, a Guggenheim Fellowship, and nominations for the National Book Award and the Pulitzer Prize; meanwhile, his interest in music led him to reporting and editing for *Rolling Stone* magazine. Ferris' interests in science and music converged when he produced the Voyager record, an interstellar calling card of human civilization, containing photographs, audio files, and music. His eleven books include *Seeing in the Dark* and *Coming of Age in the Milky Way*. He is a regular contributor to *The New Yorker* and *The New York Review of Books*, and has been published in over fifty periodicals. Ferris wrote and narrated two television specials: *The Creation of the Universe* and *Life Beyond Earth*. He has taught five disciplines at four universities, and is currently Professor Emeritus at the University of California, Berkeley. For thrills, Ferris tests high-performance Italian and German sports and grand-touring cars.

CI   I feel like a dilettante because you are a journalist and I'm playing one. Let's start with your career. Did you cut your teeth at *Rolling Stone*?

TF   I started in 1968 as a general assignment reporter for United Press International in New York, and soon after became a feature writer for the old *New York Post* – when it was a shabby liberal tabloid, as opposed to the shabby conservative tabloid it has since become. After working for three years as a reporter, I quit to write my first book, supporting myself as a freelance writer. I was freelancing for *Rolling Stone*; it and *The New Yorker* were the two magazines I most admired, and *Rolling Stone* was more approachable for a young, unknown writer.

     Just as I was discovering that I didn't yet know how to write a book, I was offered a job as the New York bureau chief of *Rolling Stone*. There I learned how to write long-form, 5000 to 10 000-word magazine pieces, which I find to be the hardest form of nonfiction. Once you master long-form nonfiction you can write chapters, which brings the writing of books potentially within reach. Thanks to this experience, when I left *Rolling Stone* I was able to start writing books.

CI   Do you still have a paternalistic interest in *Rolling Stone*?

TF   I do, although I haven't been following the magazine closely. Jan Wenner and I were brought a bit closer by the death of Hunter Thompson, who was a close friend of us both – and I still read the magazine irregularly. My son is a musician, so I got him a lifetime subscription – for $99.

CI   Your topic coverage for them was very broad. When did you start to home in on science?

TF   Science had always been important to me, and I wrote about science at UPI and at the *New York Post* when I could. I did a cosmology piece for *Rolling Stone* that got a lot of notice in the rock-and-roll world. It's kind of a trade secret, but many of the rock stars of that era – Mark Knopfler, Keith Richards, Bob Dylan – are avid readers, and the response of rock musicians to the *Rolling Stone* piece suggested that perhaps there was an audience for science writing, so that I could perhaps write exclusively about the things that mattered most to me. Back then there were relatively few science writers, and it was often asserted that the general public didn't much care to read about science.

CI   The book of yours that always struck me as a labor of love was *Coming of Age in the Milky Way*. It was so rich, so sweeping. Was writing that book an odyssey?

TF   That book just about broke me down. It was a young man's project, initially, and when you're young you're hot to do what's never been done. In the course of more than a decade of researching *Coming of Age,* I got a lesson in *why* it hadn't been done before. I was foundering at the end, but a Guggenheim grant helped me make it to shore. Writing a book can take a lot out of you physically, since the writing is both sedentary and apt to invoke high levels of stress. In the late stages of the book I got into the habit of hitting the gym in the mornings. If you want to be a professional writer, you'll do well to keep yourself in good physical shape.

CI   As you get into it, and the sheer scope becomes visible, do you ever have the feeling of being in too deep?

TF   Sure, and it's terrifying. While writing my first book, *The Red Limit,* I often vowed I would never write another book if I could just finish this one. Fortunately, our memory for pain is poor. By the time a book comes out, you tend to forget all the anxiety and exhaustion involved in writing it. You go on a book tour, everyone tells you how good it is, and you start to think that you actually write with felicity. While in that state of dangerous delusion, you sign another book contract.

CI   Women talk about this with childbearing.

TF   I don't know if the two are comparable, but the short-term memory for pain may come into the picture.

CI   Visual art is so different from writing. What did you find most interesting when you started to move into films?

TF   Making a film uses different parts of your brain; it gives you a nice counterpoint to writing for print. You're writing for the spoken voice.

CI   Do you start with a visual arc or a narrative? How do you put it together?

TF   I don't even know if it's an arc at all. Mostly I try to come up with scenes that I feel confident will work. I don't invest too much in how they are going to fit together, because generally the first way you assemble them doesn't work. If the scenes have a degree of independence, you can rearrange them and still make a coherent film. I love emotional effects attained by combining music and visuals with spoken words. That's what writing is – a written version of idealized speech, more like singing than thinking. I like my films to convey a sense that science is of a quality comparable to art, that it rewards aesthetic as well as intellectual involvement. That's why I put so much emphasis on the photographers, effects artists, and composers in the films.

CI   You've made amazing use of Brian Eno.

TF   We used Brian's music exclusively on *The Creation of the Universe*. For *Seeing in the Dark,* we have Mark Knopfler and Guy Fletcher, of Dire Straits. They're wonderful musicians and as stone-cold professional as you can find. When I asked Mark to do the title theme he said he would "attempt" it.

CI   Have there been any scientific concepts that you've found either impossible or incredibly challenging to convey visually?

TF   I recall saying, at our first production meeting for *The Creation of the Universe*, that there are essentially two approaches to presenting a difficult subject. One is to be so *un*ambitious that you know you'll succeed. The other approach is to go ahead and swing for the bleachers – admitting that your audience isn't going to get all the content, or even *most* of it, in one viewing, but hopefully making a film they'll want to see more than once. That's the approach we took, and I'm always gratified when I hear from people who've seen *Creation* multiple times, because that was how it was meant to be viewed. I encouraged my collaborators to think of it as more like a record than a film.

CI   A lot of the skill is coming up with those metaphors, those analogies, and doing it visually. That process must be fun. Do you do that alone?

TF   I'll take good metaphors anywhere I can get them, although ultimately it's my job to come up with them. All metaphors are inexact – as Robert Frost said, that's the beauty of them – but the trick is to make them no more inexact than they need to be. To paraphrase Mark Twain, the difference between the right metaphor and all the others is the difference between lightning and a lightning-bug.

CI   From a broad journalistic background, how did you get drawn to the issues of life in the universe? Was the Voyager project your first involvement?

TF   I had a wide-ranging interest in astronomy since boyhood, and the Sagan–Shklovsky book, *Intelligent Life in the Universe*, helped inform my interest in the subject of extraterrestrial life. While at *Rolling Stone*, I proposed an interview with Carl, which was published in 1973. He and I became friends. I used to stay with him when I was at Cornell, and he would stay at my apartment in New York, and we listened to a lot of music together. When he and Frank Drake came

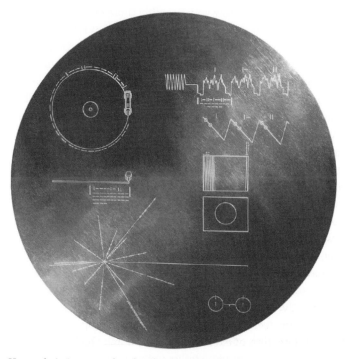

Humanity's "message in a bottle," the gold-plated analog phonograph record launched over 30 years ago on the twin Voyager spacecraft and now heading into interstellar space. The most distant human artifacts have multilingual greetings, world music selections, and images coded into the grooves. Timothy Ferris was on the small group who selected the images and music, and he wrote a book called *Murmurs of Earth* to describe the process of devising a time capsule for humanity (courtesy NASA/JPL).

up with the idea of the Voyager record, he asked me to produce it. I absorbed several of Carl's principal approaches to the question of searching for extraterrestrial life, which I think are still effective and well thought out. So it was thanks largely to Carl that I had a sound framework for thinking about the subject.

CI   Who else was involved in that project?

TF   Frank Drake had a formative role, and the full cast of characters is set out in the book, *Murmurs of Earth*. Most of the people involved in the record weren't exobiologists, but people working in music and natural sounds – field recordings, ethnomusicography, and other arcane items like Roger Payne's hydrophone recordings of whale songs.

CI   Have you had letters and emails over the years critiquing the musical selections?

TF   To some degree, but unfortunately the record was never properly released, so it's been difficult for people to evaluate it since so few have ever heard it.

CI   There's a slight irony in the fact that the object winging its way through space is an obsolete technology, even on Earth.

TF   Perhaps, but if I were making the record again today I might well use the same technology. A metal analog disk is like Sumerian cuneiform script. We know that it will endure, whereas we don't know the lifetime of the optical dots on a DVD. Some of the people involved have since said, "If we'd had new technology, we could have put much more on it," but it's not necessarily the case that you'd get a better record by having five times as much music or by having five times as many photographs. Limitations can create a superior result in any creative project, and we can vouch for the billion-year projected survival time of the record.

CI   Freeman Dyson argues that, on purely thermodynamic and physical grounds, analog computing is more powerful and more energy efficient than digital. Analog probably rules, cosmically.

TF   One of the fascinating things about analog records is that you never know how much data are in the grooves. With a CD you always know that exactly, down to the bit, which means that there's an overt limit to what you can extract from the recording. Digital is a dance of seven veils without the veils.

CI   With your awareness of astrobiology, what are the most exciting research results right now?

TF   One exciting development is the detection of extrasolar planets, and the imminent prospect of obtaining spectra of their atmospheres to look for signs of life. The first detection of extraterrestrial life may be in reflected starlight passing through the atmosphere of some unseen planet many light years away. That would transform the field – dividing it, in a BC and AD manner.

Unlike SETI, extrasolar planet observations require that scientists make relatively few assumptions about life and how it evolves. Using spectroscopic methods, scientists ought to be able to detect evidence of a wide range of life forms on other planets. That's encouraging, particularly for those of us who feel that while life is probably commonplace in the universe, it's difficult to know how often intelligence arises, and how long it lasts when it does appear.

CI   Do you think the first evidence will come from a distant star or our Solar System?

TF   I don't know. Life on Mars is still an open question. Lewis Thomas wrote years ago in *The New York Times,* when we started to see the first images of Mars close up, that Mars may have life, but if so, we've never seen a planet that has life and looks so desolate. But the Earth prior to the emergence of life from the seas would have looked just as desolate in many ways.

CI   And we're down to planets a few times the size of Earth.

TF   That's a terrifically exciting prospect. The history of the Earth gives us grounds for optimism that similar planets might give birth to life, since life appeared on Earth so early in its history. There are always arguments against looking for life in any particular way, but the deciding issue is phenomenological: if you don't look, you're not going to find it. We tried *not* looking for tens of thousands of years, and sure enough we didn't find extraterrestrial life. The rationale for looking is not some perfect argument that it *must* be out there, it's that otherwise we're not going to find it.

CI   But there's presumably still a lot to learn from scouring our own planet.

TF   That's the second area that really excites me: the tremendously enlarged phase space within which we now examine life here on Earth. No one knows how far down into the Earth organisms prevail; those roots can be very deep. Some estimates have half the biomass down in solid rock. There are living organisms floating high in the atmosphere. To find terrestrial life thriving in these extreme environments does add encouragement to the search for extraterrestrial life.

CI   What do you make of the Rare Earth hypothesis, the "Goldilocks" idea that certain things about the Earth and our environment in the Solar System were "just so" to make life possible?

TF   I think the Rare Earth arguments are mostly the post hoc fallacy writ large. That doesn't mean I'm right – but again, the way to find out who's right is to keep looking. The people who think life is rare understandably get frustrated and say, "You could keep looking for life forever. If we examine ten thousand planets and they're all sterile, you'll still say, 'Ten thousand and one might pay off.'" It's true. Sometimes exploration takes place in the service of illusory goals – like Ponce de Leon's search for the fountain of youth – but it's still a good idea to explore.

CI   Suppose the universe is full of bugs, microbes, small life organisms, but bereft of large, sophisticated creatures. As far as the public is concerned, is there a sense that finding bugs out there won't be satisfying? That what we're looking for is companionship, and if the universe is full of pond scum, the public won't care?

TF   It's easy to underestimate the public. When Carl and I were at JPL for the Viking landing and the first live picture came down from the surface of Mars, people all over the world were watching live – they were up at odd hours in Europe and Japan, where those pictures were shown live on TV. In the United States, the

The barren, desert-like surface of Mars, as revealed by one of the first pictures taken by the Viking lander when it touched down on Mars on July 20, 1976. The images dashed hope of obvious life on Mars, but it is a more interesting planet than it seems at first sight. Viking worked flawlessly for six years and it was spotted by the Mars Reconnaissance Orbiter in 2006 (courtesy NASA/JPL).

landing occurred when morning shows, like *The Today Show* and *Good Morning America*, were on the air, live. But they didn't cover it. They didn't break in and say, "Here comes the first live picture from the surface of Mars." The network science people were tearing their hair out with anger and frustration because their producers said the public didn't care.

Fast-forward to the Pathfinder landing, twenty years later. Now people didn't have to rely on television gatekeepers to tell them what to see: They could go to the Pathfinder mission website. And they did: The number of hits on that website on the first day was more than the cumulative TV audiences of all three morning shows on the major TV networks combined at the time of the Viking landing. People voted their interests, and it turns out that they were much more interested in Mars exploration than the communication professionals gave them credit for.

CI You've hit on a bigger point in our culture. It seems there's a thirst for science in the popular culture that's not being met.

TF Whenever anyone – a cab driver, a bartender, someone sitting next to me on a plane – asks what I do, I tell them about the book I'm working on, and I find that even when the subject is difficult they are always interested. The idea that "Joe Sixpack" won't pay attention unless you tell him why it's important is a journalistic fallacy that has lived on through repetition. It complements the gatekeeper idea, because it implies that you and I are smarter than Joe Sixpack, so we ought to tell him what to think and why. But I've been a college professor for decades, and in my experience, college professors aren't any smarter than Joe. Throughout my career I've written books that don't dumb things down. They ask a lot of their readers, and the readers have been up to it.

When we were making *Creation of the Universe*, our executive producer was reading a magazine published for cable television professionals. At that time, cable had a penetration of under a quarter of American households, and the industry wanted to know what kind of programming their audience wanted to see. So they conducted a survey, listing about fifty different kinds of programming. The survey results showed that the number-one favorite was news, number two was sports, and so on. My producer looked all the way down the list, and there was no science. He called the magazine and said, "We do science programming, and we just wondered – how bad is it? How low was the number of viewers who wanted more science on cable?" Their reply was that there hadn't been a box to check for science; they hadn't thought of it. "That's a good idea," they said. "When we do the annual survey next year, we'll put science on there." They did, and science came in third, right behind news and sports.

CI   As to how the public might react, suppose life is detected in a pivotal enough way that the evidence is unambiguous; it becomes a news story that biology on Earth is not unique. Will that change us, in the broadest sense?

TF   The change will be profound and long-ranging. The best historical parallel I can think of is the impact upon Europe of the discovery of the classical texts of Greece and Rome. Whole universities were founded to study Plato and Aristotle, plus there was an impetus to develop the printing press, owing to a growing demand from ordinary people who wanted to read these wonderful books they'd heard about. These effects rippled through the centuries, and they helped create the modern world. The discovery of life beyond the Earth, especially intelligent extraterrestrial life, could be like that. It wouldn't be a question of what happened in the first few weeks or months, but of what happened over a period of decades and centuries.

CI   Might it give us a larger sense of responsibility and stewardship of our own part of the universe?

TF   It might very well. It takes time for these things to sink in. The realization of how thin the Earth's atmosphere is, that it's just a membrane, like the transparent membrane over the eye; that we're on one planet among many; that our planet has been through enormous changes in its history, and that we don't understand the mechanisms behind many of those changes, some of which would be at best *dis*-accommodating for us were they to occur today – these realizations are just beginning to penetrate into the general culture. It takes a while for people to incorporate them into their thinking.

CI   Let me ask about the future of life. The vast majority of species go extinct. We're on an exponential cusp of technological change, and it seems hard to predict our evolving role in the universe. Is that something you think about?

TF   The question that interested me as a boy – and still does today – is, "How do we understand the relationship between the human mind and the wider universe, of which it forms a part, but from which it also stands apart?" In other words, how best can we comprehend our place in the wider scheme of things? One of the reasons we would like to know about intelligent extraterrestrial life is that we don't know whether intelligence is a fluke or whether it typically arises, or what other species have done with it. I'm optimistic – I like people; I like what people have been able to create out of this life – and if I were asked to predict how long people will survive, that number would be more like a million years than a hundred years.

CI   Let's move on to intelligence and SETI. We share a planet with a handful of species that possess intelligence, but lack opposable thumbs or technology. Is that meaningful when we try to look for intelligence elsewhere?

TF   I make a sharp division between species that demonstrate the capacity to use an abstract symbolic language – and there is only one such species on Earth – and those that don't. By that definition, there's only one intelligent species on Earth. On the other hand, if you don't respect others you don't respect yourself, and I would encourage humans to keep that in mind in our dealings with animals. We can be a lot more decent in the way we interact with other living things. The concept that they have rights is not a foolish notion, even though it's often laughed at; universal suffrage was laughed at a century ago. I don't think we'll be giving animals the vote, exactly, but there's a lot of room to treat life more equitably – based not on its intelligence but on the fact that it's diverse and wonderful, and that we're dependent on it.

CI   Is SETI burdened with anthropocentric assumptions?

TF   A number of different strategies have been attempted with SETI, and it may be that we've only just scratched the surface. Still, it's discouraging not to receive a signal. If my idea of interstellar networks holds up, signal detection ought to be easier than looking for individual broadcasters, but I don't know how to measure that against the results to date. I don't think there's anything wrong with using radio, plus we also now have optical SETI as well. Given the many uncertainties in the SETI enterprise, I think it's appropriate that it has been a private enterprise for a while now, and that graduate students are not being encouraged to bet their careers on a SETI success. There's little way of knowing how long SETI is going to go on without a result, so keeping its annual funding modest makes more sense than launching a big-budget "War on Cosmic Loneliness."

CI   That's the only substantive scientific criticism I've heard – that interpreting a null result is exceptionally difficult.

TF   SETI is more exploration than science, as Philip Morrison used to say. As long as somebody is passionate enough to want to keep funding the search, then the search can go on. It doesn't hurt anybody. People can criticize SETI all they like, but if SETI finds a signal one day, few will much care how good anybody's prior reasoning was.

CI   Is it possible that biology itself is just a transitional phase towards machine intelligence or pure technology? It's a scary prospect for many people.

TF   Two computers are sold every second worldwide and we're wiring them together into networks as fast as we can. It's fertile ground in which to construct paranoid fantasies.

On a more serious note, there are both conservatives and leftists eager to restrict scientific research based on someone's fearful projection of what might be done with it – that, for instance, intelligent computers will take over the world, or stem cell research lead to poor people being grown on farms in order to provide body parts for the wealthy. A healthy antidote to such thinking is to look back in history and ask, "What avenues of scientific research would you prefer had been blocked, so that we wouldn't have some of our current knowledge?" When I ask people this question, most come up with "the bomb." But would they like to have suppressed Einstein's 1905 relativity paper, which not even Einstein thought could be used to develop an atomic bomb? And even if they could have done that, do they think the world would be better as a result if we didn't know how the Sun shines? Smart machines would change the world, certainly, but I don't fear that they'd take it over.

CI   So you think we should ride the wild beast and enjoy our technological progress?

TF   Technology can certainly produce problems, and already has: Eighty percent of the world's energy comes from fossil fuels, the burning of which threatens global climate change. But science and technology has also saved hundreds of millions of lives, and improved billions more. As far as scientific research is concerned, everyone investigating nature should be free to do so as he or she wishes, subject only to common-law protections of human rights. It is folly to try to restrict scientific inquiry based upon a science-fiction scenario of how something might go awry in the future. A vague fear of the future is not an adequate basis upon which to talk humanity into remaining more ignorant than we would be otherwise.

CI   We have plenty of problems to solve on this planet, but it seems like the question "Are we alone?" sets a deep psychological hook in most people. Why is that?

TF   It could answer Socrates' question, "Who am I?" Who are we? As long as we're the only intelligent species we know anything about, it's going to be difficult for us to understand ourselves – and when we don't understand ourselves, we fall prey to delusions. It's why civilization is in *cities*; that's what the word "civilization" means. It is in cities that different kinds of people have bumped up against one another and come to know themselves better, by virtue of the similarities and dissimilarities they find with one another. That's been a tremendous asset for us, but we're still all humans; all forms of life we encounter are just variations of the same form. Once we have another form of intelligent life to compare ourselves with, we will understand ourselves better. I believe that's the fountainhead, the appeal of this ancient, mysterious question.

2

Iris Fry

Iris Fry has a unique perspective on astrobiology because of her scholarly work on the history of thinking about life and its origin on this planet. She teaches the history and the philosophy of biology in the Department of Humanities and Arts at the Technion–Israel Institute of Technology. Initially trained in chemistry and biochemistry at the Hebrew University in Jerusalem, she subsequently studied philosophy and the history and philosophy of science at Haifa University and Tel Aviv University. Her book, *The Origin of Life: Mystery or Scientific Problem?*, was published in Israel in 1997. Her second book, *The Emergence of Life on Earth: A Historical and Scientific Overview*, was published in 2000 by Rutgers University Press. She is also studying the historical, philosophical and social aspects of the relationship between science and religion.

CI   In your book, *The Emergence of Life on Earth*, you discuss first the history of ideas about the origin of life and then recent scientific theories and current issues. You deal also with the philosophical aspects of the subject. What is the "origin" of your combined perspectives on the origin of life? How did you get into your field?

IF   I started as a biochemist. After obtaining a BSc degree in chemistry and an MSc degree in biochemistry at the Hebrew University in Jerusalem, I worked there as a research assistant. Yet something was missing, and I decided to go back to school to study philosophy. I wrote a Master's thesis on the concept of purpose or teleology in the philosophy of Immanuel Kant, associated with his conception of the organism. Studying Kant brought me back to chemistry and biology in a historical and philosophical context. I then completed a PhD in the history and philosophy of science at Tel Aviv University. In my dissertation I examined the ideas of Lawrence Joseph Henderson, the early twentieth-century Harvard biochemist, on the interaction between life and the physical and chemical environment on Earth.

So it was Kant and Henderson and their scientific and philosophical ideas at their corresponding historical periods that led me to the field of the origin of life. My education in chemistry, biochemistry, philosophy, and the history of science was helpful in dealing not only with the history of the field, but also with its present scientific state.

CI   It's almost perfect convergence, a "perfect storm" of subject matter.

IF   That's how I got started. Then I gave a series of radio lectures on the origin of life in the framework of a "Broadcast University" on one of our radio stations in Israel. Out of that grew a little book in Hebrew which was later extended into a book in English, *The Emergence of Life on Earth: A Historical and Scientific Overview*. Among other topics, this book also dealt with the search for extraterrestrial life and examined conflicts between scientists and creationists over the emergence of life.

CI   The methodology of the historical sciences, such as geology, evolutionary biology, and also astronomy, must be different from experimental science. Does that shape our understanding of what happened on Earth so long ago?

IF   Historical sciences differ from empirical sciences in that you cannot reconstruct exactly things that happened billions of years ago. This is a basic concern when people discuss evolution: "You weren't there. How do you know?" It is certainly the case when we consider the period shortly after the formation of Earth over 4 billion years ago, a time frame within which, according to many scientists, life first emerged. Due to huge geological upheavals on the early Earth, it is difficult to locate original rocks from this era or to find in them traces of life. Nevertheless, science is also making progress on this front.

No one working on the origin of life would claim that she or he could reconstruct exactly what happened 4 billion years ago. People are trying to use as much data as possible to reconstruct the conditions on the early Earth that could have allowed a path to life. If life emerges in the lab under these conditions, it will be good enough. It will show that it *could* have happened this way.

Contrary to a common mistaken view about the "scientific method" based on experiments and direct observations, the experimental sciences, no less than the historical sciences, are based, by necessity, on inference. We can't see electrons or black holes directly but we can infer their existence based on the

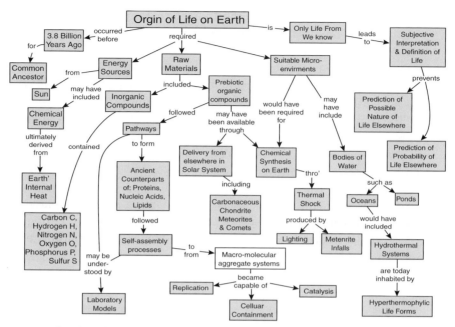

Flowchart for concepts in the investigation of the origin of life on Earth. The research involves a combination of looking at the historical record of life as early as possible in the planet's history, trying to reproduce the formation of a cell or a replicating molecule that is subject to evolution in the lab, and understanding the astronomical and geological context for the formation process (courtesy NASA/CMEX Mars).

accumulation of consistent data. In this sense, inferring a *possible* origin-of-life scenario from lab experiments is no less scientific than hypotheses in other scientific areas.

One of the bones of contention among biological evolutionists and researchers in the fields dealing with the history of life is the random versus deterministic nature of this history. The late paleontologist Steven Jay Gould suggested a thought experiment in which "the tape of life" is rewound and run again. Gould claimed that history is characterized by "contingency", the fact that every historical event depends on all the past events that led to it. Change one event on the way and the outcome will be completely different. Since so many historical events are unique, there is no way to predict the outcome and the tape of life, run again, will produce a different history altogether.

CI   What did Gould say about the emergence of life? Did he see it as "contingent"?

IF   Interestingly, he thought that the origin of life was virtually inevitable, given the chemical composition of the early Earth environment and the physical principles that govern self-organization.

CI   The origin of life became a subject for scientific research in the twentieth century. What were the ideas that had to be overcome for this to happen? Can you talk about "spontaneous generation"?

IF   For most of human history, naturalists and laypersons alike believed that whole organisms are repeatedly formed before our eyes, not only by sexual generation from parents but also from inanimate matter, triggered by the heat of the Sun and wet conditions. We find references to this belief in the mythologies of all ancient cultures. In different versions it was considered self-evident for centuries, lasting until almost the end of the nineteenth century.

The concept of spontaneous generation went through stages. As knowledge of biological phenomena grew, the organisms that were supposed to be generated spontaneously became smaller and smaller. First people spoke about insects, birds, fish, and mice, but after the end of the seventeenth century it was basically microorganisms. The belief in spontaneous generation became a problem when scientists began to focus on the ancient Earth. For this to happen, the conception of the world as static had to be dismantled.

CI   Did spontaneous generation not encompass the idea of evolution at all? Had people believed that organisms were being formed all the time, and always been generated spontaneously?

IF   Right. No evolution at all. The idea of time came into science through geology. The idea of change in the Earth itself, the realization that the planet itself had changed, became established in the eighteenth century.

CI   Wasn't Louis Pasteur a pivotal player in the spontaneous-generation drama?

IF   Indeed. In the eighteen sixties, Pasteur, the renowned French microbiologist, carried out a set of impressive experiments in which he clearly showed that microorganisms were not spontaneously generated in organic solutions undergoing fermentation, such as blood, urine, and vegetable extracts. When these liquids were efficiently boiled, sealed and isolated from the air, no single microbe was to be found in them. Pasteur concluded that, again based on his experiments, fermentation and decomposition in organic solutions resulted from contamination by microbes originating in the atmosphere.

CI   Was that why Darwin was almost mute on the issue of the origin of life?

IF   The origin of life was hardly mentioned by Darwin in *On the Origin of Species* or in his other published works. He did discuss the question in several private letters to close friends. Darwin clearly regarded the issue as too complex and beyond the scientific knowledge of the time. He also realized that having enough trouble on his hands with the theory of evolution, he should better refrain from being too explicit on the question of the origins of life. At the same time, there is a famous letter to his friend, the botanist Joseph Hooker, in which he contemplates the daring hypothesis about "a warm little pond" on the ancient Earth in which proteins could have formed from chemical compounds under the influence of heat, light, electricity, and so on.

CI   What was the role of the panspermia idea in thinking about the origin of life?

IF   Several developments took place at the turn of the twentieth century. Although Darwin did not discuss publicly the origin of life, the philosophical significance of his theory clearly indicated that life could have evolved and originated naturally without need for divine intervention. Scientists interested in the origin of life faced a dilemma between Pasteur and Darwin – between a divine creation of life and its natural emergence. In addition, progress in genetics and biochemistry early in the twentieth century revealed the complexity of the cell. L. J. Henderson, who I mentioned earlier, says in his 1912 book that because the cell is so complex, most biologists would prefer the Genesis story to the idea that life could emerge naturally from matter. The panspermia hypothesis was a reaction to this situation.

Various theories of panspermia claimed that life on Earth originated from seeds of life reaching our planet from space, either on meteorites, or as spores pushed by solar radiation. These ideas were entertained by some of the renowned scientists of the time, such as Lord Kelvin, Hermann von Helmholtz, and Svante Arrhenius. Panspermia was regarded as an option, based on the common cosmology of the time – the belief that the universe and matter were eternal. These people claimed that not only matter, but also life was eternal in the universe, wandering from planet to planet. There was therefore no need to explain how life originated from matter; it was always there. So panspermia "explained away" the big dilemma surrounding the origin of life from matter.

CI   Doesn't that notion of eternal life run headlong into the idea of evolution? It's a Victorian idea that we're ever improving, but if the time frame is eternal, what is the meaning of evolution?

IF   Most people who supported panspermia, like Lord Kelvin, who had a big debate with Darwin, didn't accept evolution, though Helmholtz, the German physiologist and physicist, suggested that once simple life landed on Earth, evolution could proceed unimpeded.

Panspermia "died" when the eternal cosmology was discarded and the hazards to life in outer space were realized. The term "panspermia" has become popular again recently. Astronomers and astrobiologists speak about panspermia when they refer to the possible transfer of life from Mars to Earth or vice versa by the ejection of meteorites at the early stages of the Solar System. From a historical point of view, this usage is inaccurate. Nowadays, no scientist believes in the eternity of life, and scientists realize that life originated from matter either here, on Mars, or on another planet.

CI   The panspermia episode seems to reflect a pessimistic outlook on science. How did attitudes eventually change later in the twentieth century?

IF   Indeed many scientists early in the century preferred not to touch the subject of the origin of life. A breakthrough to alter this attitude required new scientific data and a philosophical change in the conception of matter–life relationships. This was achieved by the contributions of a few pioneers, especially the Russian biochemist, Alexander Oparin, and the British biochemist and geneticist, J. B. S. Haldane.

Oparin, in a 1924 booklet published in the Soviet Union, and Haldane, in a much shorter article published in 1929 in London, suggested specific conditions on the primordial Earth that were conducive to the emergence of life. Furthermore, they both understood the need for an evolutionary process, including many chemical intermediates and a sort of primitive natural selection to evolve from the simple to the complex and organized. Oparin extended and specified his theory in a book, *The Origin of Life*, published in the Soviet Union in 1936.

CI   What about the role of biochemistry and genetics in this breakthrough?

IF   Interestingly, these new scientific fields, by revealing the complexity of the cell, turned the origin of life into a sort of taboo. But combined with an evolutionary outlook, Oparin and Haldane could use the new data in a constructive manner. Oparin, being a biochemist, focused mainly on the cell as a system of interacting enzymes and on life as an organized metabolic system. There was no knowledge yet of how enzymes worked, but more and more enzymatic proteins were isolated and their role as catalysts of specific metabolic reactions was being defined. In contrast, Haldane was inspired by the discovery of viruses and by the similarity of their behavior and what was known of genes. No one yet knew what genes were made of, or about nucleic acids as the hereditary material, but for Haldane the defining characteristic of life was reproduction. However, he and Oparin realized that the first step on the way to life was the synthesis and accumulation on the early Earth of organic molecules, the building blocks of life. Oparin relied on new astronomical data and concluded in 1936 that the early atmosphere was "reducing," that is, unlike the present atmosphere, rich in hydrogen compounds. This composition allowed an easy synthesis of organic molecules. Haldane also claimed in 1929 that there was no oxygen in the early atmosphere. He coined the term "soup" for the primordial ocean in which organic material was dissolved.

Their claims became known as the Oparin–Haldane hypothesis. The "hypothesis" referred to the reducing atmosphere, to the formation from primordial soup, and to the synthesis of organic building blocks and organic polymers under the influence of various sources of energy. Oparin then went on to assume the formation of metabolizing bubbles from protein-like polymers that could grow, divide and then evolve into more complex systems. Haldane, on the other hand, suggested the development of organic polymers that could reproduce themselves with the aid of building blocks abundant in the soup. They were responsible for establishing the two major research traditions in the field that became explicit only with the advent of molecular biology in the late fifties and sixties.

CI   Was it recognized at the time that they'd moved beyond speculation to a testable hypothesis?

IF   Not in the twenties and thirties, and not for quite a while later. Nobel laureate Harold Urey read Oparin's book in 1951. He had independently arrived at the idea that the ancient atmosphere was reducing and was also interested to know what it meant for the origin of life. His young PhD student, Stanley Miller, convinced his mentor to let him check the Oparin–Haldane hypothesis in the lab. Miller built a glass apparatus in which he simulated the primordial reducing atmosphere, the

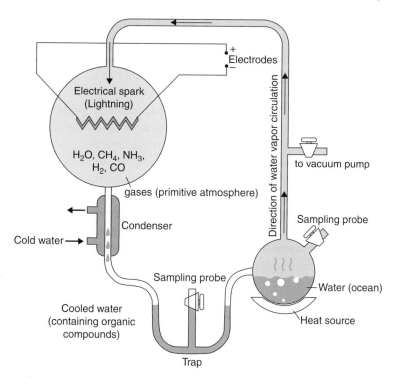

In the classic experiment on the origin of life carried out by Stanley Miller and Harold Urey in 1952, gases thought to be present on the early Earth were placed in a sealed vessel with water, and energy in the form of an electrical discharge or UV radiation was added. Within days, amino acids and other organic molecules had formed in the vessel (courtesy Wikipedia Foundation and Yassine Mrabet).

ocean, and lightning in the form of electrical discharges. So it wasn't until 1952 to 1953 that ancient-Earth conditions were reconstructed in the lab. To the delight of Urey and Miller, and the scientific community, the products of the experiment included several amino acids, the building blocks of proteins, and other organic substances. The results caused a big stir with the public when daily newspapers claimed erroneously that Miller and Urey had synthesized life in the lab.

CI   The Miller–Urey experiments are rightly famous, and yet, relative to the goal of creating a basic life form, it's like being in the foothills of a large mountain range.

IF   Certainly. They didn't presume to create life and they didn't. They conducted the first experiment in prebiotic chemistry and they created the necessary conditions for whatever came later. They showed that you can simulate certain physical, chemical conditions and get the building blocks of life. That work had tremendous significance for establishing the origin of life as a scientific field of research subject to empirical investigation.

Miller and Urey's experiment was followed by many variations. One of the most crucial, done by Juan Oró under prebiotic conditions, demonstrated the synthesis of adenine, a major component of DNA and RNA. NASA was an early and major funder of origin-of-life research. Without their funding, it couldn't have proceeded.

CI   Nearly sixty years after Miller–Urey, the experimental problems are still difficult. How do we connect the dots from simple chemicals to the first cell? Are answers, or even the means to ask the questions, within reach?

IF   It's extremely difficult. The researchers disagree on many issues. The ideas of a primordial reducing atmosphere and of the soup are called into question. There are two traditions, dating to Oparin and Haldane and their different conceptions of life. Is life first and foremost a replicating molecule, or is life a multimolecular, multifunctional, metabolic entity? Most researchers regard the "genetic" tradition as the only possible one. The major reason is the ease with which a population of replicating polymers can undergo molecular evolution through natural selection, which is crucial for the emergence of more complex systems. Evolution based on this "genetic" mechanism is evident in every living organism.

For many researchers, the major genetic system in the emergence of life was composed solely of RNA molecules. It was found in the nineteen eighties that some RNA molecules can function in existing cells both as genetic material and a catalyst, an RNA enzyme. This distinctive feature of RNA is a huge advantage compared to proteins that function today only as catalysts and DNA that functions as genetic material. The discovery of RNA's dual function led to the hypothesis of the "RNA World," that is a set of chemical systems composed exclusively of RNA molecules. A lot of work is being done in the field to find out how such an RNA World could have evolved on the ancient Earth.

This issue has proved very difficult to resolve. The "metabolists" therefore claim that RNA was a latecomer. They suggest instead a primitive metabolic system at the beginning of the emergence process. However, the crucial question for this camp is whether a population of metabolic systems without any genetic polymers could have evolved through natural selection, using a different mechanism than the one known to us today.

CI   Despite these uncertainties, is origin-of-life research considered a mature field?

IF   It is mature in the sense that the difficulties involved in answering the questions are much clearer now than a decade ago and are candidly acknowledged. When can an answer be reached? I'm not sure. There's still a huge amount of work to do. But a lot of ground has been covered. We know more about the conditions on the ancient Earth than ever before. It might be that life didn't emerge in the soup but on the bottom of the sea, near hydrothermal vents, that is, hot springs that spew chemicals from within the Earth's crust. We can connect the early Earth to the delivery of organic molecules to Earth and to our knowledge of geology. By using techniques of molecular evolution we can trace back the characteristics of the universal common ancestor at the root of the tree of life. So, by going from top to bottom and from bottom to top, we are narrowing our ignorance.

CI   Say we succeed in making a plausible scenario for how life emerged on Earth, and then we're faced with a new set of environments on exoplanets that are each different from the other, and from the Earth. Can we predict what we'll find on other worlds?

IF   There is recent talk about the possibility of "weird life" in other environments, made of very different chemistries and using solvents other than water. There are calls for a less "Earth-centric" conception of life when we look for extraterrestrial life. On the other hand, a representative of the alternative view, Nobel laureate and biochemist Christian de Duve, claims that the constraints of physics and chemistry are so strong that life in any other place would be basically the same. He believes that it's not accidental that a limited number of organic compounds formed under the physical and chemical conditions on Earth, and these organic compounds reacted among themselves and created life. Even though the details might be different – not exactly the same amino acids or the same nucleotides – many people would say that life elsewhere will be based on organic chemistry, on carbon and water.

CI   Most astronomers and planetary scientists think that microbial life is widespread in the universe, and may indeed have the same biochemical basis, but there is a large degree of contingency associated with higher organisms and intelligence.

IF   This thesis was promoted in the popular book *Rare Earth*, by the geologist Peter Ward and the astronomer Donald Brownlee. The two presented several empirical arguments for the uniqueness of Earth as a suitable location for the evolution of complex, especially intelligent organisms, and they emphasized the contingent nature of evolution. De Duve notes the strong selective advantage of intelligence and other people call attention to the development of nonhuman intelligence on Earth in other mammalian species, such as whales, dolphins, and the great apes. So it's an open question. We need data, another sample of life, or an intelligent signal captured by the SETI project, to decide one way or the other.

    Philosophically, however, it is clear that the science of astrobiology presupposes the Copernican principle – the idea that Earth is not uniquely chosen or endowed with characteristics not found elsewhere in the universe. No less important is the Darwinian supposition that life originated and evolved on Earth through natural means and might do so on other planets given the appropriate conditions.

CI   If we find the first example of life elsewhere – whether it's an ancient and extinct life form on Mars, extant life on Europa, or a metabolism-induced alteration of the atmosphere of a distant Earth-like planet – I assume that will change everything, philosophically and epistemologically?

IF   Today, based on the achievements of science on all fronts, including evolutionary biology, we're justified in the assumption that life can emerge naturally whenever physical and chemical conditions are appropriate. I see this postulate as part of a general, well-substantiated world view. For me, the possibility of the emergence of life somewhere else is strong. I would be elated by such a discovery, but not surprised, because it will not in any way contradict my basic outlook.

3

Steven Dick

After earning his Bachelor's in astrophysics, Steven Dick went across the street at Indiana University, Bloomington, and did a PhD in the history and philosophy of science, with a focus on the history of the search for intelligent life. Despite advice that his subject offered neither scientific evidence nor scholarship, Dick's thesis material became the basis for two books, *Plurality of Worlds: The Origins of the Extraterrestrial Life Debate from Democritus to Kant*, and *The Biological Universe: The Twentieth Century Extraterrestrial Life Debate and the Limits of Science*. Steven Dick worked at the US Naval Observatory and authored *Sky and Ocean Joined: The U.S. Naval Observatory, 1830–2000*, among other books. He has served as chairman of the Historical Astronomy Division of the American Astronomical Society, and as the President of both the History of Astronomy Commission of the International Astronomy Union and the Philosophical Society of Washington. Dick is a member of the International Academy of Astronautics, and its SETI Permanent Study Group. Until his recent retirement, he was NASA's Chief Historian.

CI   How did you get the history bug?

SD   I'm from Indiana, and I got my Bachelor's degree in astrophysics from Indiana University in Bloomington. Right across the street was one of the few, and one of the best, history and philosophy of science departments in the country. I decided to go there for grad school.

When I was learning all those concepts in astrophysics, I always wondered how we came to know what we knew. The history-of-science people dealt with that. In grad school, I specialized in the history of astronomy. I had been influenced by Shklovsky and Sagan's 1966 book, *Intelligent Life in the Universe*, which was the bible for extraterrestrial-life people. When I went into grad school, casting around for a subject, I thought, "I wonder what's been done with the history of the idea of extraterrestrial life?" It turned out absolutely *nothing* had been done.

CI   That's strange. It's a subject that fascinates a lot of scientists.

SD   It was almost taboo through the sixties and the seventies, and the history of it even more so. I was interested in the entire span from the ancient Greeks to the present, but I didn't know what I was getting into. When I proposed a dissertation on the history of the extraterrestrial-life debate, they told me, "There are two problems from the perspective of the History of Science Department. First of all, it's not science, and secondly, it has no history that's intellectually substantial."

I had to switch advisors. There was a medievalist there, Ed Grant, who knew that there was a medieval plurality-of-worlds tradition, what they called a *plures mundi* in Latin. Medieval scholastics commented on Aristotle, who had written a book called *De Caelo* or "On the Heavens," where he had an argument for only one world. He was using the word "world" in a different way than we do now, as cosmos. Ed Grant ended up being my dissertation supervisor, and I had a chapter on the medieval scholastics in that dissertation. I also went back to the ancient Greeks, and then up to the middle of the eighteenth century.

CI   You really should keep quiet about that, if you're a closet medievalist!

SD   [Laughs] Right. I had proposed to do the whole history. But after spending four years on it, I was only up to the middle of the eighteenth century.

CI   Giving lie to the people who said there was nothing.

SD   Exactly. I have no regrets about having continued, and in some ways I haven't finished. I wrote a book on the twentieth century called *The Biological Universe*, which was abridged and updated in *Life on Other Worlds*. My dissertation was published by Cambridge University Press and titled *Plurality of Worlds*. I think it was the first dissertation published out of the department.

CI   Have others caught up with you on recognizing the richness of this area?

SD   I think so. Other scholars have taken up the task and looked at new questions and offshoots. We showed that it was a scholarly, reputable subject. This debate was not just discussed by weird people, it was taken up by people like Kepler – well, he was somewhat weird – but it was taken up by people who were well known, and it was associated with different world views. It was associated with the ancient atomist world views of Democritus, Leucippus, Lucretius, and Epicurus.

The medieval tradition argued against other worlds, although the commentators gradually decided that it might be possible.

The heliocentric theory made it possible for the Earth to be a planet and for other planets to be potential Earths. Then Cartesian cosmology argued that there were other planetary systems outside of our own, which carried it beyond the Solar System, and that was superseded by the Newtonian tradition, which didn't *prove* that there were such systems, but it allowed them. We're now playing out that theme, finding out how solar systems are formed according to Newtonian laws. I showed that these questions were tied to real scientific traditions and that the concept of extraterrestrial life was part of the natural philosophy of the time.

CI   There were a lot of ideas floating around in ancient philosophy. What were the most prescient strands of Greek thought, as far as the possibility of other worlds?

SD   As you say, there were a lot of random ideas, like the idea that the Moon might be inhabited. The ancient atomist tradition held that the world was made of atoms, and the extraterrestrial-life debate was connected to that tradition. There were two world views among the ancient Greeks. One derives from Aristotle. The other is the ancient atomist tradition, which argued that there were an infinite number of atoms: this world is finite, and since all the atoms haven't been used up in this world, therefore there must be an infinite number of other worlds, some of them inhabited and some of them not.

CI   It's a strange chain of logic. How did they arrive at that idea?

SD   It's especially strange when you consider that they were talking about more than one cosmos. They were talking about the world as a cosmos, and in fact that's the Greek word, *kosmos*.

CI   So this is the multiverse?

SD   In a way it is, since you can't empirically verify it. They're saying that everything you can see, from the Earth to the fixed stars, is part of our *kosmos*, yet that is finite. There are an infinite number of atoms, so there must be other *kosmoi*.

CI   Was life on our planet a piece of that infinitude of places?

SD   They said that some of the worlds might have life, some might not. The idea of an infinite number of inhabited worlds was based on their idea of the physical system of the world. It wasn't a random idea. That was one Greek tradition, and the other tradition was Aristotle. Aristotle believed in a hierarchical *kosmos*, with the Earth at the center of the geocentric system and the outer sphere of fixed stars. His idea that there could be only one *kosmos* was based on his physics. You had four elements – earth, air, fire, and water – and earth naturally moved toward the center, and fire moved away from the center. You couldn't have more than one center of the world, or everything would get confused! [Laughs] Therefore there was only one world.

CI   Was it a natural consequence of the Copernican revolution for people to jump to the idea of "many worlds"?

SD   No. Copernicus himself never said a word about it. Of course, there was religion to consider there. But others, like Kepler and Galileo, followed the Copernican idea that if the Earth was no longer at the center, and it was just one of the planets, why shouldn't other planets be like the Earth? As soon as the telescope was available, they looked at the Moon and saw craters, and got the idea that it might be Earth-like. Galileo thought the Moon might be inhabited, although the beings there would be beyond our imagination. Kepler went so far as to say that there might be life on the moons of Jupiter, and even among the stars. During the long time since then that tradition has been developing – all the planetary sciences are working from that Copernican question: how similar are the planets to the Earth?

CI   With the benefit of hindsight, the recognition of the Moon as a geological place seems so powerful that it would lead to speculating about other worlds. Is there any evidence for that?

SD   Certainly not with Copernicus. It took a while. When Copernicus came out with his theory of the universe in 1543, it wasn't immediately accepted. It took fifty or a hundred years, depending on where you were. By the early seventeenth century, people got the idea that there might be life out there. Kepler, in his book called *Somnium*, or "The Dream," speculated there might be life on the Moon, because some craters were so perfectly circular that he thought they might be artificial. In a way, Kepler foreshadowed the "canals of Mars" idea. Some people consider *Somnium* the earliest science fiction.

CI   In this long period between the cementing of the Copernican idea and the beginnings of exploration of the Solar System, how did the "many-worlds" idea take root?

SD   All kinds of threads are involved. Let's start with the Newtonian tradition, which is where I ended my first book. There's no implication that planetary systems or inhabited planets are necessary from a physical point of view. Newton said that the laws of nature existed and he was criticized for taking God out of nature, because you don't need God any more – the planets move according to natural law. As compensation for that, the natural-theology tradition developed, which said that a universe full of life would be greater than an empty universe. That was a theological idea, and it had some force during the seventeenth and eighteenth centuries. Other people raised scriptural objections, because the Bible didn't say anything like that.

    The philosophical argument, the idea of the uniformity of nature, didn't develop into a substantial, provable idea until the nineteenth century, after spectroscopy was developed and astronomers saw that all the elements that exist here exist out there. They started to get some empirical data to support the idea that things out there could be the same as they are here. It's still a big jump to get to life from that. But the basic commonsense argument is very powerful. You look out there at all the stars, and you say, "Why should we be the only ones?"

    When did this become a scientific question? It depends how you define science. If you define science as natural philosophy, the way it was defined in those

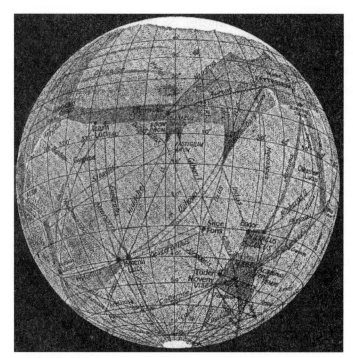

Percival Lowell's drawing of Mars at its closest approach represented markings on the surface as more linear than they are as imaged by modern space probes. His observations played into the idea of a Martian civilization trying to stay alive by using canals to transport water from polar to equatorial regions. At the time, nobody else had access to a telescope powerful enough to replicate the observations.

early days, going back to the Greeks and the Middle Ages, then this was part of a scientific tradition. But if you're talking in modern terms, then you have to wait until the twentieth century, when people started to do telescopic observations of planets in detail, and spectroscopic observations to see if they had atmospheres. At the end of the nineteenth century, Lowell had claimed the existence of canals on Mars. What *is* good evidence? That question is valid today, even when you're talking about the results of the Viking Landers, or the Allan Hills Mars meteorite ALH 84001 – even with a Mars rock in their hands and all the analytical tools of modern science at their fingertips, scientists *still* couldn't decide for a long time whether or not it had microfossils.

CI   Did Lowell and his canals cast a shadow over the subject for a while?

SD   Lowell fired the imaginations of a lot of people. The data were spurious, but a lot of people knew about Lowell and thought that even if the canals were not real, at least Mars might be Earth-like.

CI   It gets people thinking about how you would scientifically test the hypothesis. How would you decide a set of features were artificial or natural? How would you decide when a planet's too dry for biology?

SD   Exactly.

CI   Mars has always been a focus for thinking about astrobiology. Viking dashed a lot of people's hopes again, decades after Lowell.

SD   That's right. No spacecraft went to Mars after Viking for twenty years.

CI   How did astrobiology develop in those early decades of the space program?

SD   Despite the Viking Lander results, there was still a lot of research on the origins of life. The NASA exobiology program started in the early sixties, just a few years after the formation of the agency, because they were going after Mars and they knew they needed some empirical data about the possibility of life. That was when they focused on lab studies of the origins of life, prebiotic synthesis, and Miller–Urey experiments.

CI   Were those things too controversial to be funded any other way?

SD   Yes. The exobiology program developed a reputation for being forward-thinking in funding controversial topics. Viking was an 800-pound gorilla. The whole thing cost a billion dollars and it had the three life-science experiments. Two Viking spacecraft landed in 1976, and the results were in some ways ambiguous. Gil Levin, one of the Principal Investigators, still claims that he found life. Even after Viking, they could still do origin-of-life experiments and frontier research, like James Lovelock's work on Gaia, and Lynn Margulis's work on endosymbiosis. NASA funded those projects when the NSF wouldn't touch them.

CI   What were the major elements that turned astrobiology into a mature field?

SD   In the eighties, researchers realized that the Earth's primitive atmosphere might not have been reducing and the Miller–Urey experiment was called into question. Of course, Miller and the guys at Scripps still argue that it was relevant – if not for the overall atmosphere then for microclimates. Another new idea was exogenous delivery: the idea that life could have come from outer space. It was so difficult to cook up here on Earth that maybe it came on a comet, where we know there are complex organic molecules, although there's a large leap in going from organic molecules to life.

       We certainly had a watershed in the mid nineties, when we got the Mars rock and planetary systems. I was reading page proofs of *The Biological Universe* when 51 Peg was announced. I have a whole chapter on planetary systems, where I talk about how many times people claimed that they'd found other planets. I added a footnote: "Here's another announcement about a planet." But it was a real dawn.

CI   History doesn't have neat boundaries, but was there a point at which scientists started to flip to the expectation that there would be life in a lot of other places, rather than the expectation that life was rare or unique?

SD   It's hard to gauge. Certainly the late forties and early fifties was a turning point, as the tidal theory for the formation of planets lost favor and the nebular hypothesis came back into favor. By the time NASA was founded in 1958, speculation was rife about life on other worlds. As time has gone on, people have become more open to the study of exobiology. But I'm not sure how you would gauge if more of them actually think life is out there. And you've got to define what you mean by life. The universe may be full of bacteria – but that's a long way from intelligence and the SETI programs.

CI   NASA's support of programs like SETI has been controversial. Why is that?

SD   It's an interesting dichotomy. In October 1993, the federal funding for SETI was cut off when Senator Bryan, Democrat from Nevada, put in his amendment and canceled NASA's program. It had only been operational for about a year, when they were using the Arecibo dish and the Deep Space Network out in California. NASA now supports astrobiology at a level of $50 million a year. The SETI Institute is one of the NAI teams, but it's not funded for SETI itself.

CI   Have you encountered skepticism about SETI from other scientists?

SD   Yes, especially among biologists – people like Ernst Mayr of Harvard University, who died a few years ago at the age of a hundred. He believed it's too unlikely to get intelligent life; you're not going to get anything like us. I think it's possible that there's intelligence out there, but the SETI people are too sanguine in thinking we can actually communicate with them. [Laughs]

CI   Science fiction has an interesting role in the history of astrobiology, because it has sometimes led the way in ideas. Do you agree?

SD   It's fired the imaginations of a lot of people. Many NASA and SETI people with whom I've done oral history interviews were influenced to come into the field because of science fiction. It goes all the way back to H. G. Wells and *The War of the Worlds*. Good science-fiction writers, Arthur C. Clarke in particular, led the way with their sophisticated ideas. It's interesting to compare Asimov with Clarke. Isaac Asimov's most famous work is the Foundation series; there's not a single alien in it. His universe is populated by humans who migrated from the home planet. Asimov had almost no extraterrestrials in all of his writing, whereas Clarke had extraterrestrials in almost everything he wrote.

CI   Science fiction written by people who understand science is useful in teaching scientists how to think outside of the box.

SD   Sure. Scientists have written far-out science fiction, like Robert Forward. He was a well-known scientist. His *Dragon's Egg* is about life on a neutron star. [Laughs]

CI   Didn't he develop the theories of solar sails?

SD   He may have. He was involved in gravitational-wave work with Joe Weber. And of course Sagan did *Contact*.

CI   Thinking about life in the universe can slip quickly into metaphors with religion. In Spielberg movies it's direct; the analogies are literal. Sagan captured the spiritual awareness that stems from a scientific understanding of our place in the cosmos. How does astrobiology couple to metaphysical issues of how we view our place in the universe?

SD   It depends. If you're talking about microbial life, it doesn't play much into theology or religion at all. But if you're talking about intelligent life, then you've got the question of why the Earth should be central in any way. It calls into question the whole of Christian religion. Why should salvation only happen on this Earth? If you accept life on other worlds, you have theological and scriptural problems. There has been some good science fiction written on that. Maria Doria Russell's *The Sparrow* plays out some of these scenarios in interesting ways.

CI   What's your view of popular culture and the way it sees astrobiology?

SD   There's not much recognition about what's been discovered in this field. I think of exobiology and the whole idea of life in the universe as a world view. A lot of people have that world view, not so much based on the scientific evidence, but on the general philosophical idea that there are so many stars out there, and we shouldn't be the only intelligence in the universe. Pop-culture arenas like science fiction and the UFO debate are ways of working out that world view. I call it a type of cosmology, in the sense that cosmologies are world views. That's why the title of my book is *The Biological Universe*, to distinguish it from the physical universe. Is the outcome of cosmic evolution just planets, stars, and galaxies, or is it commonly life, mind, and intelligence? Those are two different world views.

CI   Maybe it's the last stage in the Copernican revolution. We're just unexceptional biological entities.

SD   It's been looked at that way, as a completion of the Copernican revolution. In a paper a couple of years ago I pointed out that there's another possible world view, what I call the post-biological universe. We may be too parochial in thinking that there are biologicals out there like us. Those last two or three terms in the Drake equation represent cultural evolution. Nobody ever takes cultural evolution into account in terms of what the extraterrestrials are going to be like. Cultural evolution dominates biological evolution on Earth and it needs to be taken into account. One possible scenario is that since any civilization that *could* improve its intelligence, *would* improve its intelligence, you may well have a universe full of artificial intelligence.

CI   Doesn't that bring you straight back to the Fermi paradox?

SD   In a way, although it's a slightly different view on the Fermi paradox. SETI people haven't thought this out seriously. How does your strategy for SETI change if you've got machines or artificial intelligence out there, rather than biologicals? For starters, they don't have to be on a planet around a Sun-like star – they could be anywhere, which doesn't help in the search. SETI people need to think out of the box and take cultural evolution into account. The weak part of that argument is that it's based on a current trend with artificial intelligence, which leads you into arguments about strong AI versus weak AI. There may be other directions that cultural evolution can take. I don't see what any civilization would consider more important than its intelligence, or its emotions. It raises interesting questions.

CI   It certainly resets how SETI should be conceived. Thinking about the next ten years, where do you anticipate the next big news in astrobiology?

SD   With the Kepler spacecraft we're going to find thousands of planets, including Earths. Finding out whether there's life on them or not will probably take longer than a decade. The sure thing is that there will be more planetary systems, and more Earth-like planets. But in terms of life, that's going to be harder. SETI is probably a long shot for getting a positive result in the next decade.

CI   I want to finish by asking about your job. For a long time, you were NASA's Chief Historian. That's a pretty cool title; was it created for you?

SD   Oh no, NASA goes back to 1958. The history office was founded in 1959 when they realized what they were doing was going to be historic. I'm the fourth or fifth Chief Historian.

CI   What were the duties?

SD   It varied. We would contract out books to be researched and written, histories related to NASA. That took a lot of time, because we'd have to go through procurement and put out requests for proposals, get proposals in, evaluate them, monitor them, and see the final products through the printing process. We also did conferences. The History Office is the NASA liaison to the humanities and social sciences. Nobody else does social science, so we reached out to the history community and the humanities. I gave a talk at the American Anthropological Association on anthropology's role in SETI.

    In addition to the books and conferences, we would answer a lot of inquiries. We also did research and writing ourselves; there is a staff of seven at headquarters. The other big thing is the NASA Historical Reference Collection. My neighbor was a Chief Archivist who had two thousand cubic feet of records.

4

Ann Druyan

Ann Druyan is an author and media producer who has long been committed to sharing the joy of astronomy with a wide audience. While married to Carl Sagan, she worked with him and wrote for the hugely influential *Cosmos* TV series and companion book, and she was a producer of the film *Contact*. She was Creative Director for the NASA Voyager Interstellar Record, humanity's first "message in a bottle" beyond the Solar System, which has a shelf life of about a billion years. For ten years she was Secretary of the Federation of American Scientists, part of her commitment to warning against abuses of science and technology. Her vocal support of scientific rationalism was recognized when she was made a Fellow of the Committee for the Scientific Investigation of Claims of the Paranormal. She has written several popular science books, two widely seen planetarium shows, and a novel. Druyan is the CEO of Cosmos Studios, which has been involved in solar sail development, as well as a range of media and humanitarian projects.

CI   Do you ever imagine what Carl Sagan's reaction would have been to recent progress in astrobiology?

AD   Yes – Carl would have been in a perpetual state of excitement. He would have enjoyed the unbelievable plethora of research that has been coming at us for the past few years. It's been so dramatic.

CI   I imagine he would have appreciated the biannual astrobiology meetings because the community is still small enough that everybody goes to every talk. It doesn't follow that unfortunate trend in science where everyone gets so specialized that they can't operate outside of their own box.

AD   Carl was such a pioneer in that he believed in being so completely, fearlessly, interdisciplinary. He viewed things holistically, integrating not just science, but literature, math, and everything else. That's a great way to begin to explore the dimensions of these overarching questions in astrobiology.

CI   You've hit on something at the core of being a scientist – what it means to venture outside of your expertise into unfamiliar territories. There will always be unanswerable questions in science and you have to be okay with that. That vagueness is something the public seems to have difficulty accepting.

AD   Yes. Intolerance of uncertainty is one of the reasons we've had this horrifying retreat to religion and mysticism. There's a craving for certainty and an intolerance for ambiguity that runs very deep. One of the massive failures of education has been the failure to inculcate in the young the respect and appreciation for what we *don't* know.

CI   And that's part of the fun of being a scientist – having this delicious sense of all these things we don't know, but either have, or are creating, the tools to find out.

AD   Exactly. The scientific methodology is so powerful. And instead of a society that teaches and shows respect for this, we have a society that seems to be driven by such an extreme hunger for answers that even when those answers have proven to be demonstrably false, they're still acceptable in some way. That's what we're up against.

CI   Since you produced *Cosmos*, you must know how strongly it motivated many people in my generation to become scientists.

AD   It was incredibly inspiring, particularly because it demonstrated the crossroads where science, ethics, history, and culture meet. Instead of a compartmentalized view of science, it was a completely integrated view of the impact of what we *know* on what we *do*. It also looked at what we *think* we know when we actually don't, and how dangerous that is. If we don't have that error-correcting mechanism, which the scientific method is so effective at employing, we make misinformed decisions with disastrous consequences.

CI   You have such a passion for science, it's difficult to believe that you didn't pursue a career in science. I've heard there's a story behind that?

AD   [Laughs] One day in my junior high school math class, when we were learning the concept of pi, I had a kind of religious experience at the moment my teacher explained the relationship of the radius to the circle. I impulsively raised my hand and blurted out, "Do you mean this is true of every circle in the universe?" I said it because pi seemed to me a universal signature, or a moment of decryption. She looked at me for a moment and her face grew angry. Then in a harsh voice she said, "Don't ask stupid questions."

CI   Wow. Ouch.

AD   I plummeted from a state of ecstatic revelation to complete humiliation. In those days, when I was twelve or thirteen, I was known as someone who cried easily, and I fled the classroom in tears. From that moment on I developed this terror of math class, which was my effective derailment from a life in science. I was immensely fortunate to meet Tim Ferris and Carl Sagan; they made me develop that part of myself.

I had, however, already developed an interest in the history of science – because it didn't require math. I was fascinated by the pre-Socratic philosophers and the idea that you couldn't use God as an explanation. I thought it was like leaving the ocean for the land.

CI   I know that you're developing a big project for science curricula. Having suffered that terrible humiliation as a young person must have driven that in some way. To me, everyone's born curious and inquisitive.

AD   We're all natural scientists. Carl wrote about how essential the gift of pattern recognition is. That's really what science is about, with the scientific method formalizing the rules – a way to not lie to ourselves as much as we've been known to do. If that's the essence of it, why is it that school science is so horribly boring? I have a teenage son, so I have some experience perusing his textbooks, and not only are they tedious, they're impenetrable. I've struggled with my son through homework assignments and I can't even tell what they're after! I've come to the conclusion that the central problem is that we compartmentalize science.

Now this is a leap – and I'm willing to find out I'm wrong – but I think the reason we don't teach our children science from day one, so they see it as a way of thinking and looking at everything, is because we don't want to look at everything scientifically ourselves. Spiritually, we teach our children a pre-Copernican view. To accept completely what science is telling us about the number of worlds in the Milky Way and the number of galaxies in the universe is to reject those illusions of centrality that are absolutely key to our spiritual beliefs. We have a society where people admit that science is going on, and they admit that their DNA is being examined – but it remains a complete abstraction, because fundamentally our world view is not only geocentric, but insanely anthropocentric.

CI   In terms of methodology, perhaps the ideal in education is the Socratic method, but that kind of intense questioning and skepticism is uncomfortable and challenging to the educator. Thinking out of the box is subversive.

AD   That's another dimension of what I'm saying. Education is partially about social control, but it's also about maintaining the agreed-upon fiction that the universe was created for us.

CI   How can we do better?

AD   If I was creating the curricula for a science class, on the first day of school the teacher would say something like this: "Shh ... I'm about to tell you a deep secret about the magical place you find yourself in." That would be at the core of the curriculum; an induction into mystery from the beginning. Not forty-five minutes of science, during which someone who is every bit as uncomfortable with science as the rest of us is expected to dole out some teaspoon of reality that's completely separate from the rest of the whole.

Religious people understand the context of community and social organization and use it to their advantage. We need to harness the power of the charismatic teacher who inducts us into the mysteries.

CI   And the power of ritual.

AD   Exactly. The problem is that we're reeling from post-Copernican stress syndrome and we haven't developed any of those things. Everyone thinks that in order to have an uplifting, revelatory experience, you have to lie to yourself; you have to make something up. As if the construct we're able to create could in any way rival thirteen and a half billion years of cosmic evolution! It's a ridiculous notion; yet we prefer to keep our rickety machinery because it's predictable. You can see the gears turning and how it comes from the previous iteration of this fantasy. The question is, why do we want the lie more than we want reality?

CI   Science is usually presented in textbooks in an austere and unengaging fashion. The story of atoms is just as amazing as any Harry Potter story – and it's true!

AD   Yes! After *Cosmos*, Steve Soter and Carl and I worked on a project called *Nucleus*. One of the stories in *Nucleus* was "A Tale of Two Atoms." It traced two atoms from the origin of the universe to now. They end up becoming a nuclear weapon somewhere. But the basic idea was to tell a story of the great adventure of two atoms; to travel with them across the great sea of time. That's not being done either, because we try to return to our old touchstones of safety, and reassure with sentimentality that we're important. That's the mythology of everything in the entertainment business and in politics.

CI   *Cosmos* changed the way science was presented in the media. Does it still influence people?

AD   Yes. One of the reasons it was such a big success was that Carl was completely high on the joy of nature. As they say – when you're in love you want to tell the world. That passion is infectious. Even though he took an enormous amount of abuse from the scientific community, he wasn't afraid to show that what we've been able to discover about nature has a tremendously spiritual component to it. In other words, though you do have to be absolutely rigorous and unflinching in terms of applying scientific methodology, once you get the fruits of the methodology, you can go nuts with joy about what it means.

CI   You mentioned that he took some professional abuse. That's surprising.

AD   It was interesting, because it was never, ever to his face. We always say "Those cutthroat academics – they can be so vicious." But in my experience, whenever

Carl Sagan, from a publicity still for the *Cosmos* TV series. The series and the accompanying book were the most successful ventures into popular science in the media in history. Ann Druyan was a producer of the series and a coauthor of the book. First broadcast in 1980, the series has been seen in 60 countries by over 600 million people (courtesy Ann Druyan).

anyone met Carl they were always affable and complimentary. It was stuff like the National Academy.

CI   Right, he was nominated but turned down. For me, even if you completely set aside his popularization, his books, and his movies, and based the case on his academic credentials and his research, it shouldn't have been an issue.

AD   I certainly agree. He was a team player for science. He would have been the first to say he had a better life than most people. He had the most realized life of any person I've ever heard about, so no real complaints.

CI   It would be a shame if popularizing science was seen as being somehow less intellectually robust. Does that perception persist?

AD   My impression is that, now, the penalties for doing the kind of stuff that Carl did have diminished. It's become much more acceptable in the scientific community.

CI   Because of him, in part.

AD   Yes. Carl's particular niche remains untenanted. That is, there's no single person who is a household name around the world; who stands not only for science, but also for the ethical and political ramifications of science and high technology, and as a voice against the powerful. I would remind you of Carl's campaign against the nonsense of "Star Wars," and against the nuclear arms race. He didn't have a publicist, but he was constantly striving to awaken people to these issues.

Jane Goodall was a great teacher, someone admired the world over and a strong voice for conservation. I think of Richard Dawkins as a voice against superstition and religion, but it's not quite the same. We were once in the state of Tamil Nadu in India, in a village so rural it didn't have a hotel, and people still recognized Carl. We couldn't go anyplace on Earth without people not only recognizing Carl, but having the same thing to say, which was: "Thank you for opening up the universe to me. I didn't think I could understand these things until you explained them. Because of you I became a science teacher, or a researcher, or went back to school."

CI   We need scientist citizens – Benjamin Franklin types – people who can speak the common language and know the technical issues. Let me move to astrobiology and talk about *Contact*, because that was yet another project that resonated with a wide audience, both the book and the film. What draws you to film as a way of expressing science?

AD   Film is where we go to worship and have transcendental experiences. Film is striking to me in that we have the capacity to create a completely immersive and convincing similitude for the universe. It's astonishing how rarely we use that capability to convey to big audiences the wonder of cosmic evolution and how much more often it's used to show car crashes and explosions, and tragic, impoverished fantasies of extraterrestrials – all in the absence of any real knowledge. They seem like transparent projections of our terror of reptiles, and nature in general. Those first few minutes of *Contact* are my favorite part of the whole movie, because they give a glimpse, an inkling, of the vastness of the universe and its magnificence.

CI   Making that movie must have been a very creative undertaking.

AD   Yes, that particular part of the movie comes from Carl standing with his tiny dictation machine pacing up and down in our house. It was probably 1985, when he and I cowrote the treatment for the film. We had a great experience with Bob Zemekis and Jodie Foster. Everyone involved was very respectful, and certainly when it coincided with Carl's illness, they were tremendously kind. That vision of the message traveling is another great teaching tool; it seems to have inspired a lot of people.

CI   It's the ultimate hook because it draws you in so immediately. Do you think it's possible, even in a popular entertainment medium, not only to use images in a scientifically inspired movie to engage and immerse, but also to reflect?

AD   Yes. Here we are in this time after Apollo, and Voyager with the pale blue dot image – just at the point you would think our concept of the world would have changed in some revolutionary way, and we'd have the planetary perspective Carl spoke about. And yet, most countries conduct themselves is if they had no concept of the size of the universe and the rarity and preciousness of life.

CI   I see your point.

AD   This is a grandiose theme, but it's one I'm constantly thinking about – how do we take these insights of science to heart before it's too late?

CI   I have one more question about *Contact*. The conventional wisdom among the scientific community is that Ellie Arroway's character is Jill Tarter. Is that right?

AD   In a way. Carl and I wrote the outline for the treatment of *Contact* in 1981, a long time ago. We did it because in the course of doing *Cosmos*, I had come across a historical figure named Hypatia.

CI   She's the one torn apart by the mob – the mathematician?

AD Yes. I was really struck by her. It was at a time when people were asking – in a smartass way – where all the female Leonardo da Vinci's of the world were: "If women are as smart as men, what the hell is going on?" Carl was sympathetic to the difficulties of being a woman with intellectual curiosity – wanting to be a scientist and not being able to. He was also inspired by Hypatia. Before we knew anything else, we decided to consciously write a movie about a female scientist who would have the equivalent intellectual and physical journey of a hero – she would be the heroine. That's all we knew. Since we were both lifelong admirers of Eleanor Roosevelt, we named her in part after Eleanor Roosevelt and in part after Voltaire, whose actual name was Arouet. She became Eleanor Arroway.

CI That's a great concoction!

AD At that time, I didn't know Jill Tarter. I've since come to know and love her, but she had no relationship to the formation of Eleanor Arroway until decades after the treatment was written. She only came into the picture when Jodie Foster asked us who she should get to know or study for her part. At that point, late in the process – the late nineties – we said Jill Tarter and Carolyn Porco. Jill is certainly a worthy model for anybody, but she had nothing to do with the creation of Eleanor Arroway as she exists in the early manuscripts of the novel and the movie.

CI Astrobiology is in an interesting phase, because many people believe we're within a decade of discovering biomarkers on another planet. Do you think that knowledge would change us culturally if we found strong evidence of microbial life somewhere else?

AD There's no telling, for a number of reasons. Most people in the United States would say contact has already been made. So, the question is – what are the lines of communication between science and the larger culture? They're not that good, because there's so much pseudoscience that blurs the line between reality and entertainment. That's not just in science; it's in politics and everywhere else, too. So the question is, how much of a sense of reality do we have as a culture? It's uneven at best.

CI So most people don't have the context in which to know whether they should or shouldn't be surprised by the discovery?

AD Yes. It's a tragedy. We have the ability to communicate information as we've never had before, but it's sadly underutilized. We don't teach critical thinking, so a lot of people don't have the ability to discern between what's happened and what's imagined.

We're undergoing a real change in consciousness, and the way it happens is difficult to chart. It's like the big earthquake in San Francisco in 1992, which scientists are now suggesting was an aftershock of the 1903 quake. There are seismic patterns that are so long-term in our culture that we can't see what's going on. It's not like you see the Apollo image of the Earth and suddenly realize that we have to love one another and take care of each other because we're on a tiny little planet. It takes a long time for this to permeate our consciousness.

I think it's happening – I'm ridiculously optimistic and frequently wrong – but I really believe that the old, authoritarian, absolutist, religious doctrines are on the ropes.

CI   That's not how it feels right now.

AD   But think of all of the science fiction you've read, that you see in movies, in television, think of all of it. How many depictions of the distant future include the gods of our time or the religious figures of the present? Maybe one or two – not many. How many times in our fantasies of going to other worlds, or of beings from other worlds coming here, do they tell us that Allah or Jesus is their God? Never. It never happens.

Right now we have this renaissance of intense traditional Christianity and Islam, but I don't think we're taking this stuff with us into the future. That's a profoundly hopeful sign, because the spasm of fundamentalism that's plaguing us now is a sign of insecurity on the part of the believers. They have to impose their belief system on us because it won't stand on its own strength; they doubt its truth. If they really believed it was true, they wouldn't need to impose it upon anyone. God would take care of it. There's some darkness before the dawn. I have a lot of hope.

CI   The embedding of science fiction and religion in popular culture is so deep that it acts like wallpaper. When I remind students that *E.T.* was a direct metaphor for the Christ story, they're surprised. They didn't notice. There are a lot of movies and works of science fiction that embed religious elements and make the alien the repository of all our fears and longings. This is our childhood; we just have to grow out of it.

AD   We don't know how to dream in any other language. We've been given a language of myth, which has its confines in certain forms with which we have to work. *E.T.* was a turn for the better, because it didn't have that insane, intense xenophobia, and it was a nice projection, a sentimental, friendly thing. It didn't posit a hostile cosmos. My own feeling about this – because everyone's free to project and imagine – is that if you go to the trouble to traverse the vast distances between the stars, then you probably have the advanced skill set that kind of technology requires. You've probably solved the issue of protein substitutes. It's not like you're coming here to eat us. I find that such a failure of the imagination.

CI   You write about spirituality in an interesting way. It's obviously not embodied in any religion, nor is it the New Age concept that lends itself to caricature and often couples to a lot of nonscientific thinking. What do you mean when you talk about spirituality, either in your work or your thinking about life in the universe?

AD   It's a concomitant of primate existence, maybe even of mammalian existence; that sense of wanting to feel a connectedness. The origin of science is Heraclitus saying, "Not I, but the universe says it: all is one." Those early stirrings are what attracted me to science in the first place. This vision of connectedness – and when I say connectedness, I don't mean it sentimentally. I mean in the way of the origin of life and the relatedness of all living things, and what we think we know about it at this point on this planet.

To me, the scientific method is a form of highly disciplined worship. It's like saying, "I know I'm imperfect and that I have a tendency to project and to lie to myself and to everyone else. So if I can create a machine that will keep me from that, I'll be forced, despite my preference for certain ideas or beliefs, to confront the universe as it is, not as I dream it will be." If I can find a machine that, over time, will winnow out that self-deception, it will be a form of humility much greater than any other form of worship I've ever encountered. It's saying, "I can't get the absolute truth, but maybe I will get this approximation of reality through science, and that's the universe."

When a man says he loves another person, but he loves her for what he wants her to be, is that real love? As real as the love you feel for who someone is? I think that analogy is translatable to the most general vision. That's why to me, science is constantly misused. The only remedy for that is many more informed decision-makers who can factor an ethical framework into what we do.

Instead we've created the opposite situation, where few of us understand how science works and most of us are likely to be intimidated by the powerful. There are few who can speak independently and have what Carl called a "bologna-detection" machine to know when we're being lied to. If we learned that skill in school, it would be great. Of course, then you couldn't tell your children we never die, and you couldn't tell them a lot of other things that a lot of people, for some reason, want to keep telling their children. That's one of the social forces that hinder that kind of change in how we educate our children.

CI   This psychological force you allude to – the post-Copernican stress syndrome – may get worse if we find that we're not unique as biological entities. That will be another blow to our self-esteem. Maybe astronomy is difficult to embrace because it's not consoling to think that we're made of star barf or that the universe emerged from a quantum fluctuation 13 billion years ago. It's quite a discipline to continue that honesty and that quest toward truth when it doesn't alleviate your existential condition.

AD   When you're little, there's an appropriate psychological stage when you think you're the center of the universe. By every definition of maturity, learning that you're not the center of the universe is adulthood. We're living transparently at childhood's end, and I don't know enough about the process to know how long it takes. We're living in a moment where we still cling to the delusion that we are the center of the universe, as we may have done when we were two years old.

CI   In our civilization, where science and technology have given us the potential to do harm but have raised a billion people out of poverty in the past century, we take it all for granted. We're not embracing science as a culture. Why is that?

AD   It's because we want to pretend that the universe was made for us and that we will never die. That is an important part of this denial that keeps us from taking science to heart.

One of the greatest statements ever made was by Karl Marx, who hasn't been given a lot of credit because of the bad things that came out of his influence. He said, "'One law for science, another for life' is, a priori, a lie." That's the big stumbling block to me. We want to have a separate law for life. We still cling to the notion of a separate creation. Yet even the people who understand that Genesis was probably state-of-the-art Babylonian science of thousands of years ago, and nothing more than that, don't want to believe that we weren't created separately from the rest of nature, and that we're not the crown of creation. We don't want to believe that we're just like the other living things on the planet and should be studied in the same way, because we're afraid to discover that we're not special. This notion of specialness is deep. As long as we keep clinging to it, we'll never be able to have a society that can use its science with wisdom.

CI   If we detect microbial life beyond Earth, it may be mundane to people because they've seen too many aliens in the movies and on television. Is there a sense in which the kinship of biology in the universe will be meaningful for us and help us to grow up?

AD   I hope it will. It's my dream for my children and their children that we'll be able to make the transition.

5

Pinky Nelson

George "Pinky" Nelson has spent over four hundred hours in space and the next twenty years in the public eye. After studying physics at Harvey Mudd College, he earned his Master's degree and PhD in astronomy from the University of Washington, where he saw the NASA flyer that changed his life. While waiting for his turn in orbit, Nelson flew in the prime chase plane during the first Shuttle mission and took pictures during its launch and return. On his first space flight, STS-41C *Challenger*, Nelson tested the Manned Maneuvering Unit (MMU), having the extraordinary experience of free flying in space. Nelson also flew on STS-61 *Columbia* and STS-26 *Discovery*. His NASA resume is impressive – Nelson earned the Exceptional Engineering Achievement Medal, the Exceptional Service Medal, and the AIAA Haley Space Flight Award, to name a few – but his recent work in education has also had a very wide impact. Nelson directed Project 2061, an educational directive aimed at making radical reforms in science, mathematics, and technology education to prepare students for the twenty-first century. He believes the entire educational system needs to be reworked from the ground up.

CI   You're an astronaut, but so much more. What's your story?

PN   I grew up in rural Minnesota. I'm from Lake Woebegone, so I was above average from the start. [Laughs] I always had lots of interests. Had I been better at it, I would have gone into baseball. I played outfield at the collegiate level but didn't have the arm to go further. But I really liked learning and reading, and I had a reasonable aptitude for mathematics.

    Sputnik was launched when I was in second grade, so I was swept up in that. Even before that I'd wanted to be an astronomer. I wanted to get out of Dodge and do something different. I could only afford to apply to two colleges, so I chose Dartmouth and Harvey Mudd. I wanted a relatively small school that was good in science and math, where I could play sports. I was poor so I had a good chance at scholarships. I was offered the same deal at both places: the choice came down to playing football at Dartmouth or playing baseball at Harvey Mudd. I decided that I was too small to play football and would probably end up in the hospital, so I went to Harvey Mudd.

    Harvey Mudd had a program run by this neat couple who taught kids to fly. The premise was that smart kids at Harvey Mudd were taking a theoretically based curriculum, and learning how to fly could allow them to apply what they were doing in the real world. During my sophomore and junior years at Harvey Mudd I got my private pilot's license with instrument ratings.

CI   You obviously took to that like a duck to water.

PN   I liked the practical side of it, having to do something with my hands and my head at the same time.

CI   Were you one of the kids who had little telescopes and traded up, or did you get into astronomy from math and physics?

PN   I had a telescope when I was a kid, a 2.5-inch refractor my great uncle bought me. I spent lots of time outside looking at the sky. I'm one of the few astronomers who knows the constellations. [Laughs] I can find my way around the sky. I did that mostly in the summer – you don't go outside much in Minnesota winters.

    I took all the astronomy courses Harvey Mudd had to offer. When it came time to go to grad school, I was recruited to Washington by George Wallerstein, who flew down in his Cessna 180. Harvey Mudd was a great preparation for grad school because I had a solid physics background.

CI   For most people, grad school is a lot of fun and a lot of work, but you immediately went into the astronaut program, so you must have had other things on your mind.

PN   I was focused on becoming an academic astronomer; that was my career path. I spent two tours down at Sacramento Peak working on solar physics. I did some work with Jacques Beckers, and worked on convection and granulation. Then I spent a year in Europe: half a year at Utrecht working on radiation-driven winds, then Gottingen for six months for my thesis. After that I got a postdoc with John Castor at Boulder, working on radiation-driven winds. My first day as a postdoc in

Boulder was the day the astronaut selections were made – so I started a new job and quit in the same day. [Laughs]

CI   What level of hope and intention did you have when you put the application in, given that you had a career track already planned out?

PN   I was perfectly happy being an academic astronomer. I had been a big fan of the space program during the sixties and saved all the pictures from *Time Magazine*. Having been a pilot, it was an exciting prospect. I thought I had a reasonable chance, and I knew that I was intellectually and physically up to the task.

CI   Has anyone who was selected ever said no?

PN   I don't think so. Brian O'Leary quit after being there for a while; he didn't like flying, ultimately. The interviews were a lot of fun. It's a week-long physical; you talk to shrinks and get poked and prodded, and then you go through a selection board – a group of astronauts and other folks who ask you a standard set of questions. Ed Gibson had just come back from Skylab and was pumped about solar physics, and was interested in the research I'd done on granulation, so that helped. George Abby, the head of the flight crew operations, the one in charge of selecting the astronauts, was from Seattle and a big baseball fan – so that helped a lot! [Laughs] My group was half scientists and half test pilots. There were only three people selected out of our group: me and Sally Ride and Jeff Hoffman, who all had astronomy backgrounds.

CI   Is the astronaut corps a unified group, or is there division between the scientists and engineers and the people who came up through the military?

PN   There wasn't any division, at least in our group. We were the first group of shuttle astronauts and there was *so* much work to do to get the shuttle flying that we all just had to jump in and do it. It was the first group with women, so the six women had to prove themselves, which they did amazingly well. The scientists had to do the same thing; we jumped in and did whatever was necessary. I wanted to do space walks, so I volunteered to work on the space suit. That was my first assignment: during the day I worked on the space suit and at night I worked on the malfunction book. I spent nights in the simulators going through malfunction procedures, so I probably got more simulator time than almost anybody – but I was doing really boring stuff. They'd fail something, and I'd take out the book and go through the procedure and diagnose what was wrong, and then make sure the procedure worked and make suggestions – and then do the next one.

CI   It was six years from when you signed up to when you first went up. Are the work and the training engrossing enough that you don't obsess about whether or when you're going to get up there?

PN   Well, you do. [Laughs] Nobody flew for the first three years, and the first four flights were already assigned, so we knew it was going to be a while down the road until anybody from our group flew. I was assigned to the Solar Max flight, which was a cool mission. It ended up being the eleventh flight.

CI   I have to ask the boneheaded question. What's it like up there?

George "Pinky" Nelson (far right) and the other crew of the Discovery Orbiter in preparation for the STS-26 mission, which would launch the NASA Tracking and Delay Satellite TDRS-3. The launch, on September 29, 1988, was the first after the Space Shuttle fleet was grounded after the 1986 loss of the Challenger Orbiter (courtesy NASA/George Nelson).

PN   [Laughs] I've never been asked that before! Here's my standard answer, which I've given a zillion times: there are three parts of being in space that are unique and different. One is the psychological aspect of being part of a well-prepared, well-trained team that includes not just the folks on board but everybody on the ground and in mission control. It's a great feeling to play on a great team.

The unique part about flying in orbit is zero gravity. It's cool to float around and experience microgravity; it's something not everybody gets to do. There are all kinds of games you can play and things you can observe and do. I have footage somewhere: you can rotate an asymmetric top around its intermediate moment of inertia axis and it's unstable, so it will flop, and you can do that really slowly in microgravity. I really enjoyed microgravity.

The third part that's unique is looking out the window. It's an incredible visual scene to be going five miles per second over the ground and look out and watch the Earth go by, see the sky, see both the northern and southern hemispheres.

CI   I imagine the people who go up are so well-trained and self-motivated that their time gets divided and they *don't* get to stare out the window and soak it in.

PN   You do. And you plan for that. I learned from Story Musgrave. In addition to the checklist you strap to your leg on the launch, he had a sheet of paper where he'd written a list of things he wanted to make sure to do while he was up there; things like, "Can I see the stars during the daytime?" I did that too, made a list of things I was curious about or that I wanted to be sure to pay attention to. We got a lot of time, because ground controlled us for sixteen hours a day and we had an eight-hour sleep period. If you weren't doing an EVA, which was really exhausting, you didn't get tired because in microgravity you don't even have to hold yourself up. We didn't sleep much; we'd get mentally tired but we could relax during the sleep period. We spent a lot of time up on the flight deck as a crew, looking out the window, talking, and enjoying the experience.

CI   You were one of the first untethered Americans in space. That must have been an amazing feeling – you're an exposed organism in a place where human organisms have never been and aren't supposed to be. What does that feel like?

PN   I can't believe they let me do that. I was on a flight with four test pilots, and for some reason Crip chose me to fly the MMU. The training was intense, so I was incredibly well trained to fly that particular machine. I had been the expert on the space suit, so I was comfortable in the suit itself. None of that technical work got in the way of the experience. I had trained myself to stop and revert to being a human being and say, "Wow, I'm going to take this all in." Flying the machine itself was incredible – and the feeling of taking off my tethers and stepping off of the payload bay.

CI   For which simulation could never really prepare you.

PN   Not at all. I was supposed to do a test to make sure all the controls were working. My test flight was to fly up to the back windows and look at Crip and Dick Scoby and Terry Hart looking out windows at me, all green with envy. [Laughs] And then I flew out of the payload bay and prepared to fly over to the satellite. It was an amazing journey. The Space Shuttle was pointed tail-to-the-Earth and bottom-forward, so I was going to fly retrograde out of the payload bay toward the satellite. I had done some rough calculations to figure out the right speed to fly over the satellite, to use the least amount of fuel to get there in a reasonable time and do what I needed to do before it got dark. I went right at sunrise. Because of the orbital dynamics, if you go too fast, you change orbits, so you have to go slowly. We'd calculated the targeting and how to point and when to do course corrections, so the mechanics lived in my fingertips; I didn't have to think about it.

I had trained myself to take my hands off the controls about halfway over and stop and look around – do a yaw back around about halfway so I could see the Space Shuttle, and then yaw back toward the satellite. I wanted to use as little propellant as I could. What an amazing sight! Here I was in my own little spaceship, about fifty yards away from the Space Shuttle, which looked like this huge spaceship, with the Earth at my feet going by at five miles a second. When the Shuttle used its control rockets there were 800-pound thrusters shooting big green flames. It was impressive. I was like a thirty-three-year-old kid thinking, "Holy smokes, I can't *believe* they let me do this!"

CI   Is there a downside where you think, "I'm never going to get this feeling again?"

PN   Maybe. But that's not the way I think. My thinking was, "This is probably the most unique thing I'm ever going to do – how cool is that?" Not that it's going to

George "Pinky" Nelson and a crewmate on an EVA during the STS 41-C mission of the Challenger Orbiter in 1984. The astronauts made critical repairs to the Solar Maximum satellite, successfully deployed the Long Duration Exposure Facility, and tested the Manned Maneuvering Units (courtesy NASA/George Nelson).

be downhill from here on out. If you want to live an interesting life, you ought to do some interesting things, and this is certainly one of them.

CI   You've been involved in aspects of spaceflight where the human presence was critical. Solar Max was a good example. What's your take on humans in space?

PN   I think humans are essential in space more from the psychological perspective than from the mechanical one. It's not cost-effective to send humans to space because you have to bring the life support system. But right now, humans are incredibly capable compared to robots. For something like servicing the Space Telescope, since we can carry a life support system into space, we might as well use it. The telescope has a higher risk if you design it *not* to be serviceable, and a much higher initial cost if you design it to be serviced by robots, because robots don't have as much capability as humans. The reason we fly people in space is because it's people who are doing the exploring. We want to be there, and we're willing to pay the price to do it.

CI   The achievements of space are spectacular, but we have wavered; our progress has been uneven. Why is that?

PN   That's the nature of human progress. Look back to the last great exploration era, when they were charting the new worlds, that progress was uneven too. There was often a long gap between sailings; but nonetheless, Western European culture spread over the new world. Over the next century or two, that's going to happen in the Solar System.

CI   Do you feel the public is committed to our future in space, or do you meet people who say we've got too many problems on Earth to waste our energy on space?

PN   I get both of those opinions. People my age see the investment because they were so excited by Apollo. The younger generation sees it less. There's so much happening because of the advance of technology that's mind-boggling and a little overwhelming – it's possible that they'll vicariously participate in so much. I don't sense this urge from the younger generation to send people to other planets.

CI   You've had the rare perspective of looking down on the planet. It's parallel to the environmental sense of the Earth – of looking at a planet with no visible borders. It gives you that extra perspective.

PN   If there's *one* common perspective that everyone who flies in space – test pilot or cosmonaut – comes back with, it's this notion of being a better steward to the planet. That's something the younger generation can appreciate. I wrote a letter to the Secretary of Energy trying to generate interest in a program with all the universities to think about environmental cleanup as an Apollo-type program, using the nuclear facilities as a model. If we show the world it can be done, and develop the technologies necessary to do it, it would be exciting, and the world needs to have it done.

CI   The space program has a strange history, born out of superpower rivalry. Do you think the box is now opening wide on commercial ventures into space?

PN   I hope so. It's a difficult task – there's a big difference in going Mach 3 like Burt Rutan did and going Mach 25 to get into orbit. That's one of the few areas where Buzz Aldrin and I agree: tourism might someday fund the space program. There are some technological hurdles we have to overcome – like how to keep paying customers from puking for twenty-four hours – but I'm excited about it. I loved the X Prize competition, I thought it was so cool that those guys were doing that technology on the cheap. I hope the government will stay out of their way at the minimum, and help them as much as they can at the maximum.

CI   Paying customers will have to acknowledge the level of risk involved. You guys knew exactly what it involved.

PN   It is a risk. The next generation of spacecraft will be even safer. But it's never going to be trivial to go from zero to Mach 25.

CI   If space becomes all about tourism and entertainment, is that backing off from the grand vision?

PN   Not a bit. I'm all for it.

CI   You sound confident that we're not going to turn our backs on space. We will eventually do more Solar System exploration.

PN   We may have to learn to speak Chinese. [Laughs] But somebody will do it.

CI   Is this an imperative for more than just one set of apes on one terrestrial planet in one part of the Galaxy?

PN   The two most important discoveries in this century will be making something alive in a test tube to show that it can happen, and then finding something alive that's had a genesis independent of Earth.

CI   Astrobiology can drive the space program. It will be hard to gather evidence from Mars, or figure out what's going on in Europa's oceans. Those are big, expensive missions.

PN   But that's one of the real, scientific justifications for doing those missions.

CI   Do you think people are interested in the "are we alone" question?

PN   A lot of people think we already know the answer. I tell audiences, "You're not going to believe me, but NASA has no secrets." [Laughs] A lot of people think NASA already knows there's life in the universe, and isn't telling.

CI   You get that question?

PN   All the time.

CI   Science fiction inspired many scientists and some astronauts too. Did it play a role in your thinking?

PN   Not much. My taste in reading fiction is more towards "real" novels. I was always a big fan of Kurt Vonnegut, and I read *Dune*, and some classic science-fiction books, but they weren't my source of inspiration. I like reality better.

CI   The space program is only fifty or sixty years old. If you were asked to speculate a hundred years downstream, how established in space will we be?

PN   A hundred years is a long time. Provided we can keep from blowing ourselves up or poisoning the planet, provided we can continue to have an economy that has resources to spend on exploration – which I think is a fairly good bet – I think we'll be on Mars. Mars isn't a pleasant place to live, so we're not going to sell condos there. I think we'll be even further out, on the asteroids and looking at the moons of Jupiter.

CI   You compared space exploration to the exploration of the Earth, which was an itch that had to be scratched for hundreds of years.

PN   I could be completely wrong, but that imperative seems to be there. The need to explore is wired into us, and the drivers that make things happen are unique individuals – Columbus and Magellan and Von Braun – who are able to harness the day-to-day, short-term worries of people and governments and societies to do the long-term work of exploration. I don't see that happening right now, but I'm confident that it will.

CI   People justify the space program and a lot of science by saying that it needs to have a practical or economic benefit, and that always helps when you're trying to fund it. But isn't exploration like science, in being curiosity-driven?

PN   Yes. A lot of times it isn't explicit, it's implicit. Von Braun wanted to explore the Solar System, go to the Moon. That was in the back of his mind, but what was in the front was: I need to build rockets, and rockets are useful.

CI   Kids live increasing fractions of their lives in virtual reality, and that projective capability is what we will presumably do more and more in space. We'll be doing sophisticated forms of remote sensing and telepresence.

PN   I have no problem with that. Other than the finite speed of light – it makes things hard to control from back on Earth.

CI   My last obvious question is: where did get your nickname?

PN   [Laughs] I got it the day I was born, from my evil father.

CI   [Laughs] It didn't have to stick, though.

PN   I was never called anything else the whole time I was growing up.

6

Neil deGrasse Tyson

    Neil deGrasse Tyson is the director of the American Museum of Natural History's Hayden Planetarium, a *New York Times* bestselling author, *People Magazine's* "Sexiest Astrophysicist Alive," a regular guest on Comedy Central's *Daily Show* and *Colbert Report*, and the man who killed Pluto. Tyson has written several books, including *Death by Black Hole and Other Cosmic Quandaries* and *The Sky Is Not the Limit: Adventures of an Urban Astrophysicist*. He was born the week NASA was founded, and raised in Manhattan's Skyview Apartments, so maybe it was always his fate to be drawn to astronomy. Tyson's charisma and accessibility have made him the public voice of all things astronomical, from books and articles to TV appearances and his regular spot on PBS's *NOVA scienceNOW*. He served on the President's "Moon, Mars, and Beyond" commission in 2004, and has been awarded nine honorary doctorates and the NASA Distinguished Public Service Medal. His professional life is largely dedicated to sharing his knowledge with others, and his sense of the excitement of astronomy is truly infectious.

CI   I want to get your take on astrobiology, as an observer and a synthesizer. Astrobiology is a field where you need both synthesis and thinking out of the box.

NT   Yes, both of those. My first encounter with people not thinking out of the box was the Mars rock, ALH 84001. I was on the *Charlie Rose* TV show a couple of days after that story broke. Among other people on the program, there was a biologist piped in by video camera. We were going over the evidence – which one had to agree seemed compelling. That spatial coincidence of reduced iron and oxidized iron would not happen in chemical equilibrium. Of course, life is always out of equilibrium. We have reduced iron and oxidized iron coexisting in our body: veins and arteries. The biologist cited Carl Sagan's mantra, "Extraordinary claims require extraordinary evidence," and this was not extraordinary evidence.

My reply was, "It's not that the possibility of life in the universe is extraordinary, because we're in the universe." We're not inventing something wildly new and different from what we've already observed. Also, life is the *simplest* explanation for what's going on in that rock. Without life, you have to cobble together an environmental story. For my money, life was not the extraordinary explanation, it was the simple one.

We ended with that famous image of the little worm-looking thing, which was not presented as primary evidence in the original paper. But it was the only photo the press had, so they led with it. It was tiny, about a hundred nanometers long, but it was intriguing. The biologist said, "That can't possibly be life!" I asked why not, and he said, "It's one tenth the size of the smallest life on Earth." I'm waiting for him to give his reply, but that was it! So it's smaller than the smallest life on Earth, but the point is, this rock is from *Mars* ... [Laughs]

CI   Mars is dry and cold and nasty. You're going to be as small as you can get.

NT   I found his argument completely unconvincing. Then I realized what our problem is. In physics, we have laws. If you come to me with a solution to a problem, I might say, "You're full of it, go away," because you're violating laws of physics for which exceptions are *out of the question*. If someone's going to discover a new law of physics, it's going to be because they're hanging out at the edges of what we've already explored, not because they did something on the tabletop. No one's going to take an inclined plane and some wooden blocks and a metal ball and come up with a new law of physics.

But in biology, there's no absolute theory of life other than evolution as an organizing principle. They can't say, "That can't be so, because it violates such-and-such theory, which has been tested and is without exception." Biology camps spring up around certain philosophical approaches. We don't get that in physics – if we do, it's fringe, like cold fusion or steady-state cosmology. But that was because we had the data wrong, and we admit it. Biologists celebrate what they call "biodiversity," but they only have a sample of one. We are all genetically linked, so we have no diversity at all.

CI You alluded to something that's a tension in astrobiology. On the one hand, there's the expectation that chemistry is universal and we shouldn't be surprised if the universe is littered with biology. Then there's the notion that we're special and life elsewhere is extraordinary, and needs extraordinary evidence to prove it.

NT Astronomers tend to be much more accommodating of unusual ideas. Here's why: every few years, if not every decade, we find *something* in the universe that just sucker punches us.

CI We get hit in the head by quasars, gamma ray bursts …

NT … black holes, dark energy, dark matter, hot big bang. There's something that knocks us over, and we just have to say, "Well, were your observation methods sound, is your logic good? Okay, that's how the universe is. Let's move on."

CI That hasn't exactly made us humble.

NT I think we're more humble about things we don't know. You don't have to scratch an astronomer too hard for them to say, "We admit that we don't know what ninety-six percent of the universe is made of."

CI That hasn't changed since Newton's famous quote about turning over pebbles while the vast sea of knowledge lay undiscovered.

NT In terms of thinking out of the box, there are a few other arguments. For example, people bucking for silicon-based life. It's intriguing, given the commonality of chemical function between carbon and silicon – except that there's more carbon in the universe than silicon. But we're not going to find a place that has more silicon than carbon, and carbon has much more interesting chemistry. If you're going to use that creative energy, invest it in another way. Let's look at the kinds of access to energy life might have. How can it tap that energy? Is it geothermal, solar, volcanic, plate tectonics, convection, water rolling down a hill? Be creative on different ways to tap energy, because we know from physical principles that energy makes things happen.

CI Right. And we might not need a star.

NT We don't need a star. That's a nice way to step out of the usual paradigm. We now know that the Solar System probably began with about thirty planets and kicked most of them out – how about the ones that got kicked out? They could have an energy source if they've crusted on top and are melting on the bottom, and we know from undersea life that you don't need light. The universe or Galaxy could be teeming with vagabond planets with life thriving under some crusty surface. Such planets can be kept warm at the bottom from decaying radioactive elements, or leftover heat from its formation.

   People shouldn't be so open-minded that their brains spill out and they lose sight of the actual laws of physics that matter and guide us. Take the size range from the smallest to the largest primate. There's a monkey that fits in a teacup and a 300-pound gorilla. That's extraordinary; it's a factor of a hundred in the same family. All life on Earth has common DNA, yet we've got people, lobsters, oak trees, and jellyfish. I lay awake at night wondering how much stranger we can get, going to another planet, than the difference between the most diverse life forms on Earth.

CI   Since life started with parasitic and symbiotic processes, and then colonies – and multicellularity is just an extension of that – what happens if symbiosis operates at the macro-organism scale?

NT   Isn't that what ants do when they build an anthill?

CI   I mean with higher organisms. Dolphins use holographic mutual imaging – they have a communal sense, as well as an individual sense. So you take the ratio of a human or primate brain compared to an individual ant brain and then scale by a second factor that's the square of the number of organisms to encapsulate the cooperative capability. It's a huge gain. We don't know what they might do.

NT   What's the difference between that and building the pyramids? Everyone works under one plan, one set of instructions; there's coherence to the mission and the directive.

CI   And we know we couldn't build the pyramids right now.

NT   [Laughs] That's like the old line, "If we can go to the Moon, why can't we do X?" Well, actually we *can't* go back to the Moon right now. I'm skeptical about finding intelligent life. I think if intelligence were evolutionarily compelling, it would have shown up much more often in the fossil record.

CI   So you're not using the Fermi argument and the fact that we haven't been visited – you're just relying on what we know about Earth?

NT   Let's benchmark ourselves. Look at how many millions of species we have – and whatever the number, ten times that are now extinct. We can probably agree that none of them had the ability to do complex math, even simple math, if that's what we define as intelligence. If intelligence was compelling in terms of evolution, it should have shown up many times, the way sight has shown up, and locomotion. Look at all the ways animals get around. Snakes have no arms, no legs; other animals have four legs, or a hundred, ten, six, eight. There's hydraulic propulsion in the jellyfish. Everything's got a way to get from A to B. The eye has evolved separately multiple times, from the octopus to the fly to the human eye. Attributes that matter for survival show up all over the place; intelligence does not. If you throw your dart into the cosmos, and invoke that statistic for the likelihood of intelligence, it drops to near zero.

CI   It's not a Darwinian attribute?

NT   [Laughs] It may even hurt you – you could get smart enough to know how to kill yourself. What gets me are the dinosaurs. The day I realized this fact, I couldn't go to sleep: the dinosaurs were around for 300 million years before they went extinct! We've been here for about a million years, but the 65 million years since the dinosaurs disappeared would have been just an eon tacked on their life span.

CI   Astrobiologists talk about timing issues. Technology has put us on the path of exponential change, which you don't see elsewhere in biology, illustrated by your example of the dinosaurs. We try to project to creatures that may be much more advanced than us, since they may exist somewhere in the Galaxy. Is this really a special time?

NT   It's easy to talk about living in special times, but the only thing special about *this* time is that …

CI   … it's *our* special time.

NT   [Laughs] It's an illusion. I have an annual report on the science results from 1850. The preface from the editor says, "At the pace that science is advancing, we'll have to split it into two halves of the year. The Foucault pendulum is *amazing*." The volume from 1869 talks about Neptune: "Isn't it amazing how this was discovered in our time?" They're apoplectic over how advanced they are, and we look back on that and say, are you kidding? You don't know what advanced *is*. Everyone gets to say that in an exponential growth of science and technology.

CI   Is induction useless in setting expectations for advanced life in the universe?

NT   We can't make empirical statements. The history of people attempting to do that says that we should just stop.

CI   [Laughs] The good thing is that you won't be embarrassed in your own time.

NT   One of my favorite books on the subject is called *Celestial Worlds Discovered*, printed by Huygens in 1696. He speculates on what life might be like on other planets. It's so hilarious that it's charming. He says, "There's a separation of land and water on other planets the way there is on Earth. And whatever those life forms are, if they are endowed with curiosity, they might want to build a ship to sail those seas. How would they move around? Well, planets have atmospheres, so there's surely movement of air from the heating of the surface. They would make sails. How would they control the sails? They would need rope, so they would surely grow hemp plants." [Laughs] So he predicts sailing ships with ropes on another planet, whatever these creatures are. He *can't* think outside that box.

CI   The first UFO sightings in Jules Verne's time were flying galleons, because that's all they could imagine. Once we could make sleek, shiny metal, we saw sleek, shiny metal things.

NT   [Laughs] Right. It's charming and unimaginative, and I'm certain we are as guilty of that today.

CI   You've had a role as an advisor for the US space program. A lot of people are impatient. They don't realize how hard it is – space travel is incredibly expensive. Is it our destiny as a society, as a culture, to explore space?

NT   Setting star travel aside for a moment, planetary travel is as hard today as sailing the open seas was to the cultures of five hundred years ago. About a dozen people crossed the ocean and came back alive five hundred years ago. That's how many have walked on the Moon. The effort and investment required by the nations that committed to it was enormous. Most of South America speaks Spanish and not Italian, because Columbus got funding from Spain. Space is an equivalent frontier in cost and in resources, in human capital and financial capital. It was never a matter of "If people just want to do it, let's do it!" That's never been the case; that's our sanitized reflection.

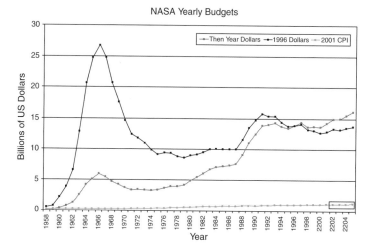

History of NASA's budget. The upper curve is in 1996 dollars, the middle curve is then year dollars, and the lower curve shows the gain in the Consumer Price Index. The big peak in space program funding was for Apollo as a direct result of the Cold War and the US–Soviet superpower rivalry. It is likely that private investment in space will eclipse government investment in a few decades. For perspective, the amount spent on manned spaceflight is less than 1% of the amount spent on the military (courtesy Benjamin Heasley and the Wikipedia Foundation).

CI   It makes exploration sound a little nobler.

NT   It takes a driver that goes far beyond the act of simply *wanting* to do it. Human nature is remarkably consistent from civilization to civilization and from era to era. There are certain things we've always done – we have spent an unlimited amount of our resources on waging war.

CI   It's a bit depressing that the glory of the Moon landings was an epiphenomenon of the quest for military dominance. And that appetite is undiminished.

NT   Undiminished. We have spent an unlimited amount of our resources in search of greater wealth. Economic return is where we get Lewis and Clark, and Magellan, all the great voyages. We've also praised entities that were perceived to be more powerful than we are. The pyramids, cathedral building – all in praise of power, royalty, and deity. That's less of a factor today than it used to be; most powerful nations are secular, which was not the case five hundred years ago. Those are the three most significant drivers in the history of the human culture. And they are responsible for every great – "great" as in scale – enterprise that humans have ever undertaken. But the biggest driver of them all is war, from which we get the Great Wall of China, the Apollo project, the Manhattan project, and many more.

So if we're going to go to Mars – which costs a lot of money – we had better put a check in one of those boxes, because if we can't, we're not going. The biggest challenge in getting humans to Mars is that expensive projects take a long time. They have to survive economic cycles, political cycles, changes in leadership. Political will goes with the breeze; it doesn't last. But the will to endure, the will to become rich and powerful, does.

CI   The worst-case scenario might be that we mess up the planet so badly that we actually *have* to go.

NT   There's a simpler argument. If China says they want to build military bases on Mars, we'll be at Mars in a year. [Laughs]

CI   Do actual *people* have to do this? With nanotech, we could send out intelligent motes, swarm them through the Solar System, and send back information. We could probably do interstellar exploration in fifty years that way, but does the buy-in only come when we send people?

NT   Yes. I wish that were not true. We put on our scientist hats and we can't believe how much money we spend putting people in orbit who want to come back! No one gives ticker-tape parades for robots; nobody names elementary schools after robots. We should not deny this fact. I see the public reaction. We hold events here with astronauts. People wait in line to get an astronaut's autograph. They don't care what the astronaut's name is; it's the *fact* that he or she's an astronaut.

CI   That's powerful – they are an emblem.

NT   They're emissaries of a frontier, in a way that robots are not.

CI   You were on the President's "Moon, Mars, and Beyond" commission. How did that group deal with the debate over manned versus unmanned spaceflight?

NT   The original presidential guiding document didn't vote one way or another, it said to investigate both and see what would be sensible – use one or the other or a combination, from one mission to the next. It was clear, clean, and simple. We took that to heart and tried to make sure that the proposed plan would be vetted by other experts, so that we could make the judgment as to the best combination of people and robots.

CI   What about the privatization of space – the idea of tourism?

NT   I think that's the single greatest potential driver of our presence in space. If we can't make that happen, I have relatively low hopes for our future in space. The capacity to launch into space exceeds the current demand. Satellites cost vastly more to make than they do to launch. Companies make a simple calculation. If they drop the launch cost in half, that's only 5 percent of their budget, because 90 percent was the R&D to make the satellite in the first place. So we need a marketplace where launch costs drop not by half, but by 90 or 95 percent. If the vehicle can carry people, I think there's an unlimited demand. When we do that, we have a free market driving vehicle safety and vehicle design. And we get to ride the fruits of entrepreneurial thinking.

CI   And it will benefit science indirectly.

NT   Yes, because now they develop patents, so now the government pays *them* to fly the astronauts. Companies start making money off that, and get to keep all their patents and then make something else with them. The new Mars plan will promote institutional utilization of resources, which is a long way of saying we would exploit the soils where we're going. Let's say I need a way of extracting water out of the soil on Mars. I do that for Mars, and now you can take that to the Sahara desert and produce water in arid places on Earth.

CI   A last question. If the cost came down reasonably, would you go into space?

NT   I would, though it's never been a life goal. I came of age in the sixties, but the astronauts were never my role models – I didn't look much like them, and they all spoke like mission control, and they had crew cuts at a time when the Broadway musical "Hair" was number one; there was a disconnect for me. I knew that going into orbit around Earth was not *space* in any grand way. The Moon was a little better, but I was already interested in the Galaxy and large-scale structure and the universe. So it's not been a goal, but it'd be fun to be weightless in the vomit comet.

CI   You were a city kid, and even with a small telescope and knowledge, you could venture to these outrageous distances that we'll *never* get to.

NT   Yes. People always asked, "Do you want to be an astronaut?" when they heard I wanted to do astrophysics. It's the assumption that anything "up" is the same. But two hundred miles up is not the same as 13 billion light years up.

7

Steven Benner

To Steven Benner, questioning the fundamentals is as essential as pursuing the unknown. He began his quest with Bachelor's and Master's degrees in molecular biophysics and biochemistry from Yale University, and a PhD in chemistry from Harvard. The recipient of numerous honors and fellowships, Benner has been a professor at the Swiss Federal Institute of Technology, Harvard University, and the University of Florida. He established several companies and organizations to ask the big questions about genetics and biology at the molecular level, including FfAME (The Foundation for Applied Molecular Evolution), EraGen Biosciences, FireBird Biomolecular, and most recently, TWIST (The Westheimer Institute of Science and Technology). Benner has done pioneering experiments that alter the basic toolkit of biochemistry to see what the effects might be on life processes. This lets him speak with authority on the plausibility and possibility of alternative biochemistries elsewhere in the universe.

CI   Tell me about the paradox in the question of life.

SB   The paradox is a matter of origins, because the chemistry is not enough. When
      you take organic molecules and cook them normally, or put energy into them,
      you don't get organized matter or life emerging – you get tar or asphalt. People
      say, "Stanley Miller did an experiment, and from that we've concluded that life
      is the intrinsic outcome of organic chemicals reacting as they do," but it's not.
      Stanley Miller showed that if you sparked energy through organic material, you
      got asphalt or tar, from which you could extract compounds or components that
      were biologically interesting, but otherwise it was just tar.

CI   If we try to do a "life in a bottle" experiment, we don't get too far. How much
      of an obstacle to our understanding of the nature of life is the fact that we can't
      connect the dots between simple chemical ingredients and cells?

SB   There are two different origins questions. We can ask: what are the exact steps
      through which life arose? That's difficult to answer. The other side of the ques-
      tion is: how do we reconcile the chemical structure and molecular physiology of
      the life that we know with what we know about organic reactivity not occurring
      subject to Darwinian mechanisms?

      We're trying to answer that second question. People complained that the NSF
      was funding incremental research and not creative, forward-looking research. So
      I called Jack Szostak and said, "Is artificial life a big enough problem for these
      people?" Jack said, "Sure," so we have Harvard, Florida, and Scripps with Jerry
      Joyce and his group as the partners in an NSF Center. We have people in the
      laboratory trying to get self-replicating systems, and self-replicating systems that
      replicate imperfectly, and then systems where the imperfections are themselves
      heritable. That's what we need chemically to get a Darwinian process going. We
      encounter many difficulties when we try to do it, because normal chemicals do
      not behave this way. DNA is very special and RNA is even more special, and if we
      change the structures of either of them only modestly, we destroy their ability
      to do rule-based templating and rule-based molecular recognition, which is the
      basis of evolution.

      The central issue is that everything that we know about organic chemistry in
      the laboratory going back two hundred years shows that we don't get Darwinian
      behavior out of chemical systems spontaneously, or even when we try. It's not
      as simple as saying that because we've got the general outline and life is the
      intrinsic outcome of chemical reactivity, that we're just left with a historical
      question and we shouldn't be all that worried if we don't solve the intermediate
      steps. We're up against a couple of problems. Take the "RNA World" hypothesis,
      which is based on rather strong evidence that there was an episode of life on
      Earth that used RNA as the only genetically encoded component of biological
      catalysis. Maybe that was the first form of life, but to get that, we need to get
      RNA to appear out of a nonbiological environment. If there's water around, it's
      very difficult.

CI   What do you make of fine-tuning or anthropic arguments about biology?

SB   There is one anthropic principle that is indisputable. If the universe had physical laws that made life impossible, we would not be here talking about it. Water is a good example. People are confusing the notion that life is adapted to water with the anthropic idea that water is adapted to life. We cannot get from one to the other. Life is clearly adapting to its environment; we see a signature of adaptation that's extremely strong. We can't expect to see a strong signal underneath that says it's the other way around.

For another example, we have a prebiotic way of making ribose and RNA, and this is the major breakthrough that was published in *Science* in 2004. We're showing that minerals containing boron magically stabilize ribose. Previously, we've shown that ribose is pretty much the only sugar that supports genetics and the general structural constraints of DNA. For some accidental reason, the sugar that's stabilized by boric-containing minerals happens to be the very sugar that best supports DNA structures and genetics. One might say, "My God, that's an argument for design." On the other hand, if we look at the solar nuclear synthesis of boron, it's one of the elements that doesn't do very well based on fundamental physical laws. With the neutron cross-section capture for boron, you shouldn't get neutrons anywhere near boron if you want to keep it around. The Solar System abundance of boron is quite low, down near scandium. We can start talking about these counterfactuals. We can marvel at how wonderful water is, because it's what supports the life we know, but the minute you put any level of thought into this problem, you can see how you or I might have done it a lot better if we had been in charge.

Then there's the problem of ice floating. When water freezes, it floats. That is an unusual property of a liquid, and people have written books about how that shows that water is ideally suited for life. However, looking at the history of the Earth for the last 50 million years, when ice floats, it's white. White reflects light, and because of ice floating, it doesn't damp perturbations in the energy input to the Earth – it amplifies them. One of our big problems right now is that the Earth has ice ages, and no one has a clear idea how it gets out of the cycle of ice ages. Antarctica started to frost over about 30 million years ago, and consequently the albedo of the Earth went up, so it frosted more. Frankly, I would rather not have ice float, or perhaps have water ice be black.

CI   Water expanding when it freezes doesn't help cells, because they burst.

SB   Exactly. We can combine all of the counterfactuals. We ask, "What do you mean, ice floats?" What we mean is that familiar, terrestrial ice floats, but ice at lower temperatures and at higher pressures is denser. On a rocky planet five times the mass of Earth, the most likely ice to form would not float. People talk about how wonderful water is because of its large liquid range. But that only means we have a large liquid range on Earth at atmospheric temperature and pressure. Water does not have any liquid range on Mars at the equator; it's like carbon dioxide on Earth. We don't consider carbon dioxide having a large liquid

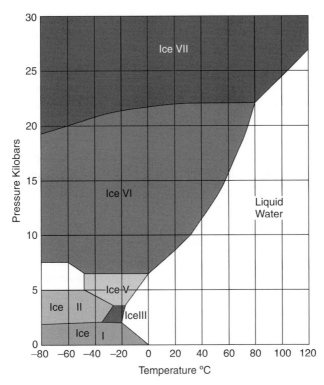

It has been argued that the special properties of water – the fact that it floats when frozen and is a good solvent – are favorable for life. But in general, as this phase diagram shows, ice has a number of crystalline forms (called polymorphs) at high pressure, and only the familiar Ice I is less dense than water. The high-pressure forms are presumed to exist in non-terrestrial settings. As far as life is concerned, water has negative aspects because it acts against the formation or stability of a number of important biochemical ingredients (courtesy Steve Dutch, University of Wisconsin-Green Bay).

range on Earth because it doesn't have a liquid state at the temperature and pressure we're used to. On Venus there is supercritical carbon dioxide, which is fluid.

You can't go far into the discussion before becoming totally anthropocentric, terracentric, and biased by what you find normal based on your own experiences. Liquid ammonia actually has a larger temperature range under reasonable pressure. With liquid water and ammonia, there is a terribly large liquid range that goes down to temperatures we expect to find on Titan.

CI   I agree with you, anthropic ideas are overplayed. Let me ask you about your own research – you're a hard-core chemist with a lab and you also do bioinformatics and biotechnology. What attracted you to these areas of research?

SB   The core paradigm of research has been expended and exhausted. There's not really a lot left to do. The problems that natural chemists worked on in the nineteenth century, and mechanistic organic chemists worked on in the fifties and sixties, are gone – the periodic table being one of them. So chemists moved into researching complex biological systems. They discovered very quickly that there are many interesting mechanistic and structural questions that were not solved as of 1975, but a whole bunch of physical technologies like X-ray crystallography and chemical methods were being brought to bear on these problems. Coming up as a student, I was able to talk about the rate of flipping of rings in proteins at the nanosecond timescale, or the positions of all the atoms to within an angstrom resolution using crystallography. We can measure the rate constants of reactions, and there's a whole field of enzyme kinetics, which I was trained in.

      One of the important questions to ask is: what is worth studying when you have this problem of physical technique outpacing the science? You can study a Picasso with an electron microscope, but what you can say about the Picasso is that it's not worth studying with an electron microscope. You may publish papers, but what you're doing is collecting information about the arcane. It has nothing to do with the art or the artist or the conception of the art, and the same principle goes for biological systems.

      By the eighties, chemistry was sufficiently well advanced that we could study the molecular systems far past the level of function. Here, function is ultimately the ability of the host organism to survive, get married and have children. We became involved in the historical side of chemistry in living systems as a research selection problem. If we understand the history of the biochemical or biomolecular system, we have a chance of knowing what in that system is actually important to biology, and what is not.

      In biology, some features of the molecular structures are random, arising from historical accidents or fixed for no good reason. Some of the structure is going to be vestigial and there will be information about past selection pressures that have vanished without changing the physiology of the organism, because it's too hard to keep the system coherent. We need to look at the molecular system and try to decide which features are adaptive and which are detrimental before we commit heavy resources to the analysis. That's how we got started, but once we ask which molecular features are important for fitness, it goes on forever.

CI   The interplay between chance and necessity is a large underlying theme of your research. How does that affect your attempts to pin down the various possibilities you just listed?

SB   It makes it harder to receive funding. We got funded by collecting more systems biology sequence data. We have collected a lot of fundamental chemical data. The human genome is nothing more than a statement about how carbon, hydrogen, nitrogen, oxygen, and phosphorus atoms are bonded in a genome. We've collected all this *data*, but little has ever emerged that resembles understanding.

      I'm not surprised or disappointed. We knew from nineteenth-century chemistry that we could isolate natural products and determine their structure, but

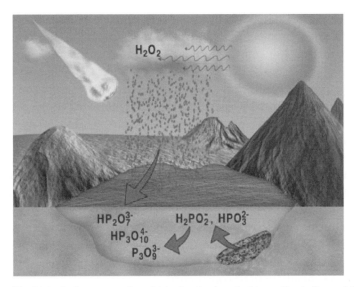

The historical sequence by which simple chemical ingredients formed the first living cells is not known. But one important step was priming the Hadean Earth with condensed phosphates, which are necessary precursors of RNA and DNA. The source of much phosphate material was meteor and comet impacts during the early bombardment of the planet (courtesy NASA/NAI).

that the biological function was not embedded in the structure in a transparent way. So why would we expect it to be transparently embedded if the natural product now happens to be a gene? The history of the genetic sequence is much more transparent in the DNA sequence than the structure of cholesterol is. I can't look at the structure of cholesterol and tell you much, other than that it's soluble in oil and not soluble in water. I really can't tell you the history of cholesterol from it. But if you give me cholesterol synth-base, an enzyme, the sequence, and the families of those, its history is accessible right away. We use that method as we try to interpret function in these biological systems. We resurrect proteins from two- or three-billion-year-old bacteria, which tell us that the protein is optimally active at 65 °C; so it probably lived at that temperature. Much of what we're trying to do is tie the historical narrative to the function.

CI Let me ask about the narrative. If you had a hundred Earths that started with the same chemical conditions and energy input from the Sun, how often would you expect to see the same chemical basis for life? How unique is life's narrative?

SB We go back and forth on this question from year to year. Given an Earth-like planet, given the general instability of the nucleic acids that are necessary for the emergence of life, given that we've tried to make about two hundred alternative versions of these molecules without finding something better than RNA, and given that the RNA seems to fall out of an interaction of organics known in the cosmos with minerals known on the planet in environments that help us solve the water problem, my view is that if we found independent life emerging on planets like "Klingon," it would still be based on ribose. It's interesting that I'm saying that, because that's not necessarily what I would have said a couple of years ago.

CI   Does that also apply to the specificity of the amino acids? Or the proteins that are used on Earth as a function of the almost infinite possible set?

SB   No, I would not say that about amino acids. If you interact with a Vulcan, I could easily imagine him having different amino acids. I'm perfectly prepared to believe he will have different bases on the nucleotides. I do believe the backbone will be a sugar, like ribose, and that he will have emerged by a similar process on an Earth-like planet where there's water. If we start talking about Titan and the organisms that might live in the supercritical hydrogen–helium mixtures of Neptune, it would be very different. For Earth-like planets of Earth-like mass plus or minus twenty percent, and a similar position around a similar star, I would expect the chemistry to be constrained.

CI   How strongly should astrobiologists be guided by the range of extremophiles?

SB   None of the extremophiles or extreme environments on Earth is in any sense extreme from the perspective of the cosmos or the Solar System. Let's talk about pH. DNA molecules, the ones we know on Earth, don't really work if we lower the pH from 7 to 5, and they don't work if we raise the pH from 7 to 10. The reason for that is that hydrogen and its positions are very important for how DNA works. If we start changing the pH, we start changing where the hydrogen atoms are.

People talk at length about Rio Tinto, where the pH is 1, and how wonderful it is that the organism has adapted, but the organism hasn't really adapted. The organism has set up a pump so that it furiously pumps the protons out of the cell, and that's how organisms in Rio Tinto survive. That's not a way to get life to emerge. I'm not going to expect life based on DNA to emerge in Rio Tinto with a pump that takes an ambient pH of 1 and decreases the proton concentrations by a factor of a million. That's something we find when an advanced life form, which evolved or emerged at a pH of 7, has moved into the Rio Tinto by accident and had to survive there.

Most extremophiles on Earth aren't actually extreme in the cosmos. We don't ever get to a point where an extremophile lives without water, we never get out of the liquid phase of water, we never go to a high or low pH, and we never have a very high concentration of salt. At the end of the day, I'm absolutely convinced that organisms on Earth are living in environments that are basically normal. Jonathan Lunine speculates about whether or not life is possible in water and ammonia at 112 K, and the answer is probably yes, but that's a real extreme environment where pH is going to be 15 to 20 – way above the pH range we know on Earth. There's no way that studying extremophiles on Earth is going to prepare us to construct a device that would detect life in that sort of environment once we get there.

CI   So astronomers should probably have a liberal definition of the habitable zone.

SB   Yes. The habitable zone concept is also terribly Earth-centric. Dave Stevenson of Caltech has pointed out that life could live off nuclear energy left over in the form of geothermal energy. It's radioactive decay of radioisotopes that were generated in the last cycle of supernovae. Once you've agreed to live off of that, there's really no constraint on where you live. Stevenson's idea was that in the early formation of planets, many rocky planets were ejected from the Solar System due to gravitational tugs and pulls. There's no reason you need a star to live if you're a microorganism living off of geothermal energy. When you're five miles down in the ocean or even five miles down in the subsurface soil or rock, you don't really care whether there's a star there or not, and you don't really know. It's an interesting problem. The ability of life to live in places where the energy source is quite different from ours on Earth means we could very easily have life scattered throughout the Galaxy on wandering, rogue planets without stars.

CI   Let me return to the specificity of life. What is the potential range of information-storing molecules for any biology? Can that be approached experimentally?

SB   Absolutely – that has been ten years of my life. We have to ask these "what if" and "why not" questions. When I said ribose is the only sugar that can support genetics in a DNA-like structure, it's because the community has made about two hundred variants of ribose, put them into the backbone, and concluded that almost all of them work *worse*. Most of them don't actually work at all, which is why we say that ribose is essential. This is why I said that if we went to Vulcan we'd expect to have the same kind of chemistry.

You can criticize it by saying, "Steve, you didn't think of the Vulcan. You haven't done everything, or anything close to everything." You can argue that there is a different chemistry we haven't thought of, so we can't draw the hard conclusion that because there is *no* other genetic form that supports the origin of life like ribose, life everywhere in the Galaxy that exists on an Earth-like planet must be built from ribose. But the facts are that there are other structures for genetic backbones and we've made them. One feature we've argued for as a universal is a repeating charge in the backbone, which is something found in natural DNA. Again, we have a good reason to argue that it's universal, because we've tried to make nucleic acids, genetics, and DNA without the charge, and failed.

CI   What about mineral or clay-type mechanisms as templates for information?

SB   We're trying. The A–T base pair is too weak, so we change the A–T base pair and we send chemists in the laboratory to make a better one, and try to get it to work better. Right now we're exploring the realm of chemical possibilities. We say, "Damn the prebiotic history," because we don't care at this point whether we can actually find evidence that the alternative structure we've thrown together existed naturally at some point in the history of the Earth. We're trying, but it doesn't work. Jack Szostak has a paper in *Science* about making membranes, so parts of the story are available. Mostly, we'd be happy if we could just get life from the standard clays we know about on the Earth, ignoring the question of whether prebiotic Earth actually had these specific clays. We'll worry about that later; right now, we can't get any clay to get RNA to have kids.

CI   Where does this leave really "out of the box" ideas about non-carbon life?

SB   I don't know. I published this paper about silicon-based life, and William Baines had an article in 2004 in which he talked about it. It's an interesting question. For example, if our principal solvent were liquid di-nitrogen II, there are some advantages to silicon compounds in that solvent. The disadvantages of silicon compounds in water go away the minute we get rid of water as a solvent. This is the kind of thing we can imagine, and I'm not ruling it out. Once we get past silicon, it's difficult to think of things that will form networks. Anybody can view an episode of *Star Trek* and see a life form that is weird, and you could argue that they can't be ruled out. There's silicon-based life and the Crystalline Entity, which is a mineral-based. There's also the Calamarain, which is a pure energy life form.

When I wrote an article for *Current Opinions* in which I had to define life, I defined it as any self-replicating chemical system that goes through evolution. The restriction to a chemical system is not because we'd rule it out if we were to come across a robotic system that was self-replicating, or a *Star Trek* character like Q, who moves in and out of the cosmos without much of a need for matter or energy. The reason why I don't include those in my definition of life is because I don't believe they exist, and I don't think they're possible. We're constantly viewing these things based on our view of what *is* and *is not*. I don't happen to think that it's possible for an organism to live in the continuum – although I find the premise hilarious – and I think that Q is one of *Star Trek*'s more entertaining organisms. Scientists are not going to suffer their definition of life to include Q.

CI   You've talked about the sculpting of life by its larger environment. The Gaia idea implies that life is symbiotic with its global environment. It's always been hard for me to understand how that works at the microscopic level of reactions, and in the Darwinian molding of organisms on the small scale. Does it make sense?

SB   Yes, it makes sense. The Gaia hypothesis is maybe two or three steps further. There's no question that life and the planet are intimately connected. We can't talk about one without the other. We're sitting on a planet that had its atmosphere poisoned by the emergence of oxygen. There's no question that geology has been greatly influenced by the emergence of photosynthesis, which generates oxygen – even down to the erosion of iron-containing rocks. We can't ignore that process as we think about the history of life. Conversely, the Earth fought back, and we can't think about life without thinking about how Earth and the cosmos are trying to kill it. Gaia goes further. It's more mystical. It's more of an intimate connection.

CI   It implies an equilibrium established at a global level. I don't understand how the small-scale equilibria that life forms establish with their immediate environment play into that, when the range of environments across the planet is so large.

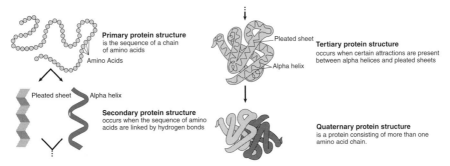

The four levels of protein structure that code for functions of all life on Earth. Terrestrial life uses about 10 000 from an almost infinite set of possible proteins, and computational methods are unable to predict the behavior of any particular structure (courtesy National Human Genome Research Institute).

SB   It's worse than that – ecologists don't know how it's possible to avoid the excesses of predator–prey relationships. It's not good for a predator to eat all the prey and drive it to extinction, because that's selfish suicide. But the mechanisms to prevent that are not terribly clear. People assume negative feedback that acts as a damper, but human society is certainly not dampening its effect on the ecosystem. Europeans came into North America, took over, and, as far as we can tell, hunted many of the animals to extinction. A lot of the big game seems to have gone because we overfed on them, and there's no mechanism in biological systems to prevent that from happening.

    This is one of the interesting aspects about ecology, which is embedded in astrobiology. People are trying to ask, "Are these mass extinctions cosmogenic? Did we just have a bad day in the Solar System or the Galaxy, or was there an interaction between the life here and the planet that was constrained by chemistry and amplified and caused things to go bad?" Some people would argue that humans are an example of that, and time will tell. The Australian Aborigines burned a relatively lush climate to flush game out and then the whole continent was driven to a desert, which wasn't the right way to do it.

CI   How useful is computational chemistry as a tool, for example, the simulation of autocatalytic networks? Does the work need to be done in the wet lab, or are purely computational techniques powerful enough to yield insights?

SB   I'll give you the party line, but there are people in the business who don't agree. My view is that we cannot compute in a way that predicts or retrodicts quantities as simple as the freezing point or boiling point of water, the solubility of sodium chloride in water, or the packing of small organic molecules into crystals. No one can do it. The theory is inadequate. Lots of people are trying to predict the three-dimensional structures of proteins by number-crunching methods, and I'd argue it's a ridiculous thing to be trying to do.

CI   If you can't do those simple things, then you're overreaching.

SB   Of course. For a protein structure, in addition to being an *N*-body problem, we also have to deal with the microscopic elements of interaction, including pluses bonding to minuses, the solubility of salt, the solution in a strong-interacting, high-dielectric solvent like water, the freezing point of water, and the packing of organic molecules into crystals. In my view, we don't have a prayer of ever getting anything serious out of pure number crunching. Now, the minute we're willing to do something creative with computers, we can make progress. We rely on computers in great detail to analyze the evolutionary history of proteins, and I'd hate to have to do it by hand. So as a tool, when we apply it with ideas, I think computation becomes very valuable.

CI   The last thing I wanted to ask you about gets straight into astrobiology. I know you're involved in discussing the bio-signatures that future Mars missions might detect. What's the best way to approach that, given that we don't know exactly what we're going to find?

SB   I'm a little frustrated, because NASA had a mantra to follow the water – which I think is what we have to do, because we can't expand our knowledge beyond that. Mars is a rocky planet like Earth, and it's not very likely that anything other than water would be the physiological solvent. The Opportunity landing site was ideal. We couldn't have wished for anything better. We expect to see borate minerals there because it's a lot like Death Valley. We expect to see ribose. We expect to see certain compounds associated with the oxidated degeneration, the degradation, of meteoritic stuff that falls in from the cosmos. There are all sorts of things we expect to see.

When I was on a panel for one of these Mars missions, we wanted one instrument. It was a laser Raman Spectrometer, which Steve Squyres had on the payload, and it was approved. One thing I learned about NASA is that you have to be at the last meeting, and if you think you're at the last meeting, they'll schedule another one after it. I had to trek all the way out to Pasadena, so it was a chore. That laser-Raman wasn't on the rover, and their alpha particle element detector couldn't get any elements lighter than chlorine. We had the geologists design instruments to detect iron and bromine. That instrument doesn't detect anything light, which includes hydrogen, carbon, lithium, boron, oxygen, nitrogen, and all the things we want to detect that are relevant to biology.

The bottom line with going to Mars – as it is with any exploratory science – is not to get too creative. We don't need to be thinking about silicon-based life and all that. What we're talking about is just getting back to Opportunity's site with the correct instruments, seeing what's there, and doing it step-by-step.

CI   What are your expectations for Europa if we actually send a mission there?

SB   It wouldn't surprise me in the slightest if there was life in the oceans under the icepack. There is presumably energy from radioactive decay. We probably don't need a lot, and we probably don't want too much.

8

William Bains

William Bains was trained in biochemistry, earning degrees from the universities of Oxford, Warwick, and Stanford. He held a lectureship at the University of Bath before leaving academia to evaluate new technology opportunities in biotech and pharmaceuticals. He led the science team at Merlin Ventures, a company created to fund biotech start-ups in the UK. He cofounded Amedis Pharmaceuticals in 1999, where he helped improve the research and production of new drugs, and in 2002 he founded Delta G Ltd, which explores a "systems biology" approach to diseases of energy metabolism. He has authored numerous papers, patents, and three books. He received the Toshiba Year of Invention Prize in 1992 and he was elected a member of the Human Genome Organisation in 1994. Bains has been a visiting professor at Imperial College in London. He took up the classical guitar as an adult and aspires to play Tarrega's *Recuerdos de l'Alhambra*. He identifies himself as a biotech entrepreneur.

CI   How did you end up hanging out with the bad crowd of astrobiologists?

WB   [Laughs] My background is in molecular biology. I did that as an academic until 1988, when I joined a technical consulting company called PA Consulting Group. This was at the start of the first biotech era in the UK. They were advising banking and marketing companies on getting into biotechnology. A lot of that was technology search and evaluation, and they also had a small development lab, primarily for developing engineering products with a biotech focus, like medical diagnostics devices. That kept my interest in science quite broad, and I developed an interest in the industrial application of molecular biology.

In 1996, the company moved towards management consulting, which I didn't find terribly interesting. I moved into a small venture capital group who were creating and funding biotech start-ups. This was a radical and unusual idea for Europe, and still is. It was a company of creation, doing new things with exciting science. A few years later, I put together ideas for a company I wanted to set up and run. I left the VC group in 2000 and became Chief Scientific Officer of that company.

CI   You've had a foot in academia, but also been part of this freewheeling world where there's a bottom line. How well do they work together?

WB   Sometimes they work together well. Other times, the interaction, creativity, and freedom of the academic world prevent people from focusing on the products, timelines, and deliverables of the commercial world. Quite often there's a real mismatch: if an experiment produces a negative result, that's really interesting to an academic. In the commercial world, negative results are only valuable in that they tell you what *not* to do. A lot of people find it difficult to switch from one to the other.

CI   Some academic scientists wander away from the commercial projects if they feel the experiments use brute force. Drug testing works through a huge parameter space, crudely and systematically. That doesn't seem intellectually interesting to a research scientist.

WB   That's entirely right. It's also expensive, and universities usually don't have the resources. It's hard to fund academic groups to work on new drug projects, even those requiring substantial intellectual input and not a brute-force approach. The biologists say it's a problem of chemistry; the chemists complain that it's not an *interesting* problem. That's what triggered my start-up, Amedis Pharmaceuticals. We developed software to predict the properties of drug molecules, which didn't work terribly well, for a variety of reasons.

We also explored candidate drug molecules using organic silicon chemistry. A few years before, I'd wondered: why not make drugs with silicon atoms? Silicon chemistry is similar to, but subtly different from, carbon chemistry, so you get subtly modulated properties. Making them is harder because it's not mainstream organic chemistry. It requires unique technology.

I decided to form a company to explore that aspect of chemistry. A small number of chemists around the world are interested in making biologically functional molecules with silicon atoms. Every new molecule we came along with was an interesting academic project, and testing those molecules and comparing them with equivalent carbon-based ones was intellectually stimulating for biologists. So the business opportunity tied in with the academic opportunity, and the academics we worked with were great about fitting in with business timescales. Our German organic silicon chemist, Professor Reinhold Tacke, would promise to produce three grams of a compound within a week, and to be sure, six days and twelve hours later, he'd phone saying, "I've got it, where do you want me to send it?"

CI   That's very business-like, working towards that kind of deadline.

WB   On the flip side, you might try an interesting experiment, and only find out after a month whether it was going to work; quite often it wouldn't. Professor Tacke was a brilliant collaborator. He enjoyed the scientific challenges and determining the biology of the molecules he made. In that case, it worked well. I've had several other cases where it failed, largely due to either the company not understanding the academic need for novelty, intellectual invigoration, and pushing boundaries, or the academics not realizing company needs. If they say they want something by Wednesday, they want it by *Wednesday*, and on cost, and what they asked for – not something else you thought was more interesting.

That's how I got involved in silicon chemistry. I've been interested in astrobiology for ages. I've been a science-fiction reader since childhood and speculated

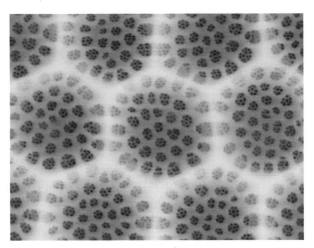

Silicon is an essential component of biochemistry and the second most abundant element in the Earth's crust. It is essential for many plants and aquatic organisms like diatoms, pictured here. In fact, the process of silification, or transporting and capturing silicic acid into silica at the deposition site, is still poorly understood, and many highly engineered silicon structures in nature have no artificial counterpart. Silicon-based biochemistry is a rich field for informing the potential range of life in the universe (courtesy M. Sumper, and Keele University).

about this casually, while lying in the proverbial bath. My interest in the biochemistry of extraterrestrial life forms was set off as a teenager by an Isaac Asimov article about the chemistry of carbon and why carbon life forms could be the only basis for biochemistry.

CI It was written in a declarative way?

WB Yes. And for good reasons. The article was called "The One And Only," in his article compilation *The Tragedy of the Moon*. It was about why life forms based on silicon are implausible. I was doing A-levels for high-school chemistry at the time and thought, "No, there must be things like silicon chains and polyphosphates and sulfates," but without any real knowledge of the chemistry. That argument stuck in my mind as over-dogmatic.

CI Surprising, even, coming from a science-fiction writer whose imagination was unbounded in other ways.

WB He'd written at least one story about a silicon-based life form. But he was a biochemist by training, so when he put his biochemist hat on, he said, "You can speculate on these things, but it's really unlikely." That stuck in the back of my head. I didn't have the tools to do anything with it. But as the Chief Scientific Officer of Amedis, I learned a lot about silicon chemistry.

I had a couple of chats about this with my German collaborator, who was a great fan of Franconian wines. We went to a terrific wine cellar in Wurzburg, where I'd go and get completely slaughtered, and we'd have scientifically wild discussions. I ended up understanding the flexibility and adaptability of silicon chemistry. The idea of looking at this from an astrobiological point of view first popped into my mind in mid 2002. For various reasons, I was being, as we say this side of the Atlantic, "encouraged to pursue other career opportunities."

CI [Laughs] Was that a positive or negative experience?

WB At the time it was extraordinarily negative. In the long term it turned out to be positive. While I was being levered out of the company I had created, I had time to play with the concept of silicon-based life. The other impetus to play with it was a paper by Norman Pace in the Proceedings of the National Academy of Sciences.

CI The one on "universal biochemistry"?

WB Yes. It was exactly like the Asimov article. But this time, with more knowledge, I read it and thought, "That *must* be wrong." It's incredibly presumptuous that a biochemist and biotechnologist who's not studied this subject at all could read a paper by someone like Norman Pace and automatically assume he's wrong. [Laughs] But I had the time to explore it. I read a few papers and did a few calculations. I was particularly interested in silicon. It was immediately clear from my knowledge of silicon chemistry that it would need to be in a nonaqueous solvent, because water would hydrolyze it. That led me into wondering what other solvents could be around, which resulted in my writing the *Astrobiology* paper called "Many Chemistries Can be Used to Build Non-living Systems."

Once I got hooked on this line of research, I wasn't going to give it up. I left the company and started my next one while pursuing this. I started out with the research and writing the papers, then ended up in the astrobiology community, rather than vice versa. It was slightly odd, reading around the chemistry, which I could grapple with reasonably well, and then trying to get into planetary physics and stellar distributions, where I was digging to the bottom depths of my technical knowledge.

CI   Did you find the community to be welcoming of different perspectives?

WB   I did find them to be open, yes. I wrote up my ideas and sent the paper off to *Astrobiology*. Mike Russell in Glasgow refereed it. He said, "You've got some interesting stuff there, William. Who knows whether it's right or not, but let's have a think about it." John Baross liked it and invited me to a workshop in Irvine, but other people thought it might be silly. Still others thought it might not be silly, who knows? It's interesting and well argued, so why not?

CI   Is that how you got to stand in front of 500 astrobiologists at their big conference in 2008?

WB   Yes. I didn't know what to expect. I could have just talked about the paper, but I wanted to do something new and extended, which I thought would take me two or three weeks to work up, but in fact took me over six months of digging stuff up and doing more calculations. At one point, I had four desktops stacked under my stairs at home, all running calculations on molecular space and diversity.

    I didn't know what people would make of it. If it was an English audience and they thought it was rubbish, there'd be a polite smattering of applause, because they weren't going to be rude about it until afterwards. But with an American audience, I didn't know whether they'd do that, or whether they'd start shouting and throwing things at me. In the end, the audience was brilliant. They were open and interested, and a lot of people wanted to talk more about it afterwards. It was terrific. Some people still thought we ought to follow the water, which means follow the carbon – that all this other stuff is cool, but it's science fiction, so let's not waste our time. They've got a sensible point. Others thought it was terrific to see people thinking about this more plausibly.

CI   There's a divergence of views. Some of the ground-based astronomers studying potential habitats are more open to unusual experiments or strange biomarkers. Those working on space missions, who get their big bucks from NASA, have a conservative approach, because they have to know exactly what they're looking for and whether or not they've found it.

WB   That's entirely reasonable, because there are limited resources and an enormous number of things you could do. With all the talk about exotic biomarkers, carbon molecules, or particular aromatic molecules, ground-based astronomers, never mind spacecraft people, say, "Guys, from here, if we pick up *water*, we're lucky." If we could land the MIT chemistry department on Titan, that'd be brilliant! But we can't, so let's look for the things we're confident are associated with life on Earth.

    If those experiments can be adapted to be a bit more general, and to look for biomarkers on a wider range, then it's worthwhile to push the people working on space missions and ask if they've thought outside the box. But they've got to build complex machines, get them across however many millions of kilometers, land them, and keep them operating. That in itself is an immensely hard task. I compare it to drug discovery and development, a field I'm familiar with. It takes 10–15 years, and hundreds of millions of dollars, to get one new drug out. The big pharmaceutical companies are naturally conservative when faced with that long timescale, that risk, and that expense.

CI   It seems like there's not much incentive in venturing out of the envelope if you're trained as a biologist; there's so much to do already. Is exobiology considered less seriously because of that?

WB   Perhaps. Biologists are aware that biology is enormously complex, and they know that if you alter one tiny aspect of a biological system, it stops working. It's not just something major, like switching carbon for silicon. If you add any one of dozens or hundreds of drug molecules at a tiny concentration to an organism, it stops working, or its function alters radically. There are multiple interactions between every component of that system. Their interactions have ramifications at different organizational levels – a cell can interact with a molecule, while a molecule can interact with an entire tissue system. There are endless feedback loops. It's hugely complex.

Biologists fall into conservative and less conservative camps, but when they've been steeped in that and know how hard it is to predict the effects of even the smallest change on a system, they get skeptical. They say, "You can speculate about silicon life swimming around in liquid nitrogen on Titan, and it sounds jolly cool, but how on Earth are you going to say anything sensible about it?"

CI   So it's maybe not an intellectual disinclination to consider it, but the difficulty of formulating a sensible question?

WB   Yes. If chemists want to make a compound out of phosphorus and nitrogen that mimics the properties of one of the bases in DNA, that's great for them. And if they want to extrapolate from that to a life form based on phosphorus instead of on carbon, they're welcome to do that. But do you have any idea how far it is from that mixture of amino acids and bases and sugars, together with a whole slew of other stuff, to the actual self-replicating biological system? It's many orders of magnitude greater in complexity. It's off the map.

CI   In the metaphor of biological landscapes, we inhabit a verdant valley, but there may be other valleys a few hilltops over with an equally excellent arrangement of metabolisms and biochemical mechanisms that make quite functional biologies.

WB   Absolutely. Some bits of biology are quite inefficient, right down to the molecular level. Our mechanism for generating energy inherently generates oxygen free radicals, and other reactive oxygen species, which cost quite a bit of metabolic energy to get rid of – and we still don't get rid of them all, which is a substantial cause of age-related damage. We live longer than mice in part because we put in a lot more effort just repairing the damage done by our basic metabolism.

I'm sure a good chemist could come up with a more efficient catalytic system. Whether it would be more efficient *and* capable of self-repair, *and* you could build

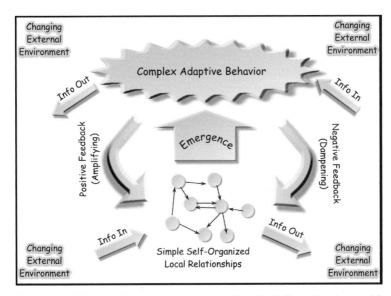

General models of biology try to capture the idea that information is stored in a dynamic and changing environment, with some processes that are damped and some that are amplified. The complexity is multiscale, such that small-scale chemical networks are part of much larger biochemical systems. The general rules that govern this behavior for biologies beyond the Earth are not known (courtesy Wikipedia).

it out of modular building blocks, is another matter. If you want to eliminate those reactive oxygen species, you wouldn't start here.

CI   Biology's good enough, it doesn't have to be perfect. Radical experiments may be mostly absent from the biological record because they were quickly erased by evolution.

WB   They'd not be viable. Humans are a good example of this. There are six and half billion of us, and any radical experiment in human physiology or anatomy or performance is almost immediately obvious. Human beings walk into the doctor. The most radical variations are a range in adult height from about 4 feet up to about 8 feet, but that scales the whole body accordingly. Occasionally you see additional or fewer digits on the limbs. There are lots of variations in color and general shape, but you don't see humans with two brains or two hearts. Two hearts would be great, but it never happens. There's an immense, interconnected complexity of life, which you can build through gradual steps from simple systems. Once you have that complex system, it's hard to completely redesign and rebuild it.

CI   What's the most sensible way to define complexity?

WB   Complexity is difficult. The information needed to describe a human being can be encapsulated in the same number of bits as the information to describe a mouse.

Fruit flies are somewhat simpler, but not a whole lot. Is it a minimum description? I don't think so, because you need to add something about the range of different components and their different behaviors in the end state. A mathematician or an information theorist would say no, it's not the simplest information you could use to describe something. But I don't think that coincides with our everyday and intuitive understanding of complexity. We think we're much more complex than rabbits, primarily because of the number of different things we can do. It's related to performance as well as static structure.

One of the things that distinguishes biological systems from nonliving systems in general is a type of complexity that I believe is encapsulated in chaos theory. Very small changes in the starting conditions can result, in a complex, nonlinear way, in large changes in the output. In living systems, large changes in the starting conditions result in almost no change in the output. Whether I'm born in Saudi Arabia or Antarctica, I'm still a human; whether I eat vegetables or meat, I'm still human. Part of complexity is the systems that allow that to happen, which feed back recursively to control themselves and their own replication.

CI   We imagine there is a set of biological worlds in the universe. We wonder how many go beyond microbial life, and how complex some of those more advanced creatures might be, after nearly 4 billion years of evolution. The metabolic and biological diversity of the early Earth was awesome back then, but not too much seems to have been added at the cellular level since.

WB   You've touched on something important: complexity depends on the level you're considering. In terms of metabolic and chemical complexity, microbial systems are at least as complex and diverse today as we are, and the overall microbial biosphere is vastly more complex and diverse than the most complex individual organisms we see. They adapt to a chemical direction and are able to do more specific chemical things, but not with larger scale, morphological structure.

CI   Does evolution have a driver towards large-scale morphological experimentation, or will it rarely go beyond a lot of small-scale microbial diversity?

WB   There hasn't been much evolution away from that, even on Earth. I don't know what fraction of the mass of the biosphere is single-cell organisms, but my guess is about 90–95 percent. You could regard plants as aberrant single-cell organisms, and animals as trivial in terms of biomass. It comes back to complexity and diversity arguments. You have distribution curves of chemical adaptability and anatomical or structural complexity or size, and when the biosphere reaches a sufficient size and energy consumption, the whole curve is high and the world gets populated by a substantial number of species, and you get things that walk around.

CI   Why do you say that metabolism is modular and limited in diversity?

WB   When Earth's primitive biochemistry found a way of putting together carbon–carbon bonds, and exploited that in a variety of contexts that became standard, the result was a model of catalysts and some basic enzyme structures that can catalyze those reactions. Those have adapted to and then extended to as many chemical contexts as are possible with that chemistry toolkit. It becomes difficult to start over and reinvent a new way of building carbon–carbon bonds.

The synthetic metabolic pathways of pyrimidines, nucleotide bases, are baroque chemistry – these huge, tangled molecules. Given the relatively small number of reaction chemistry types that life seems to have ended up with, there's no simple way of making a pyrimidine, which one could argue is the reason that pyrimidines are not part of the primitive biochemistry of life. If you look at biochemistry, the modular blocks used for transferring energy, non-carbon fragments, or nitrogen are limited. These things are recombined in different contexts to make different metabolites. The catalysts themselves are modular, a polymer, and that's probably inevitable of chemical life, because otherwise you need catalysts for everything. In order to make the catalysts, you need other catalysts, so ultimately you need catalysts that can be assembled according to predefined programs from predefined modular units.

CI   The multitude of likely habitable worlds is all going to be chemically different and have different energy sources and different contexts for biology. What are the most exciting possibilities?

WB   The Jovians and super-Jovians are unlikely to produce an environment with a suitable temperature and pressure for chemistry to happen, *and* be sufficiently stable. In our Jupiter, there's circulation of material from the middle and upper atmosphere into the deep atmosphere, where it's so hot that any complicated chemicals would break down quickly.

We haven't seen anything Earth-sized or Ganymede-sized, so we can open our minds to anything. The problem is that there's no constraint. You could imagine plants growing in liquid nitrogen, or twenty-mile deep oceans of near supercritical water. It's hard to pick one thing to search for, given that huge spectrum of possibilities. My imagination had been rather limited. But then I looked at those wonderful pictures of the surface of Titan and thought, "Rivers – brilliant! I wonder if there are fish in them!"

CI   [Laughs] The unbounded range of planet and moon environments is a real challenge: what's a problem you can bite off and work out all the details?

WB   Whatever the environment's going to be, if it's going to have life, then it's going to need solvents, and probably not a near-critical or supercritical fluid, the solvent properties of which alter dramatically when you alter the temperature and pressure even quite slightly. You're not hugely limited by available liquids, but it does limit you to cold liquids, which are going to be pretty inert, or to hot liquids which are going to be quite chemically reactive, which in turn will limit the potential chemistry – water, ammonia, sulfur, and so on.

It's worth thinking harder about carbon dioxide. In terrestrial environments we don't think of carbon dioxide as a liquid, but it liquefies at greater pressures. There is a relatively small set of opportunities you can look at, and then you go back to the astronomers and determine under what circumstances you might find a body of sufficient size to have an atmosphere, sufficiently cold that carbon dioxide can liquefy, and sufficient pressure that it will liquefy. Is that a plausible thing to look for, and what would its signature be? You feed it back to them and see if they in turn can restrict the number of options.

CI   We know there are likely to be a lot of fairly cold environments out there, the so-called cryogenic biospheres, but aren't they interesting even if you're not super-efficient or high-speed in all your reactions? You do have 12 or 13 billion years to work with.

WB   Carbon chemistry in cold environments tends to be essentially frozen, unless you excite it in some way, either with local temperature or with UV radiation. But other chemistries, which would be much too unstable at terrestrial temperatures to form sensible potential biochemistries, could react at significant rates at low temperatures. A number of spontaneous, prebiotic chemistries may be functional at a very low temperature. Then, if these other hypothetical chemistries can pass muster, what's the necessary diversity and functionality? Biochemists, molecular biologists, and myself end up saying, "Heck, the biochemistry is so complicated, how do we ever work that out?" But you can have a bash at it, throw some ideas around and see if they seem sensible. The kinetics of prebiotic chemistry is a barrier to the conceiving of life in cold environments; the kinetics of metabolism are fine, because they don't have to go into efficient catalysis that can allow other things to happen.

CI   It leaves you with a lot to work on. I hope you're going to continue to keep bad company.

WB   Oh, absolutely. Astrobiologists are a lot more fun than many of the people I meet, who tend to be investment bankers. Bankers are wonderful people, but they're not as much fun.

CI   Although you might say that they live in a much more Darwinian world.

WB   Quite.

Part II   EARTH

9

Roger Buick

Roger Buick likes nothing better than to wander in the Australian wilderness with his geologist's hammer, searching for traces of Earth's earliest forms of life. He got a BSc in zoology and geology and a PhD in geology and geophysics from the University of Western Australia, and quickly became interested in how the rock record could be read to learn about long-extinct organisms. His research has taken him to some of the world's most isolated places, like the Australian outback, the Greenland ice cap and the wilds of Northern Canada. Along the way he has been a lecturer at the University of Sydney and worked for several mining companies as an exploration geologist. Buick is a professor at the University of Washington and the senior faculty member in their PhD astrobiology program. He runs the Cooperative Facility for Isotope Research in Astrobiology, Climate, and Ecosystem Science, and has done much of his research at the NASA-funded Astrobiology Institute at the University of Washington.

CI   How did you start your higher education?

RB   I studied biology at the University of Western Australia. After that I went out and worked with a marine biologist for a year, but got sick of killing things and decided to work with things that were already dead. I did my PhD in paleontology and geology.

CI   As a kid were you always inclined towards science?

RB   Oh, yes! I always wanted to be a scientist, a natural scientist of some sort. I collected rocks and minerals, shells, butterflies, and had a fish tank. I lived in New Guinea for five years, so I had a pet crocodile, a bird of paradise, and gliding possums.

CI   What were you doing in New Guinea?

RB   My father was a librarian and set up the library at the University of Papua, New Guinea. Most of my high-school years were spent in New Guinea.

CI   How long have you been at the University of Washington?

RB   About nine years.

CI   It's a great group up there, for what you do.

RB   Oh, yeah! I had always felt like a scientific orphan until this word astrobiology was invented; I called myself a paleo-bio-geo-chemo-tectono-strato-sedimentologist who worked on the Archaean. When astrobiology came around they said, "Oh no, you're an astrobiologist," which made it a lot easier to explain to people what I actually did.

CI   Most scientists are very specialized, sometimes to the detriment of science. You're unusual in that you work in an intrinsically interdisciplinary way. What have been the pluses and minuses of that type of work in your career?

RB   It takes an awful lot longer to master an interdisciplinary area of science. But it brings insights that are much more likely to be novel, because you're bringing a unique combination of information and outlook to the area. Working somewhere in between biological and earth science is mentally stimulating. Then you have astrobiology, which involves astronomy and oceanography and aeronautical engineering and all this other stuff. I'm by no means the master of any of those other areas, but each little incremental bit of knowledge from another scientific discipline helps you see and think about things in a different way. I've been working the last couple of years with an atmospheric scientist and am now working with a postdoc on building computer box models of early atmospheres. It's something that would have been inconceivable to me five years ago.

CI   Let's start with your background in geology and paleontology. If we're asking the standard *Guinness Book of Records* question, what is the oldest rock?

RB   Well, the oldest *fragment* of rock is an individual zircon crystal that dates back to 4.4 billion years.

CI   Is that primeval?

RB   That's within 150 million years of Earth forming, so yes. That crystal actually comes from northwestern Australia. It was found in a sandstone – a metamorphosed conglomerate really – that's about 3 billion years old. The oldest *rock* is the

Acasta Gneiss in Canada. But that rock won't tell us from its formation about life because it's metamorphosed granite. Only rocks that formed at or near Earth's surface can potentially tell you anything about the early history of life. That's why we focus principally on sedimentary rocks and, to a lesser extent, volcanic rocks.

CI   I think most people don't realize just how hard it is to find old rocks ... several-billion-year-old rocks. Can you talk about how we've homed in on a rather small set of places where these valuable specimens can be found?

RB   It's not just that it's difficult to find old rocks. Anything old is difficult to find – how many people do you know that are a hundred years old? Not many. It's also difficult to find old objects that are well preserved. Old rocks tend to have been battered by the vicissitudes of existence – heating and pressure, cooking and squashing – metamorphism in other words. And most really old rocks that we find have been so metamorphosed that they're unlikely to contain any decipher-able relics of early life.

There are two or three places on Earth, mostly in the interior of continents, where rocks have been protected from plate tectonic processes for a long time and we find old rocks that are moderately well preserved. Now by "moderately well preserved," I mean by old-rock standards. Usually geologists just throw up their hands in horror and say, "These rocks are too wrecked to be able to do any-thing with." But you can do things with them if you work from first principles, are pretty cautious in your interpretations, and spend a lot of time getting all the background information absolutely right before you start trying to find life traces in them. That's basically what I've spent the last thirty years doing: getting to know one patch of rocks in the northwest of Australia so thoroughly that I can start dissecting them for signs of life.

CI   Has there been resurfacing even in these relatively well preserved regions? Or are the strata that you're interested in fairly accessible?

RB   They're reasonably accessible. In northwestern Australia the rocks are pretty well exposed. You can find old rocks that are even better exposed in South Africa on the border of Swaziland, and in the midwestern part of Greenland. In Greenland they've been recently glaciated so they crop out quite well. But northwestern Australia is a desert region, so there's not too much vegetation covering the rocks. The trouble with Australia though is that it's generally such a boring continent; everything just sat there weathering for about half a billion years. You have to be able to see through that weathering to be able to understand what the rocks were like when they were younger.

CI   Tell me about fieldwork in Western Australia. I know the population density is low compared to the rest of the world, and even to the rest of Australia. It must be pretty rugged, right?

RB   There are a couple of towns there. Also some big iron-ore deposits in the region, which means there are a couple of big mines as well. But where I work it would be fifty miles to the nearest person. It's very hot, dry, desert with hurricanes in

The Roy Hill area of the Pilbara in Western Australia, home to some of the oldest rocks on Earth. The driest parts of the Pilbara resemble the surface of Mars. Not far south of Roy Hill zircons were found that are 4.4 billion years old, dating to within 150 million years of Earth's formation (courtesy Roger Buick, University of Washington).

the summer, and plenty of humidity. In the winter it's not bad; the temperature is in the high eighties and it's quite pleasant.

CI   What type of technology do you take with you in the field to gather your samples?

RB   Just a sledgehammer – pretty low tech.

CI   Is there any dangerous wildlife you need to fend off with a sledgehammer?

RB   Snakes, camels, scorpions…

CI   Camels?!

RB   Yes. But no lions and tigers or anything like that; it's a pretty benign environment, apart from the snakes.

CI   How do you home in on the best rocks? I mean, to someone who is uninformed, the strata and rocks and outcroppings would seem to be an undifferentiated wilderness.

RB   The best strategy I've found is to map it. Geological mapping is not something that's widely practiced anymore, but I try to teach all of my students how to do it. When you create geological maps, you're forced to look at the rocks very closely and work out how they formed and how they relate to other rocks. You're also forced to cover the country, walk up every hill, and smash open every rock. That's how you find out which strata are likely to be the best for finding signs of ancient life. So, I've found mapping to be a very productive strategy.

CI   What's your ideal field team?

RB   I *like* to do it alone, but it's unwise in that sort of environment because if something happens, you fall and break a leg or something, you're dead. So usually I take someone with me, like a grad student. I spent two years up there alone mapping for a mining company and that was great fun. We had two-way radios, so if one of us didn't report in at the end of the day the rest of the mapping team would scramble and try to find them.

CI   Have you ever been in a sticky spot while you're on your own?

RB   Yes, several times. I've had vehicles blow up on me; I've had to walk about thirty-five miles out with no water – that wasn't fun. That's about the worse that's happened.

CI   I don't want to romanticize your science or make it sound mystical, but there must be at some level an art to it, at least in terms of noticing things, because there's so much that you *could* notice. Is that awareness so natural to you now that you can instantly gravitate to the most interesting outcroppings?

RB   In northwestern Australia I can, but when I go to a completely new area or different part of the world, I'm as bewildered as anyone. I recently went to Greenland for the first time, and for the first two weeks I could *not* work out what was going on. I couldn't understand the rocks, I couldn't make sense of them – it was a nightmare. I felt like a fish out of water. But I went back about two years later and things started making sense. Sometimes it helps to have a fresh pair of eyes looking at the rocks because I started seeing things that other people hadn't noticed.

CI   That transition from bewilderment to recognition and understanding is a very experiential thing. Obviously you've studied all possible geological formations and have stared at photographs of all parts of the world so, intellectually and conceptually, do you know most kinds of geologies that you might run across?

RB   Intellectually I do, but it takes a long time to train your eyes to see what's important. In different parts of the world, rocks are exposed differently. They're different colors, they show different textures, and you can't get two more dissimilar places than Greenland and the Australian outback. Australian rocks have just sat there and gradually crumbled. In Greenland, all the rocks are a grainy black color and because they've been shattered by freezing and sculpted by glaciations you get completely different surface patterns. It takes a long time to train your eyes to see through these weathering patterns.

CI   It sounds like it's an intensely experimental science – something that you couldn't convey very well in a textbook.

RB   Yeah, I've never seen a really good textbook for teaching field geology. As you say, there's an art to it; some people are naturally good at it and other people just never quite get it. You've got to be able to think in four dimensions at the same time as you're using two or three different senses to understand the rocks. You're working on scales ranging from submillimeter to many kilometers, so at any one time you're integrating several different intellectual tasks. But you're doing it without realizing it because you don't go through a conscious process when you're working.

CI   Let me ask about epistemology. In your field, the most valuable commodities – old unaltered rocks – are rare. What issues arise in interpreting the evidence?

RB   Well, the first really big concern is trying to distinguish *life* from *nonlife*. All life forms are carbon based. But when we find little one-micron spheres of carbon, how do we know whether those spheres are remnants of a microbial body or if they're a nonbiological aggregation of dead carbon?

The second difficulty is the question of how we would know independent relics of the earliest life if we found them. We assume that the process of evolution has been pretty continuous, and that very early life is going to be like primitive life on Earth now. But what if there were early failed evolutionary experiments? How would we recognize them?

Those are the two big biological issues, but then there's one more imparted by the geology: the concern of contamination. If you have a rock that's 3.5 billion years old, there's ample time in that long, *long* history for it to be contaminated with some younger biological entity. Those are the three big epistemological issues.

CI   Some scientists think life formed almost as soon as you imagine it could, given the inhospitable conditions.

RB   Everyone thinks of the origin of life as being an extremely improbable event, but if the conditions were right for it to happen once, then they might have been right for it to happen multiple times. Maybe only one strain of life ended up being the successful competitor.

CI   What's the timeframe for the formation of life?

RB   There's no shadow of a doubt that the planet was voluptuously and voluminously inhabited by diverse life forms as far back as about 3.25 billion years ago. Multiple converging lines of evidence support a wide range of metabolic styles operating at that time. At 3.5 billion years, there are still many different lines of evidence that the planet was truly inhabited.

By 3.75 billion years it starts to get difficult because we're restricted to two sets of rocks in Greenland, both of which are highly metamorphosed and deformed. It's a real effort to read anything about the history of life from them. One lot of rocks, at Isua, could well host evidence of life, but it's pretty tenuous and still open to argument. The other ones are supposed to be 3.85 billion years old, or maybe even older. But they're even more metamorphosed than the Isua rocks. I can't make anything of them – it's virtually impossible.

CI   Say a little bit about the nature of fossils. Many people know that the fossil record runs out at some point, but what type of evidence can you find that far back?

RB   There are four different sorts of fossils you can look for in the early Earth. The first are the dead bodies themselves. But that's not easy, because as far as we can tell, before about a billion years ago all life was microscopic. You can't just go to a rock and find a fossil in the field like you can with a fossil clam and say, "Ah-ha, I've found a fossil." You would have to collect rocks that *look* like they have the highest likelihood of containing body fossils and then bring them back to the lab, slice them up very thinly, and investigate them under a microscope. That's a difficult process, so not many body fossils of early microbes have been found. The oldest ones that everyone would agree on are only about 2.5 billion years old.

CI   Are they multicellular? How big are they?

RB   They're single celled! Think of a sphere of carbon, about a micron across, or a carbon tube five microns in diameter; that's the size of them. So they're exceedingly small, at the resolution limits of light microscopes. But if you *can* find them, they can tell you quite a lot. They can tell you what the organisms looked like, how they reproduced, if they happened to get fossilized in the act of reproduction. They can tell you in some cases if they're capable of movement or not. Body fossils are good to find, but I don't think we have any really convincing ones much older than about 2.5 billion years.

The next type of fossil you can go looking for is a trace fossil, which is not the actual remains of the organism itself, but something left behind as a result of the organism's activities.

CI   Does this include stromatolites?

RB   Yes, stromatolites are trace fossils. They're visible to the naked eye, but they're not the remains of the actual organism. They're like Pompeii – Pompeii contains a few body fossils of the people who built it, but what you can easily see are the buildings, not the people. Stromatolites are just like that: the "city" built by the

Cross section of a 3.47-billion-year-old stromatolite from the North Pole area of the Pilbara in Western Australia. This is one of the oldest examples of a trace fossil of a microbial colony, formed in an area that was a shallow coastal sea at the time the fossils formed. The checker squares give centimeter scales (courtesy Roger Buick, University of Washington).

organism. But this city can also tell you quite a bit about the organisms that constructed it.

CI   That's interesting. How old is the oldest stromatolite?

RB   The oldest microfossils are 2.75 or 2.5 billion years old. The oldest stromatolites
that I think everyone would accept are about 3 billion years old. But in my opinion, there are very good stromatolites that are 3.5 billion years old. They come
from Western Australia and I'm convinced that they're real trace fossils.

CI   And the other two sorts of fossils?

RB   From trace fossils, you can go down to the even more remote level of molecular
fossils. In that case, you don't get the body of the organisms, but you get a few
stray molecules from the body preserved in rock. Oil, for instance, is a classic
example of a molecular fossil of plankton. It used to be thought that hydrocarbon
molecules wouldn't survive heat and pressure and that it wouldn't be useful to go
looking for molecular fossils in really old rocks. But a few years ago we showed
that you could get hydrocarbon molecules derived from once-living organisms
in rocks as old as 2.8 billion years. These can tell you all sorts of things because
different organisms leave different molecular traces. For instance, if we were to
bury you in a rock – we'll kill you first, we'll bury you in the sediment…

CI   That's nice …

RB   … we'll heat you and squash you to a moderate degree so that all traces of
your body are destroyed. But there would be an ooze of organic molecules left
behind. From that ooze we would be able to work out that you had complex
cells with a nucleus, we'd be able to work out what sort of metabolism you had,
and we'd even be able to figure out if you had a high cholesterol level, because
cholesterol survives extremely well in geological environments. Cholesterol is
also a marker for our group of life, the eukaryotes, the organisms that have
complex cells and sex.

CI   What about bacteria?

RB   Bacteria have completely different molecular fossils. They don't produce a
diverse range of molecules like cholesterol. So if you find hydrocarbon molecules
in an old rock including cholesterol, you know that our group of life had already
evolved at that particular time. We've managed to show that our group of life, the
eukaryotes, goes back at least 2.75 billion years.

CI   When you move from a trace fossil back to the molecular level, you must lose
information. What can be learned from molecular fossils?

RB   You can't say too much about habitat, size, movement, or shape. But you can say
more about metabolism and how the organisms made their living. For instance,
cyanobacteria, which are the main photosynthesizers on the planet now – the
things that take up carbon dioxide and water and turn them into sugars and
oxygen – have a distinctive biomarker molecule. If you find that particular molecule in old rocks, and especially if you find large amounts of it, you can be
almost certain that there are oxygen photosynthesizers around, because even the
most primitive cyanobacteria have the capability of oxygenic photosynthesis. We

can find that molecule in rocks half a billion years before sedimentary rocks tell us that oxygen had started building up in the atmosphere. The ability to produce oxygen evolved well before the signs of oxygen started appearing in the geological record.

CI   Is contamination a problem in molecular fossils?

RB   Yes, because oil can flow from one rock to another. You can have recent contamination – we live in an environment where there is contamination from petroleum just about everywhere. When a diesel truck goes past your window that black smoke is blowing biological molecules over your precious sample. But the good thing is that the molecular fossils have distinctive patterns that can tell you that they're not contaminants. The oldest confirmed molecular fossils are 2.75 billion years old.

CI   So even though there are no cell walls or anything, the contamination issue can be bypassed and the biochemical tracers clearly point to living organisms. Could they be misinterpreted?

RB   Cholesterol, for instance, is strictly a biological molecule. If you find lots of cholesterol or its geological derivative in old rocks, you know it has to be biological. There is no natural process that synthesizes cholesterol from methane and carbon dioxide – it just doesn't happen. Cholesterol is a beautiful molecule. Much as you might hate it, it is a wonderful thing.

CI   What is your favorite fossil?

RB   My favorite is a 3.5-billion-year-old stromatolite. I found it during the first year of my PhD work. But what's sitting in front of me right now is an old sedimentary rock formed from evaporating seawater and it contains atomic fossils. That's going down to an even finer scale than molecular fossils. Atomic fossils are biological elements that show a distinct isotopic ratio that's different from the non-biological world. For example, the carbon that makes up your body has a different isotopic ratio than the calcium carbonate in marble.

The rock in front of me is a chunk of 3.5-billion-year-old barite from Western Australia that contains little inclusions of pyrite and sulfide. If I scratch it with a knife it will disturb the inclusions and stink of rotten eggs. Bacteria reducing sulfate to sulfide impart a signature to the isotopes of sulfur. So by sniffing this thing and also by measuring the isotopes, I can infer that there were sulfide-producing bacteria living in what was a little pond on a beach on Earth about 3.5 billion years ago.

CI   Are atomic-level tracers the primary evidence in this tantalizing zone of 3.5 to 4 billion years ago?

RB   Pretty much, apart from the stromatolites. I think there are good stromatolites at 3.5 billion years but, if you take the conservative view, the oldest are 3 billion years old. By the time we get to 3.5 or 3.75 billion years ago, we're down to a strictly atomic-level evidence of life – isotopes of carbon and sulfur indicating that biological, metabolic processes like photosynthesis or sulfate reduction were taking place.

Barite from the North Pole area of the Pilbara in Western Australia. Radioactive methods date this rock to 3.47 billion years ago. Small inclusions in the rock contain sulfur compounds that were metabolized by bacteria. The layers of the rocks were produced by sedimentation and evaporation (courtesy Roger Buick, University of Washington).

CI   How can you be sure that you know all the metabolic mechanisms that are operating? And given that the tracers become more inconsistent at the atomic level, how do you confirm biological origins when you have to worry about things like natural isotopic variations?

RB   Yes, that's very true; there are nonbiological processes that can fractionate the isotopes of biological elements. But biological fractionations are often extreme. Nonbiological processes that fractionate isotopes are usually relatively mild and are usually inconsistent in their changes from environment to environment. So we look for consistency and magnitude before we start believing that things are biological.

CI   How does the good hard evidence that you gather play into the theoretical debate over the mechanisms of the earliest life? There are a lot of ideas about whether or not metabolism came first and when the first replicating molecule appeared. What kinds of ideas do you hope to contribute to that debate?

RB   I'd like to be able to say what the earliest preserved organisms were like on Earth. And from that other scientists could extrapolate and say, "OK, maybe 3.5 billion years ago life had this kind of metabolism and was capable of living in these sorts of habitats, and had these sorts of skills." If it is indeed true that the late heavy meteorite bombardment sterilized Earth, we may have had a relatively short window for the origin of our current strain of life. So if we can get fossil evidence of life a few hundred million years after its origin, it will inform us about the origin-of-life question in general. But my guess is it's not going to work out like that because more and more we seem to be finding that life was almost modern in its sophistication, even 3 or 3.5 billion years ago.

CI   I've read that life developed an extraordinary metabolic diversity very quickly. When you consider the extremophiles, the range of habitats and range of metabolisms is amazing. How does what we're learning about early life on Earth address the issue of complexity and the developmental timescale of more sophisticated organisms? Why did it take so long to go to multicellularity?

RB   I don't know – bacteria do very well with their genetic exchange capabilities and maybe we overrate complexity just because we are complex.

CI   Also, there are organisms like stromatolites that have been successful for huge time spans without advancing beyond a certain stage. What is necessary and what is contingent in the evolution of life?

RB   That's hard to answer. It would really help if we had another strain of life evolving in parallel on Earth, or if we had evidence of an independent origin of life, or if we could compare it to life on another planet. Having just one paleontological narrative to read means that it's difficult to determine what's necessary and what's contingent.

CI   I'm reminded of the Mars rock. What's your take on the debate over life in the Allan Hills meteorite?

RB   Well, there were four lines of evidence for signs of life in that Mars meteorite, and three of them have been pretty categorically debunked. The jury is still out on the last, but the window is narrowing. I think the likelihood that the meteorite contains evidence of life is pretty low.

The same issues that face early evidence of life on Earth apply to evidence of life on early Mars. First, whether it's a uniquely biological phenomenon and second if it's truly indigenous to the rock. Contamination is a big concern. What if life on Mars was different from the life that we are used to? On a different planet, there's a much higher likelihood that life might be unfamiliar or fantastic to us. So even though we might be able to overcome problems of contamination, we might not recognize signs of life in a Mars rock because we're wearing our terrestrial-tinted glasses while looking at the fossil evidence. There's always that worry that evidence might be staring us in the face and we wouldn't recognize it.

CI   Which potentially habitable site in the Solar System is most interesting to you?

RB   Mars! Early Mars looks just like northwestern Australia 3.5 billion years ago. You know how I said everything was red? It's almost identical! I have a little toy Mars rover that I took into the field with me a couple of years ago and plunked it down in some red sand with red pebbles in the Pilbara of northwestern Australia. The picture I took of it looked just like a Mars scene! It was spectacularly similar. There was life in the Pilbara 3.5 billion years ago, and I think 3.5 to 4 billion years ago is probably the most likely place and time for life in other places of our Solar System.

CI   The "Ah-ha" moments in science are rarer than the movies and TV would have you imagine; science is more of an incremental process. But if you're fantasizing, what would be the most exciting thing for you to find that would vault your work or the evidence you work with to a different level?

RB   If the origin-of-life experiment that I'm carrying out in the basement with a graduate student actually succeeded and produced an independent origin of life that we could let evolve.

CI   It's a Miller–Urey experiment? What are you doing in your basement?

RB   We have a vat of early Archaean environment that we're letting sit there and stew to see if life could originate under plausible early environmental conditions. The original Miller–Urey experiment wasn't a very plausible early Archaean environment.

CI   Right. The fundamental problem, I guess, with all such experiments is how to mimic the huge timescales that were required.

RB   Sure, but were they really required? In the right environmental conditions, I could imagine that life might have originated pretty quickly, and not have needed hundreds of millions of years. If conditions are right, I think it can happen fast.

CI   The work you do almost sounds ideal, because you get to stay rooted in your fieldwork and add to its intellectual pursuit through many interdisciplinary strands.

RB   Yes, exactly, that's the great thing about astrobiology. It's *multi*disciplinary, not just interdisciplinary. To be an astrobiologist you need some awareness of half a dozen different disciplines.

10

Lynn Rothschild

Lynn Rothschild decided early in life that protists were cool and she has spent her career doing research on their evolution and physiological ecology. She got her undergraduate degree in biology from Yale University, a Master's degree from Indiana University and then a PhD in molecular and cellular biology from Brown University. After an NRC Fellowship at NASA Ames she stayed on and is now a Research Scientist there. She goes to Baja, California, Yellowstone National Park, and sites in New Zealand to model Precambrian ecosystems and their response to global climate-change variables, in particular ultraviolet radiation. She is a past president of the Society of Protozoologists and a Fellow of the Linnaean Society of London. Rothschild teaches astrobiology and space exploration at Stanford and Brown universities. A new realm for her is 100 000 feet up where she flies experiments on high-altitude balloons.

CI   How did you become interested in microbes?

LR   In third grade we had a unit on microscopes. I saw an amoeba for the first time, and that was all it took. I was hooked! Later, when I went to college, I wanted to study protists, but the classes that covered biological organisms at Yale were heavily oriented towards evolution. After Yale, I went to graduate school in Indiana, where molecular biology was a much bigger deal. I realized molecular biology could be an extremely important tool for evolutionary biology. I put the two together and ended up finishing my PhD at Brown University. My thesis work was on chloroplast evolution.

At a Harvard seminar, Andy Knoll came up to me and told me that NASA was interested in early evolution. Through him, I heard about postdoctoral fellowships at NASA and came to Ames Research Center at the end of 1987. I was with a group who had done a lot of work on Viking. Since that mission had been unsuccessful in detecting life, there was a general feeling of depression among the biologists there.

CI   They thought Viking was a null result? The ambiguities in the experimental data weren't taken seriously?

LR   They weren't, particularly at Ames. That doesn't mean that there was no residual interest amongst the biologists. At the same time, more and more work was being done on life in extreme environments. Around the time I was hired, the Allan Hills 84001 meteorite showed up. That sparked a lot of interest.

I'm currently a member of the Institute through the Ames team, but I'm not an employee of the Institute. I'm coeditor of the *International Journal of Astrobiology*, and ran our first three meetings in the US.

CI   Do people in the field define themselves as specialists who just happen to be working on astrobiology?

LR   Yes and no. I get paid to say I'm an astrobiologist, but if I'm giving a technical seminar in a biology department, I say that I'm an evolutionary biologist. It depends on the audience. It's almost like saying you're an American. Sure, we are all Americans, but ultimately no one's a native – not even the so-called "natives." We all come from somewhere else. The same thing is true to some extent in astrobiology. We've all migrated from another discipline.

CI   You've worked closely with planetary scientists and astronomers for a long time. Is it challenging to work in concert with people from other disciplines?

LR   Yes. There's a difference in culture, and certainly a difference in vocabulary. I was asked to give a talk at a "Geology of Mars" meeting in 2004, and after a day I noticed that all of the geologists used the word "constrain" every third sentence, which made me laugh. We *never* say that in biology, but there wasn't a single speaker who didn't use that word. There are particularities in other fields, too.

CI   Astronomers are guilty of the same habits. But since I share them, I can't think of what they might be. I hear them enough that they seem normal.

LR   I think astrobiologists are more open to people in other disciplines. Astrobiology, when it's done right, integrates data *across* fields. If you start with cosmology on day one of a conference and end up with ecology on day four, you're really having a bunch of smaller conferences while being physically registered for the same meeting. That's not astrobiology.

In my own research, I look at environmental influences on evolution. That can mean either early evolution of life on Earth, or from planet to planet. To understand the environment, I need to understand planetary science, radiation physics, and stellar evolution, and then meld that together with biology, all the way down to the molecular level.

CI   You've said that you "fell" into doing extremophile research. What has been the direction of your research in the last ten years or so?

LR   I got involved by asking what it would have been like to have lived through a day on Earth 3 billion years ago. I try to find communities of organisms that are similar to ones that were around then. That means going to an extreme environment – not because those were necessarily the conditions on the early Earth, but because it's often more difficult for nonmicrobes to live there. Essentially, I end up with a strictly microbial community.

I was interested in how photosynthesis might have happened at ten in the morning at 3 billion BC, or what the Sun might have meant to DNA back then. As it turns out, those kinds of questions are still unknown in terms of Earth today. I've gone back and forth between looking at what happens during the course of the day here, and what might have happened with different carbon dioxide and radiation levels.

CI   Is the DNA evidence of current microbes indicative of what organisms might have been like 3 billion years ago?

LR   I do as much fieldwork as I can, looking at organisms in their natural communities and environments, but we do a lot in the lab as well. The stuff in the lab isn't focused on extremophiles, but we're finding a lot of crossover. Ultraviolet radiation was probably tremendously harsh on the early Earth. It's much harsher today than we like to admit. People with zero dermapigmentosis can't go outside during the day, because they don't have DNA repair mechanisms. Imagine if the radiation flux on the Earth were orders of magnitude higher. The fact that all organisms have several backup systems for repairing damage due to radiation and oxidation means that it was an enormous issue, right from the origin of life. Radiation probably had a profound influence on evolution and on whether or not an organism could survive from planet to planet.

CI   Could the diversity of adaptation forced on organisms by extreme and varying physical conditions actually accelerate evolution?

LR   Yes, but remember that evolution is lazy. We don't do any more than we have to. Let's say an organism is living in an environment of 113 °C. It's going to have to make all sorts of changes to make sure that it doesn't denature its proteins and boil like an egg. If an organism is living at low pH, it's not as big a deal, because all it needs to do is keep the protons out. There are a handful of organisms that can live down to a pH of zero, and they all have exceedingly good proton pumps.

Some extreme environments require specialized adaptations, but others aren't that difficult once you figure out the trick.

CI   How does the cumulative insight into extremophiles over the last few decades alter the traditional notion of habitable zones?

LR   These insights expand the definition for the range of habitable environments in the Solar System and beyond. What we know about extremophiles defines the envelope for what is possible for life. It doesn't say that this is the envelope that life lives in, but if we know there's an organism on Earth that can live at 113 °C, we know there's nothing about 113 °C that would preclude life from surviving at that temperature elsewhere. This new knowledge has created a multidimensional bubble for life that continues to defy the previously set minimum requirements.

CI   Have we fully explored the boundaries of the envelope on Earth yet?

LR   We know we haven't! *Science* and *Nature* regularly update the community with new high or low temperature champions, or new high or low pH champions. Everyone wants their extremophile to make it into the *Guinness Book of World Records*. We're not at the edge of the envelope yet.

CI   What's your favorite extremophile?

LR   That's like asking who your favorite kid is!

CI   I don't know – extremophiles are too primitive to have feelings, right?

LR   Some microbes have antifreeze mechanisms that allow them to live in brine inclusions in the Antarctic ice. Those microbes are so cool!

But I also recognize that if I go out in the middle of winter in New England, the trees and birds there are alive. When I think about the emperor penguins sitting on their eggs for months on the ice, or about other animals, like polar bears, I recognize that lots of organisms are active at very low temperatures. We forget about the enormous adaptations life has taken for different environments. It isn't cheating to have fur or feathers. There isn't a rulebook for how you should live at a given temperature.

CI   One of my favorite examples is the rapid spread of early humans into incredibly inhospitable zones 10 000 to 30 000 years ago. People didn't just cling to the temperate middle of the planet. With no technology and limited shelter, they were living far into the Arctic zones. That's extraordinary human adaptation.

LR   Absolutely! Even though I'm a microbiologist, I appreciate the nonmicrobial life that makes it.

CI   The organism that withstands huge radioactive doses has always amazed me.

LR   *Deinococcus radiodurans*? Radiation destroys its DNA like that of any organism, but it has a tremendous capacity to repair the damage. It would be analogous to smashing a plate and piecing it back together. This microbe and its close relatives can live in conditions with high salt and little water, which are often correlated. While an organism has little water, it's constantly accumulating DNA damage. As it emerges from such a desiccated state, it has to fix the damage quickly. It probably developed this mechanism because it had to fix its DNA very quickly when it re-hydrated.

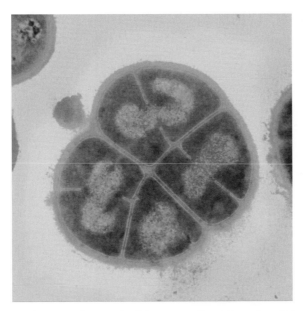

An electron micrograph of the extraordinary bacterium *Deinococcus radiodurans*. A radiation dose of 700 rads would be lethal to a human; this microbe can withstand 5 million rads. It's also tolerant to extreme UV radiation, acidity, oxidation damage, and dehydration. In each case, its strategy is to store multiple stacked copies of its DNA and utilize highly efficient repair mechanisms (courtesy Michael Daly, Uniformed Services University of the Health Sciences).

It's good to remember that a lot of organisms aren't extremophiles all the time. There are frogs that can freeze solid, but if you find one jumping around a pond in July and stick it in the freezer, it will die. To survive being frozen, the frogs start to acclimate in the fall, and switch over their biochemistry gradually. *Deinococcus* is the same way.

CI   Some of the stranger extremophiles to me are those that thrive in interior rock environments. It's hard to imagine how the water and nutrients reach there. How do these kinds of organisms adapt and survive?

LR   Actually, they're relatively comfortable because they're protected from radiation. In some cases, the humidity is higher inside these crevices than outside. They're also not exposed to the wind. The real challenge is drying out, but this tends to happen more slowly than for microbes in other environments.

CI   Larger adaptable organisms like tardigrades are fascinating. Which larger organisms are the most interesting in terms of being able to potentially hitchhike around our Solar System?

LR   Tardigrades are tremendously interesting. We've worked with them a little bit in the lab, but they're extremely difficult to culture. You can bring them out of

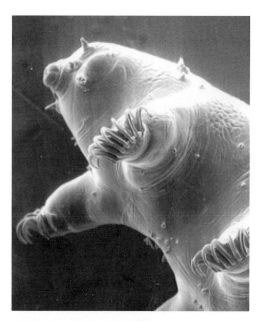

Tardigrades are water-dwelling segmented animals with eight legs. Over a thousand species are known; most are no larger than the head of a pin. They live everywhere from the high Himalayas to deep sea trenches, and can withstand temperatures near absolute zero and high doses of radiation. They can survive for decades, and perhaps much longer, in a state of suspended animation called cryptohydrosis. Many of them survived unharmed when they were launched into Earth orbit for ten days in 2007, experiencing extreme UV radiation and the near perfect vacuum of space (courtesy R. Gillis and R. Haro, University of Wisconsin-La Crosse).

desiccation and do a few experiments, but getting them to go back and forth between states is problematic.

CI   I read once that there's more DNA in a tablespoon of seawater than there is in the human genome, but since much of it comes from organisms that are difficult to culture in the lab, we don't know much about them.

LR   That type of statement gets repeated all the time – we don't know 99.9 percent of what lives here. What's happening is that people are making RNA sequences, and every time they see something new, they think it's a new organism. But because they aren't culturing them, they don't know how much natural variability there is. It may be that only a tiny fraction of new things actually belong to separate species. In other words, if they grew them in the lab, they would see the thirty sequences they just isolated all popping up in the same culture.

On the other hand, the lab where I worked in Indiana had done the work on the mating types of paramecium in the twenties and thirties. Tracy Sonneborn and his students found about a dozen species of paramecium that look identical.

However, their mating systems are completely different. For example, 1 would mate with 3 and 4, but not with 2. There are a dozen species of paramecium called cryptic species. They *look* identical, but they aren't. That's another hint that there may be more diversity than we think.

Quite honestly, it's difficult to culture organisms in the lab. It takes time and effort, and the reward is not what you would expect on a normal funding cycle. Culturing microbes isn't always easy, and most people today are not trained. When they say it's not culturable, it may just mean they threw it into two or three different media they found in a book, and it didn't work. I'm not convinced that all of these things cannot be cultured. The threshold for working on them is very low. It's easy to get the DNA and find sequences for flashy papers.

CI   Complexity is another word that's tossed around casually in conversations on biology and the evolution of life.

LR   There's no such thing as a simple organism. They're all incredibly complex.

CI   Can complexity be defined in terms of base pairs or metabolic pathways? Is it a sensible concept at all?

LR   In my mind, there's a colloquial definition and there's a scientific definition. Unfortunately, people use the word interchangeably. When we're talking about an organism being complex, I mean that it has a lot of different reactions going on. I'm not making a comment about its thermodynamic complexity. Presumably it's got a much higher level of thermodynamic complexity than a pile of mud.

I don't understand why people say that a human is more complex than a paramecium. We feel that humans must be complex because we *think*. The creationists use this type of argument. They say you can't go towards increasing complexity, so evolution couldn't take place. They confuse three or four different definitions of complexity.

CI   What's the most useful aspect of complexity for a biologist when thinking about the evolution of life on Earth?

LR   Perhaps the number of genes or proteins. The number of processes occurring within a single cell is a measure of complexity. According to that definition, things like unicellular eukaryotes are the most complex, because a single cell has to mate, eat, and excrete within a changing environment. In comparison, humans have millions of cells to deal with the same problems. On a cellular level, these single cells are more complex, but it's a difficult thing for scientists to define. Scientists are sometimes guilty of saying that trees are more complex than algae, because we see ourselves as the endpoint of evolution. In some ways we are; in other ways we're not.

CI   Why did it take so long for life to evolve to multicellular forms on Earth, despite its early start?

LR   I'm not convinced it did. Multicellularity occurred many times. There are many lines of algae and protozoa and even some bacteria that can form fruiting bodies or multicellular units. When we ask that question, we're really talking about the origin of the multicellular plants and animals – or at least the lines that became successful.

CI   And those are just a twig on the tree of life.

LR   Yes. Because multicellularity has risen so many times, I find it incredible that even those lineages didn't arise much earlier. We haven't been able to find them in the fossil record.

CI   Besides the fossil record, are there other lines of evidence, possibly biologically based, that would support this idea?

LR   Really only two points, both of which are inference rather than direct evidence. First, it should be relatively trivial to evolve multicellularity, since it involves incomplete division, and then later differentiation. Second, there are many examples of multicellularity among the protists, including especially the ciliates, chrysophytes, red algae, and green algae. Even the multicellular plants, animals, and fungi arose separately and, in the case of fungi, probably several times. Arriving at a living being is apparently much more difficult.

CI   Successful is another casually used word when talking about life on Earth. There are fairly simple organisms that have been "successful" for long periods of time.

LR   You're absolutely right. I meant successful in that they gave rise to the plants and animals that we know and love today.

CI   The inevitability of higher organisms is a tricky issue. I've noticed a culture gap between astronomers and biologists on the likelihood of life in the universe. Astronomers tend to think that not only is microbial life likely to be littered around the Galaxy and the universe, but also that higher organisms and intelligence are widespread. Evolutionary biologists have traditionally been skeptical about the prevalence of large, intelligent organisms. Where does this discrepancy arise?

LR   I suspect that we're not alone in terms of intelligence. If we are, we have a huge responsibility in taking care of our universe.

CI   We don't seem to be doing very well.

LR   We're not even doing well with our own planet.

CI   Given the organisms that can hibernate or go into stasis to survive, what are the possibilities for life hitchhiking around solar systems, ours or others?

LR   I'm part of a group that is looking at that specifically. We're studying organisms that can survive high levels of radiation and desiccation as potential models for organisms that could hitch a ride from planet to planet.

CI   Does your experience with extremophiles, and your knowledge of the way life on Earth evolved, make you optimistic about the existence of microbial life in the growing number of distant solar systems?

LR   Yes. I can't believe that life didn't occur elsewhere. It seems so easy. We don't know all the steps, but we know many of them, and the building blocks are out there. If we *are* alone, it's absolutely stunning. We have to look around and say, "Wait a minute! We're special!" In that case, it's either harder than we thought, or there's something unique about what happened on Earth.

CI   Is there any point in speculating about alternative biologies or mechanisms for life that are nontraditional – that is, biologies not based on carbon or DNA?

LR   I don't think there's any point in looking at biologies that are not based on carbon. You're on solid ground with carbon chemistries. That's what's out there. When you start to talk about something specific like DNA, it's up in the air. Steve Benner is doing a lot of very cool stuff with alternate molecules at the University of Florida. Leslie Orgel has also thought about this question for many years. He originally thought about substituting glycerols; now he has another favorite compound for a pre-DNA molecule. Limiting ourselves to particular structures of DNA and RNA is foolish, but holding to organic chemistry and water as a solvent is not so unreasonable.

CI   What projects do you want to be working on in five or ten years?

LR   The more I delve into radiation, the more interesting it becomes. Evolution is basically heritable variation and selection, and radiation is involved in both. We are looking at the origin of changes that become pathologies in humans – such as skin cancer and cataracts – which may have had a reason to exist in early evolution. Understanding how the physical environment influences evolution is a great challenge. Viruses will probably play some role in what we are doing.

My real love is still working with microbes, particularly protists. I can't imagine doing projects that don't involve microbes. In the end, most biologists have a favorite organism.

CI   Your extremophiles have spawned a burgeoning biotech industry. Have you ever been tempted leave academia and start a company?

LR   No, I haven't been tempted to do a start-up. I wouldn't be the best person for it because I love pure research. I had an advisor years ago who said that one of the cool things about having your own lab in academia is that you change the course of your research just because of something you read in a paper. You don't think something's quite right, or you pursue a new idea, because you can always go back to your lab and test it. That's amazing! I adore that academic freedom.

11

John Baross

When a research team studying deep sea vents off the coast of the Galapagos
Islands turned up thriving ecological communities, John Baross was eager to get his hands
on organism samples any way he could – including those the team had preserved in the
nearest bottle of tequila. Baross earned his PhD from the University of Washington, where
he is a professor of biological oceanography. He has had seventeen research projects with
six different teams since the NASA Astrobiology Institute was established. He specializes
in extremophiles, particularly those that survive in hydrothermal vents, which he divides
into "weird" life and "seriously weird" life. Baross was cochair of the Committee on the
Origins and Evolution of Life, and led a National Research Council study titled "The Limits
of Organic Life in Planetary Systems." He applies his research expertise in extreme environments to speculation about the potential for life on other planets, and enjoys exploring
his field's more philosophical avenues. Baross edited, with his colleague Woody Sullivan, a
recent book on astrobiology for Cambridge University Press entitled *Planets and Life: The
Emerging Science of Astrobiology*.

CI   How did you become an astrobiologist?

JB   I started out as a chemistry major wanting to go to medical school. I was part of the molecular biology revolution; I was really fascinated with biochemistry and molecular genetics. I took my first course in microbiology in my junior year and loved it. In that same time frame there was money coming into the microbiology department as part of the pre-Viking experiments. In addition to people going to Antarctic dry valleys and looking for microbes, there was a program to look at what kind of microbes contaminate the nose cones of rockets and whether or not these organisms could survive in space. At the age of twenty, I went to Lompoc with a professor and swabbed the nose cone of a Delta rocket just before it took off, and that was my first astrobiology research project.

      I wanted to learn more about environmental microbiology. In particular, I was interested in the role of viruses in the environment, because viruses at that point were studied primarily as genetic tools. I felt we needed to determine their real role in the environment, besides being a genetic tool or killing off organisms. I went to the University of Washington because they had marine biology and they let me do what I wanted. So I pursued marine microbiology – but always thinking that, somewhere along the line, this astrobiology thing was going to be important. As an undergraduate I had a sign made, "cosmo-geo-microbiology," and I put it in my office. I still have it in one of my labs.

CI   When did it become clear that the microbial life forms under study were the tip of a much larger iceberg of organisms that were difficult to culture, organisms we knew very little about?

JB   I don't think the community was persuaded until the late seventies. When I was a first-year graduate student, I was told by an eminent professor in oceanography, "We will not teach anything about microbiology because it's insignificant; those organisms are small and their numbers are small so they don't do anything in the oceans." And there I was as a first-year graduate student, wanting to do marine microbiology! It wasn't until new methods were developed that we realized they play a dominant role and are probably the most important organisms in the ocean. By the nineties, we understood that there is a high diversity of organisms we know nothing about – maybe 99 percent. They're involved in virtually every kind of geochemical cycling, including primary production. They're major carbon-dioxide-fixing organisms in the marine environment. At that same time, while I was starting my postdoc in the late seventies, submarine hydrothermal vents were discovered on the bottom of the ocean.

CI   What's the story – why were people looking, and what were they looking for?

JB   Years before the expedition was mounted to send a submersible to the bottom of the ocean, temperature anomalies were measured in water columns near areas where we believed the plates were separating, in particular off the Galapagos

Islands. We're talking about hundredths of a degree, but to measure hundredths of a degree increases in a two- or three-thousand-meter water column indicates that there's a major heat source. A group of scientists mounted an expedition based on those temperature data to search for deep-sea volcanoes, using the submersible Alvin, in 1977.

No one thought that biology would be important in these environments, because two or four thousand meters down are muddy bottoms, with sparse populations of animals that are dependent on whatever organic material from photosynthesis floats down from the surface. I was a new postdoc in microbiology, working on Antarctic stuff, and the call came back from the ship that on one of their dives they had found this incredible oasis of marine animals – huge clams, tubeworms close to a meter high.

CI You must have thought, "What were they smoking?"

JB Well, they were smoking; it was that era. [Laughs] There was a tremendous amount of excitement. The whole crew was focused on these amazing animal communities and how they were living, how they were being sustained. When the expedition ended I got hold of some of the water samples, which were preserved in various forms of alcohol – including tequila and other things. I made some of the first counts of those organisms and got hooked.

A year later, in 1978, another expedition discovered black smokers off Peru. That excited me because they were measuring 350 °C water coming out of these big smoker vents. Back in the Galapagos, there was warm water venting out of the crust – usually a few degrees above the ambient seawater, which is about 2 to 20 °C. That was warm, but not really hot. What got me excited was the idea that here, because of depth, you could actually maintain liquid water at very high temperatures – up to about 450 °C. Here's the chance to test the hypothesis that liquid water, and not temperature, might be a limit of life. At the same time, it got me extraordinarily interested in what kind of life might exist in crustal material, the subseafloor, which nobody knows anything about – it's a brand-new environment.

CI So while Viking was dampening people's expectations on Mars – the experiments were somewhat ambiguous but the cameras showed a pretty dry, arid, and dead-looking planet – you were simultaneously becoming aware of these extraordinary ecosystems on Earth. Was that when the nature and extremes of life on Earth became one of the central pillars of astrobiology?

JB There was definitely a sense of depression during the Viking period. It was an expensive experiment that failed; the normal conditions on the surface of Mars probably aren't going to support life. But along with hydrothermal vents a variety of new research methods were discovered – particularly in microbiology and the sampling of the ocean.

The interest in extreme environments came about as a result of the discovery of hydrothermal vents. In the eighties, we discovered all these incredible, interesting, bizarre, novel microorganisms. And molecular methods showed that many of the organisms found in these extreme environments represent a separate

	Hydrothermal Fluid	Seawater
Temperature (°C)	360–365	2
Acidity (at 25°C)	3.35	7.8
Dissolved Oxygen	0	0.076
Hydrogen Sulfide (mM)	2.3–3.5	0
Sodium (mM)	537	464
Potassium (mM)	17.1	9.8
Calcium (mM)	30.8	10.2
Magnesium (mM)	0	52.7
Silica (mM)	20.75	0.2
Chloride (mM)	636	541
Sulfate (mM)	0	27.9
Manganese (μM)	680	0
Iron (μM)	5590	0.0015
Copper (μM)	98–120	0.007
Zinc (μM)	47–53	0.01

A hydrothermal vent near the mid-Atlantic ridge at 26°N, where black smoke arises from the interface between the superheated water from below the crust and cold, high-pressure water near the seafloor. In addition to the high temperature compared to seawater, the hydrothermal-vent water is extremely rich in hydrogen sulfide, silica, and metals. Entire ecosystems have been found near hydrothermal vents, living independently of the Sun's radiation (courtesy Wood Hole Oceanographic Institute).

domain of life, distinct from normal bacteria, and these organisms may represent the most ancient groups on Earth. We started making the connection between analog extreme environments, origins of life, and life on other planets.

CI   You're talking about the Archaean branch of the tree of life. That must have been controversial when it was first put out by Carl Woese.

JB   Time was not put on the genetic trees. It was the distance between certain very highly conserved gene sequences; the distance one organism has from another.

There are some conserved sequences in genes shared by humans and archaea that grow at 110 °C, and bacteria that grow at pH 2, and tomatoes, fungi, etc. That was exciting, a lot of people jumped in.

We have an evolutionary molecule to play with. By looking at that molecule we're able – for the first time ever – to compare unicellular organisms like bacteria to humans or tomatoes or fungi. It wasn't accepted initially, and occasionally still isn't accepted by people who work with metazoans and higher organisms. They feel that, in some cases, structure may be more important than these molecular clocks. Many researchers have tried to put time onto these genetic trees – not only using the gene that Carl Woese used, but a variety of other genes. We've matched these trees with paleontological data and geological data, and placed different organisms in the different time frames. Before about 2.7 billion years ago, it gets more difficult.

CI   When you were a junior professor and not established, did your oceanographic colleagues look at you strangely because of the "astro" part of what you did?

JB   I didn't call myself an astrobiologist in the eighties, and I wasn't working directly on that topic. It was clearly something we were thinking about. As soon as there was any kind of culture in which you could actually call yourself an astrobiologist, the community became extremely divided, as it is today. Many of my close colleagues think astrobiology is nothing, that it has nothing to give. I call it the science of optimism, because we're going after something we know may not exist, and even if it does exist it may take more than our lifetime just to find it. A lot of people who enter astrobiology are attracted to the marriage between philosophical issues and astronomical issues.

CI   It seems that astrobiology has established the *expectation* of microbial life in a lot of other habitable places beyond the Solar System.

JB   Absolutely. Each year gives me more confidence that there is *at least* microbial life out there; to me it's not even an issue. How do we detect it? How do we get it, and how do we discover specific habitats on these planets where it may exist? Are there separate origins of life, or different evolutionary pathways? Even within the realm of carbon-based life, are there other options besides the terrestrial-life option? Those are the most pressing issues. I chaired a task force at the National Academy of Sciences on what we call weird life, or the limits of organic life in the universe. We don't want to miss out on finding life by being too Earth-centric with our detection methods.

CI   What's the evidence that the Archaean organisms, life's earliest organisms, were extremophiles? Are existing extremophiles similar to their ancestral versions?

JB   We don't know for sure that the most deeply rooted organisms on our trees of life are the earliest organisms. They give us clues – in terms of their metabolism, the way they derive energy, what they eat – and those clues can tell us something about those early processes among organisms. But it's hard to extrapolate to the earliest life forms.

We have a common ancestor. All existing organisms have the same genetic code, the same way of using a code to translate gene messages into proteins, a limited number of ways in which you can derive energy – from either light or chemistry. There's biochemical unity. This means that before the separation of the three major domains of life, there was a genetic pool, probably of high diversity, with lots of experiments going on within the evolutionary context. That eventually selected out mechanisms – genes – that were the best, and created an ancestral pool of organisms. That's what we're most interested in: how we got this common ancestral pool of genes, and how it developed into our unity of biochemistry. Why this one version and not other alternatives?

In looking at existing organisms, many of them deeply rooted on the tree, we're trying to figure out what some of those ancestral genes were like, and perhaps how the genetic code was formed. How did genetic material get from one organism into another to homogenize this diversity? Understanding that requires a better understanding of the origins of viruses, of how organisms exchange genetic material, of how to make a large genome or a large chromosome in a relatively short time, of how cells fuse together, and of symbiotic associations.

CI   What was the metabolic diversity of life's earliest organisms? Many of them did not rely on photosynthesis. Did they use chemical energy?

JB   I claim that a hydrogen-based ecosystem was the early driving energy source. That means organisms that can make methane as hydrogen reduces carbon dioxide. There are other groups that use hydrogen and carbon dioxide coupled with sulfur. There appears to be a diversity of pathways for reducing that carbon dioxide to other organic compounds. There may have been a wide range of ways to reduce carbon dioxide with hydrogen at some point, but the ones we've been studying are primarily the pathways that make methane, which is also considered an ancient pathway.

A second one is what we call anoxygenic photosynthetic microorganisms. They photosynthesize but in the absence of oxygen, and they don't make oxygen; so rather than split water, they split hydrogen sulfide, which was very plentiful in the early stages of ocean chemistry. What's interesting about these organisms is that in the process of reducing carbon dioxide, they oxygenize hydrogen sulfide, so you end up with oxidized forms of sulfur including sulfates. We do see sulfates before 3 billion years ago, so we know there was some process making and reducing sulfates microbiologically in a very ancient system. The two types of metabolisms I've just described can also absorb at wavelengths closer to the infrared, so they can be out of the UV penetration range in the ocean. I feel that where there's hydrogen on any other planet, along with carbon, there's a key energy source.

CI   Hydrogen's so abundant. It shows how primitive humans are, because we're still trying to get to a hydrogen economy. Microbes figured it out way before us.

JB   Yes. I see evidence for metabolic pathways and other catalytic systems without any proteins on metal–mineral surfaces. Catalytic reactions occur once metallic compounds start forming, not dissimilar to the way that planets are formed. Little organisms – essentially little planets – are formed in much the same way early on, with spontaneous creation of structure and of energy-yielding pathways in the absence of any sophisticated catalytic protein, perhaps in the absence of any kind of information macromolecule. I think there's something inevitable about creating some kind of carbon-based life, based on the physics of how elements come together and how reactions occur on various metallic minerals.

CI   Are there modern analogs, or living relics, of these primitive metabolisms?

JB   Absolutely. One that's being studied is a pathway called the reductive TCA cycle that we and other respiring organisms use to derive energy from oxidation reactions that occur in our mitochondria, for example. You feed organic material into the organism and it carries out these various oxidation–reduction reactions – it produces carbon dioxide and it produces energy as ATP. If you take that cycle in reverse, then you're pulling down carbon dioxide and reducing it into organic material and you're using energy. As it turns out, the reductive TCA cycle exists in most organisms, and it's thought to be the most ancient of metabolic cycles. There are groups now finding that we can almost replicate the whole cycle, without enzymes, on pyrite and other minerals. We can form these intermediate organic compounds for life and also derive energy for those reactions directly from minerals. There's a lot of interest in mineral catalysis.

CI   From what you know about the likely metabolisms and Earth conditions, where was the most likely place for life to start?

JB   I think the best place for life to have originated would have been in the sub-sea-floor associated with hydrothermal activity. You can generate energy – catalytic surfaces with minerals in the Earth's subsurface that may have produced the organic compounds and condensed them into larger molecules. We may have to think a little outside the box. Some people are looking at metabolic pathways in the absence of protein enzymes using hydrothermal vents. Others are looking at other catalytic functions that may reduce nitrogen gas into ammonia for life, and they're looking at hydrothermal models.

The big problem is making nucleic acid. Gradients in deep-sea hydrothermals would have been a very plausible habitat for the earliest microbial ecosystems because of the abundance of energy and the abundance of carbon. Everything is there, and at the same time they haven't yet evolved mechanisms to protect themselves from ultraviolet radiation, particularly in the absence of an ozone layer, which wasn't around when early life evolved.

CI   But if life began in that kind of environment, how would it be sustained and propagate? These suboceanic environments are little ecosystems or worlds on their own, but aren't they transient? How would biology become global?

JB   First of all, there's no comparison between the early Earth and what it looks like today. Tectonics would be much stronger and hydrothermalism would be robust. We're not looking at major plates moving around, we're looking at a jigsaw puzzle, with the whole crust hydrothermally active and lots of subducting crust. You still have a lot of heat and you haven't formed large plates. Hydrothermalism was universal on the ocean floor. Those environments were not transient.

Secondly, when you form new crust, called ultramasic rock, it's very high in iron and magnesium, usually in the form of silicates. It reacts with water to produce hydrogen, and in the process produces heat. It's an exothermic reaction. It's one we've just recently discovered in the mid-Atlantic ridge, and we realize that also would have been rampant. So we have magma-driven systems, and new rock, ultramasic rock, interacting with seawater that's also producing heat and lots of hydrogen. Hydrothermalism was universal on the ocean floor. If a particular vent site clogged up, there were others popping up all around it. That would enhance chemical reactions by creating even more gradients, in terms of temperature and different quantities of minerals.

The subsurface back in the Archaean, and even to some extent today, is an open system that behaves like a chemical reaction. Seawater interacts with hot rock, and extracts nutrients and minerals; it's basically the whole periodic table and all these rocks and volatiles, and then all that remixes and creates another set of minerals and volatiles. It provides the most options for creating diverse chemistry and diverse habitats.

CI   It sounds like the places most likely to be living worlds are dynamic environments with sources of free energy. They sound very different from current Mars.

JB   If we go to Mars, where surface life is impossible, there's plenty of water; most of it is ice in the regolith. There is evidence of past volcanism that has spilled out water, and if there's still any kind of a heat source, then it's possible that there are still small pockets of heat generating the kinds of chemistry that can sustain life, and perhaps enough heat to melt through part of the buried permafrost. That's where we may have life.

Then there's the discovery of methane on Mars. We can probably rule out a biological source. If methane is being generated geophysically, then the most likely processes involve hydrothermalism. If there's water buried deep in the subsurface along with ultramasic rock, that's an exothermic reaction that generates hydrogen. The hydrogen continues to react with the metals – iron, nickel and others – and minerals to produce a variety of organic compounds, including abundant methane. We see methane abundances that can reach extremely high levels in natural systems where ultramasic rock is reacting with water. It's also an exothermic reaction; it could produce heat up to 150 °C or higher. And that's just a chemical reaction. With this energy source, we could have the warmer temperatures to release liquid water, and we could sustain a group of living organisms.

CI   Hopping to Europa, what does your knowledge of colder oceanic environments on Earth tell you might be going on there?

JB   That's a big cipher I'm very interested in, because recent models are trying to invoke hydrothermalism on Europa. There's one that claims the tidal heating and the flexing of the bottom core would create some kind of hydrothermalism, and that means a number of nutrients that I've talked about, and warm water at the

Oceanographers and biologists were surprised to find a complex web of life living in cold seeps, deep in the Gulf of Mexico. The sea worms seen here live in methane ice. As with hydrothermal vents, this is an environment where photosynthesis cannot operate, so the organisms utilize a variety of chemical energy sources (courtesy National Science Foundation).

same time. They and others have tried to imply that you could have enough heat concentration out of a robust venting system to get up to the ice and cause some melt. Those are just models. The key to Europa is not so much cold temperature, and not so much that there's no oxygen – it's whether or not nutrients are being generated. The only thing we know about Europa's chemistry is that there is probably sulfate and there appears to be plenty of carbonate, and that's based on spectral data. But what about the energy sources? We don't know anything about those.

My view is this: if it is a hydrothermically active moon and life did get established, say in the subsurface where the nutrients would be, then it's been pumping microbes out into the water column for more than 3.5 billion years. Whether or not they've adapted to grow in the low nutrients in that water column, there still would be an accumulation of organic material. It's possible we could detect that, if we could get samples of some of the brightly colored ice along some of the ridge areas – that color might be organic material. Or we could find an area of shallow ice that we could penetrate somehow and analyze some of that material.

CI   You've been pretty involved in issues of weird life. What are the most useful ways for us to relax the bounds of how we define life?

JB   I try to divide up "weird." There's "slightly weird," which means carbon-based, using a lot of the same biochemical processes but maybe different building blocks. That's one level of weird: different amino acids, maybe different bases for nucleic acids. The second is a little more weird, still carbon-based, but using a completely different molecular architecture for the cell. No central dogma of DNA–RNA–protein; it's something different. How do we imagine that? Even in a carbon-based scenario, we're still thinking about an informational macromolecule and some kind of translation of that into a product.

    Then we start getting the "seriously weird," and the first option is still a structured entity but perhaps it's not carbon-based. It could be silicate-based or silicate–carbon-based; but structurally it's radically different. Can there be carbon-based life or silicate-based life that can live in solvents other than water? That's a separate issue. We do know that a lot of enzymes work in the absence of water and organic solvents, and in some cases they behave quite well and differently. But to form, the structure of the enzyme has bound water to it; no experiment has been done in which water is completely absent. Even though water might be less than one percent, it's absolutely essential to create the three-dimensional structure that allows catalytic activity. In terms of the solvent issue, we don't have any information.

    The most radical are pure speculation, like life living in the atmosphere of Venus that's more sulfur-based. Or life in organic solvents like an ammonia ocean on Titan. We are mostly looking at carbon-based systems and other solvents; those are going to be the dominant recommendations. From a cosmic chemistry point of view, if we have any rocky planet or moon with liquid water, then carbon-based life is the way to think. You can define life any way you want, and hypothesize something living in virtually any environment you want. For now, I'm going to stick with what we can do in a carbon-based life system.

CI   Let me finish with a question about the field of astrobiology. Can you recruit good students and do you see the profession growing at the grassroots level?

JB   I see more students applying to work in astrobiology than in oceanography. Astrobiology is attracting some of the very best students, and they're attracted to a lot of the same questions that motivate senior scientists; not just a search for life elsewhere, but the philosophical issues. When you ask them, "Why do you want to study this?" their answers are very personal, very existential: "This is my way of finding out more about myself." You realize they're thinking more about why they're doing science than most science students. I really like that, because it's rekindling my interest in having those discussions again.

12

Joe Kirschvink

Joe Kirschvink has challenged conventional wisdom about our planet in a number of ways, such as his claim that the entire Earth once resembled a giant snowball, causing a crisis for biology that stimulated biodiversity. Another example is his idea that the Earth experienced a period of true polar wander, rotating about the equator, which led to the Cambrian explosion. He also courted controversy with his contention that magnetic minerals in the Allan Hills meteorite indicate life on Mars. The common thread in these ideas is the use of magnetism as a diagnostic of the turbulent history of the planet. Kirschvink was an undergraduate student at Caltech, did his PhD at Princeton, and then returned to Caltech as a faculty member. He runs a lab devoted to paleomagnetism, something few people know exists and only a handful actually study. He swims, skis, and plays with high-tech microscopes for fun; his two children have names that are the Japanese words for "magnetite" and "gemstone." He is a Fellow of the American Society for the Advancement of Science, the American Geophysical Union, and the American Academy of Arts and Sciences. Asteroid 27711 is named after him.

CI   Have you always been a rock hound?

JK   I started collecting rocks before high school. I grew up in west Phoenix. I was not raised where it was green and yucky. I was raised where you could see the rocks between the cacti.

CI   Were you turned onto rocks by a teacher, or just by your environment?

JK   I was interested in rocks and in Earth sciences. The first science course in high school was an Earth science class. It occurred to me that in Earth sciences you can study math, chemistry, and physics, and have fun with it.

CI   So it was an integrated science course?

JK   Exactly. We had a great set of teachers in high school; two of them had their PhDs. I came to Caltech as an undergraduate, unable to choose between biology and geology. So I did both.

CI   Was that possible to do in your schedule?

JK   There was no break for a double major. You did both. I took overloads every term but had a ball. Halfway through junior year, I saw a notice that undergraduates who earn 135 units or more of graduate-level classes could graduate with a Master's. I applied for graduate standing my senior year and got a Bachelor's in biology and a Master's in Earth sciences.

CI   To satisfy your joint interests, you must have wanted to do fieldwork and have a lab. Do you have both running at once?

JK   I worked in Lee Hood's lab my sophomore year, before I hooked up with Gene Shoemaker's group. I wanted to get a taste. I spent three months with his group sequencing protein from a clam shell. He was developing automatic sequencing technology. Then he started working in DNA sequencing, and now that's the way you sequence a protein, DNA first. I swore after three months in a subbasement laboratory with no windows that I would *never* study something I couldn't see. That's why I focused on field studies. Now *I've* got a subbasement laboratory studying magnetism, which is something you can't see.

CI   Is that your main area of research?

JK   That's my bread and butter: paleomagnetics and magnetic stratigraphy.

CI   What about the fieldwork? It must be fun; it gets you out of your head.

JK   Studying the rock is how we learn about the planet. In terms of planets, Earth is the only one we can bang a rock hammer on. We took a continuous core across the Permian–Triassic boundary in South Africa a few years back. We got 39 meters of almost continuous core, right across the boundary beds in the group.

CI   That's a mass extinction where the cause is still controversial?

JK   There's a big debate about whether there's an impact or not.

CI   No smoking gun? No crater?

JK   That's the debate. One structure that has been identified was looked at ten years ago as a possible impact, then rejected for good reasons. The debate still swirls.

CI   Fieldwork is a traditional apprentice system. Is that how your students learn?

JK   Yes. It works extremely well, even at the undergraduate level. I mainly work with undergraduates.

CI   Your two disciplines play into a shift in thinking about planets and biospheres. Life isn't plastered on the surface of a planet, or suspended in its air or water. The interplay between the rocks and the atmosphere and the life is so deep and profound that you can't understand one without the other two. How did that awareness emerge?

JK   My colleagues in biology look at the biosphere at time zero. Geobiology worries about time zero *plus* the last 4.5 billion years, and in a constructive, evolutionary way. You've got to worry about *how* life evolved. The interaction with the planet is a central piece of it, both ways: life interacts with a planet, the planet interacts with life.

CI   How did that connection emerge? Was it thinking about mass extinctions and their causes?

JK   The Alvarez hypothesis was one of the big factors, but that connection has always been obvious to me. I've never made a distinction. That's why I started looking at the Precambrian–Cambrian boundary back in 1974. Here's a wonderful evolutionary burst. You're going to understand it best by understanding what the world was doing at the time.

That's why I started working in paleomagnetics and magnetic stratiography. That tool would help us put the continents back together. If you're going to understand one of the most important biological radiation events on the planet, you need to understand where the continents were. Few people were working on it at the time, and it was fun and fascinating. It let me into the lab.

In high school I toyed with electrical engineering. One of my earliest memories as a kid was trying desperately to remember where I had put a coil of wire, because I was soldering something. Then I realized I had been carrying it for the last hour! [Laughs] I realized when I came to Caltech that I wasn't an engineer. I didn't know that in high school. I'm not interested in what humans can do with nature, but in how nature got to be the way it was, in biology and geology. It was a realization.

CI   Many people don't know much about magnetism beyond the field and its periodic reversals. What's revealed by more detailed paleomagnetic study?

JK   If you want to correlate an extinction boundary from the ocean realm to the terrestrial realm, or even across terrestrial realms, you're not going to be able to find a volcanic ash in China at the same level as in Peru or South Africa. Magnetic reversals are global and nearly instantaneous. That's one way to refine the precision of correlation. You also get extremely good control on where the continents were. It's a stratiography.

CI   How far back does it work as a tracer? Before 1.5 billion years, everything's so messed up by heat and pressure that it's hard even to find fossils. Are magnetic signatures preserved as you go back in time?

JK It depends. Fossils, magnetism, and geochemical signatures all have their own ability to withstand nature's erasing attempts. Magnetism is sensitive in some sediments, and not in others. We can go back easily to about 2.5 billion years. In South Africa, it's about a quarter billion years, with perfect signals. We have beautiful basalts from the Kalahari at 2.3 billion, which is the first snowball event. For the rest of it, you try it and see. I've had whole countries that don't work in magnetics. Namibia was wiped out.

CI I'll cross it off my list. [Laughs] Let me ask about epochal events in the history of life that connect to geological events, the Cambrian explosion particularly. I've seen a number of books, each with a theory about what happened. What have you brought to the table?

JK The Inertial Interchange True Polar Wander.

CI For the uninitiated, how would you summarize it?

JK The planet spins about the principal moment of inertia. The moments of inertia of a planet depend upon mass densities. On Earth, these are subductions on plumes, and they grow and die with time. Forty years ago, people like Don Anderson envisioned situations where the residual principal and intermediate moments of inertia could cross. You're spinning around nicely, but because of a plume dying, the intermediate moment increases to where they cross, and you get an instability. Anything up to 90° is limited only by the viscosity of the upper mantle, which, if you had an instantaneous mass change, could give you a 90° polar wander in 10 000 years.

CI Wow.

JK From the Cambrian explosion, we had about a 15-million-year interval of a 90° shift. That brackets the Cambrian explosion interval. We suggested there was a fundamental tectonic driving mechanism.

CI When was this?

JK About 530 million years ago, and the 10 million years before and after that. In 1988, I was on a field trip to Northern Africa, Morocco, where we'd been working on the Precambrian–Cambrian boundary carbonates in the northern flank of the Antiados Mountains. I noticed a peculiar unit that turned out to be a volcanic-ash flow, about 80 cm thick. Shit! There's an acidic, volcanic ash interbedded in our dig, where the trilobites start. At that time, age estimates for the Precambrian–Cambrian boundary were from 530 to 610 million years.

The problem with paleomagnetism is that half of our data comes from volcanic units, and half from sediments. The Cambrian data from all the continents combined was a complete, utter shambles. I realized that the geophysicists didn't understand the timescale; they got a mess. We eventually got a date out of our volcanic ash, which yielded about half a milligram of zircon, and took us three or four years. The age for this zircon population came out at 521 million years, almost 90 million years younger than the highest estimate, and 9 million years younger than the lowest.

This was a monkey wrench. It can't be that young! But it's a very good zircon date; it's an interbedded zircon. They tried everything. They threw in things that were clearly not in the same population, and got an older and older error bar. Now that we've got 15–20 different dates from it, guess what? We were right on the money.

In the early nineties, we went through the paleomagnetic database and revised it to see what would happen to the timescale. The data sorted themselves out beautifully. We ended up with clusters of points that moved 90° in, not only in Australia, but in North America as well. You could explain 90 percent of the magnetic data with one motion of the entire planet. That's a prime candidate for an inertial energy event. It was centered on the evolutionary burst.

CI   The correlation and mechanism are clear, but what's the connection to biology?

JK   One of the craziest things about this interval of time is the carbon isotopes. In this interval of time, the carbon isotopes are going back and forth, ten times larger than anything you see outside of the Permian–Triassic. They're ten times larger than the isotope anomaly across the Cretaceous–Tertiary. It's as if a volume of carbon equal to the entire biosphere is being thrown into the oceans, taken back, and then that repeats. Those carbon-isotope fluctuations are huge.

CI   What causes those fluctuations?

JK   The most robust conclusion to draw is a thermal oscillation. In the Paleocene, 55 million years ago, there's a 12-degree thermal spike corresponding to a sharp spike in the carbon isotopes. Also, in the Paleocene, there are oxygen isotopes, so you can map a temperature change. In rocks this old, we rarely, if ever, get an oxygen isotope record.

Diversity correlates with temperature. If you identify today the places on Earth with the highest number of living species, the mean temperature's higher in those places. It's the first-order rule of diversity. When you've made new body plans, you can spawn an entire ecosystem. But evolution is a bit slower. When the global temperature rises, things radiate and speciate. When the temperature cools, they don't die away, but new body plans form, and new, different groups. By thermally cycling the Cambrian, diversity is pumped up incrementally.

CI   So the diminishing parts of the cycle are not extinctions, they're just modulating it downwards?

JK   That's right – pruning and modulating. That system allows novel groups to get a hold and radiate. The other point made by my paleontological collaborators is that the radiation event was associated with the rise in carbon isotopes. That's exactly what you would expect. So here is an observation of carbon isotopes. We also have this inertial interchange event. How do you connect the two?

Organic carbon. If it's eaten by methanogens, they release methane. And if it happens at high enough pressures and low enough temperatures, the methane is a solid. It builds up a cap, no more than half-a-kilometer thick, because the geothermal gradient puts it back in the gas form. You end up with large areas today that are covered with methane.

CI   Are we talking ocean floors?

JK   Ocean floor sediments, continental slope sediments, and permafrost. This inertial interchange event took North America and big parts of South America, from the south pole up to the equator. Boom. You have a ten-million-year front of warming permafrost going right across that band, with episodic pulses. The inertial interchange event caused these thermal spikes, and evolution was poised at the point where homeobox genes had already evolved to radiate and to pulse, which enhanced that radiation further.

CI   Homeobox genes are the ones that propagate body parts and vertebrae?

JK   Exactly. They regulate development and body plans.

CI   Does life need interesting geology? Will a boring geological planet be less likely to form life?

JK   A boring geological planet probably would have a boring biosphere, maybe only bacterial-grade evolution.

CI   Is it now established that diversity is driven by geological and thermal change?

JK   Yes. The Permian's another problem. There was a big burp there, and there's a dead zone in the early Triassic before life finally radiated. In the middle Cambrian there's another negative spike going up, a big extinction event followed by a massive radiation.

CI   What about the much earlier snowball Earth several billion years ago and the emergence that followed?

JK   The Paleoprotozoic? That's the oldest and baddest. The paleomagnetic data is superb. We have two separate studies of flood dissolves that interfinger with that glaciation, and that's 11°. We know that the magnetization was acquired when these pieces were cooling in the water and flaking off. That particular snowball-Earth event has all the fingerprints of an anaerobic environment until just before it. After that, we have the best evidence for the massive operation of oxygen.

CI   When did that period begin?

JK   About 2.3 billion years ago. That's about what you would expect for a hard snowball, when the Sun is 86 percent of its present luminosity.

CI   How does the paleontology community take the idea of dramatic geological change as a driver for evolution?

JK   Twenty-five years after the Alvarez iridium–impact hypothesis, they take it more seriously. Almost all the controversy about the K–T boundary has disappeared. The extinctions in the oceanic realm are precisely linked to those layers, but there's still an argument about whether there was more than one impact in the K–T boundary. There may have been an earlier impact 300 000 years before that didn't do anything.

CI   How do you separate cause and effect, even when you have good timing?

JK   It's pretty clear the radiation of the biosphere did not cause the moment of inertia of the Earth to change. You see the Earth's polarity, you can

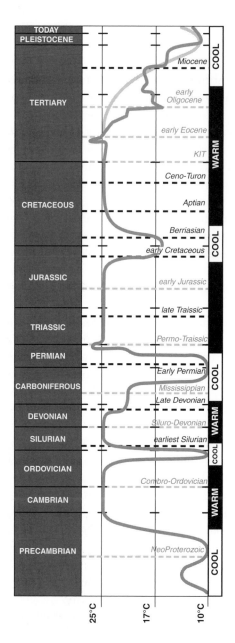

There is extensive geological and paleomagnetic evidence that Earth went through a series of extended cold phases from 630 and 780 million years ago, in the Precambrian and also much earlier, about 2.25 billion years ago. These episodes are called "snowball Earth" episodes, though there is argument over whether the ice coverage was complete. The positive feedback that causes cooling and ice, which then reflects more sunlight to cause even more cooling, is broken by volcanism and the release of the greenhouse gas, carbon dioxide (courtesy University of Bristol, and PALEOMAP Project).

reasonably infer from what the Earth is doing that you can get carbon oscillations, and modern forest ecology tells you that temperature can closely reflect diversity. It's the only parameter. Our model is the only scenario that goes right from the physical process to a prediction about what we see.

CI   The interplay of complex geology and the biosphere points back to Rare Earth arguments. When we uncover terrestrial planets in significant numbers, do we anticipate that the Earth will be unusual geologically? Are tectonics necessary? What are the necessary and sufficient conditions for life?

JK   Some of the assumptions people use bother me. Everyone equates "Earth-like planet" with an oxygen atmosphere. I think our oxygen is a freak event. It's clear that the protein systems that are central to photosynthesis – photosystem 1 and photosystem 2 – share a common ancestor. They're now together in one lineage, cyanobacteria, which became the organelle in higher plants. It's as if two photosynthetic cells became symbiotic because they had different abilities. In the process, they allowed the photosystem 2 manganese cluster to become so oxidizing that it'd bleach your hair if it had the chance – so oxidizing that it could rip that electron from water. Once they did that, it was a selective advantage second to none.

That is *not* something that you would expect to happen on any planet. From my point of view, the existence of photosystems 1 and 2 together in the same system, doing what they do, is a remarkable feature. I would not expect to see any planet with Earth's mass and composition doing it that way. Run evolution again, and it might not ever happen. If it never happens, we're stuck at a bacterial grade. It is a Rare Earth argument.

CI   So the Earth is not much help in predicting the general properties of terrestrial planets?

JK   No. I think Peter Ward and Don Brownlee may be right. I like the argument proposed by Gerard K. O'Neill at MIT back in the early seventies in his book *High Frontier*. If an intelligent civilization had begun space colonization, and if once every 500 years it successfully colonized a new star system, the entire Galaxy is settled in 15 million years. This Galaxy's been here several billion years, and nobody's here yet. Maybe bacteria are common, though I could give you reasons why even they might not be common. But if animals are extremely difficult, maybe only a dozen planets in the whole Galaxy ever got to that level. We sat around on Earth for 600 million years before anything intelligent came up and built a telescope.

CI   You're partial to magnetite. Is it true that you named one of your kids after it?

JK   There's more to the story than that. Back in 1983, my wife and I discovered single-domain magnetite in fish. It was the magnetic sense organ. We were expecting our first child, and involved in the usual parental debate: what name to pick? There was a whole set of names if the baby were a girl, but there was a problem with a boy: her family name is Kolabashi, and it's like Smith in Japan. When you have a common name like that in Japan, the quality of the given name is important. What determines the quality of the given name is the Chinese character

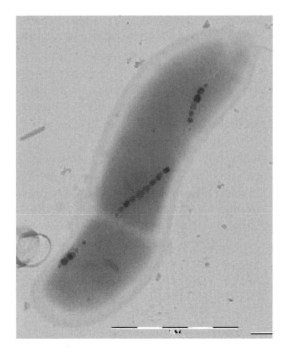

Magnetic sensing and synthesis of magnetic material is seen in many simple and advance organisms, and may represent a primeval sense. In bacteria like this, magnetite is synthesized into particles about 50 nanometers across, which are bound by membranes to form chains, allowing the bacteria to orient themselves in the Earth's magnetic field (courtesy DOE/Ames Laboratory).

that goes with it. It should have some meaning. It should in itself be a beautiful character, and it should have some relevance to where the family is and what the family is doing.

I finally said, "All right, what would magnetite be?" She scribbled the character down. When I first saw it, it looked like a couple of lightning bolts hitting the symbol for a stone. Very old, very ancient, very powerful Chinese characters. Wow! She called her mother up from the States, collect, at four in the morning. My mother-in-law went that morning to a local temple where they have a Buddhist sect that worries about names, and explained it to the monk. The monk pulled out the list of 32 things, like stroke order, and everything was perfect.

CI   This is the feng shui of names.

JK   Exactly. At that point, my in-laws would not *let* us name the firstborn child anything other than Jiseki. Then what do you do for the next child? At the time, my wife was taking a mineralogy class and she was infatuated with minerals. There's another name, equally good, for gemstone: Koseki. You must pronounce it perfectly (KHO-sek-i), because KO-sek-i is the marriage registry for a family. But KHO-sek-i is "gemstone." We were worried the kids would be made fun of, but neither one was. We're now studying meteorites, and this is why we stopped at two: the equivalent name for meteorite would be pronounced IN-seki.

CI   That's a bit of a problem.

JK   [Laughs] We stopped at two.

CI   I wanted to ask about animal sensing and magnetite.

JK   It was my idea.

CI   Do you think that's a cosmically prevalent mechanism in higher-order organisms?

JK   I think it's the primal sense.

CI   It certainly pervades the tree of life.

JK   I think the last common ancestor of all living things was probably a magnetic bacterium. We see the same magnetite bond mineralization ability across the bacterial domain. It's in all of the groups of proteobacteria. We even have an archaebacterium that's magnetized. It's in a whole slew of protists. I don't think I've seen a group of animals that it's *not* present in. It seems to be present in green plants for some reason, and we have no idea why it's there. It's clearly at the core of the magnetic-intensity sense. I don't know of any system in which you can put your finger on a structure and say it has that evolutionary ancestry. I think it's the primal sense.

CI   In the mists of the formation of the first cells, what is the selective or energetic advantage of magnetic sensing? How does it emerge?

JK   It probably emerges as a requirement for iron sequestration. Iron is ubiquitous, you need it for a ton of biological functions. So if you were in an environment where you've got lots of iron, and you occasionally move into an environment where you don't have much of it, there is a selection pressure for storing it. It's easy to store, but you want to store it in a protective form, because ferric iron can be a little bit hazardous – it can form free radicals, and you want to control that. However, once you start condensing iron, it's easy to make a little mistake and make magnetite. Simply add phosphate minerals, and boom! Once you get a little bit of a magnetic particle, there is a slight magnetic orientation, and natural selection can play on that. The rest is history. It's an easy scenario to envision.

CI   Is magnetic sensing widespread in higher organisms like animals?

JK   A lot of magnetite in human tissues is not being used as a sensory realm. There are no known sense organs in the brain itself. All the sense organs are external.

CI   Proportional to birds, is there more?

JK   The amount of magnetite in the higher-animal bodies is far more than is used for receptors. That's why we went to fish, because at the level of fish, the only magnetism is in the nasal areas. Once you start examining turtles and lizards and humans, the background tissues become magnetic. Other functions have taken over, and it's difficult to locate magnetic sensory organs. We know some of the magnetite is being used for a compass, because you can do a simple experiment: a sharp pulse to an animal with a beautiful magnetic orientation can destroy that in a second. Bunk! It takes the animal days or weeks to recuperate. Occasionally you can even induce a shift in the direction it goes. That is a unique ferromagnetic effect. The hypothesis of my thesis back in 1979, that magnetite is the basis of that magnet, has been confirmed over the last thirty years perfectly. It's gratifying to have figured out a whole new class of sensory modality. It's hard to duplicate that feeling.

CI   You spoke about having nutty theories, but you seem to enjoy working outside the envelope?

JK   I never noticed it. Biology and geology are two vastly different cultures, but there's only one science. I wrote an editorial in Japanese for the University of Tokyo newspaper some years ago, pointing out that nature doesn't care if something's called physics or chemistry. How a biological organism operates is governed by the laws of chemistry and physics as well as anything else. I don't see these artificial boundaries. The system works. How do you understand the system? Use whatever you can.

13

Andrew Knoll

Andrew Knoll dreams of the ancient, alien Earth and speculates about changes that accumulated in the planet's geochemistry, and how those changes enabled the development of biology into the diversity we see today. He got his PhD at Harvard University, and taught at Oberlin College for five years before returning to Harvard, where he is the Fisher Professor of Natural History and a Professor of Earth and Planetary Sciences. He and the members of his lab study the early history of the Earth, particularly traces of early microbial life preserved in the rock record. When the entire biogeochemical history of the Earth isn't enough, he shifts his focus to Mars; Knoll was a member of the Mars Exploration Rover (MER) science team. He has many awards and honors, including the Wollaston Medal of the Geological Society of London and the Walcott Medal of the National Academy of Sciences, of which he is also a member. He wrote a popular book on the evolution of life called *Life on a Young Planet: The First Three Billion Years of Evolution on Earth*. His favorite T-shirt would say, "Evolution is the interaction between genetic possibility and environmental opportunity."

CI   You work at the juncture of two different disciplines, biology and geology. How did you get started?

AK   When I was a kid in a fairly rural part of the Pennsylvania Dutch country, I used to collect fossils for a hobby. It never dawned on me that I could do it for a living. I had a wonderful epiphany when I went to college and realized there *was* a field where I could pursue my hobby while trying to integrate two types of science that fascinated me: geology and biology.

CI   Were your childhood stomping grounds rich in fossils?

AK   Yes. I grew up in the foothills of the Appalachians in southeastern Pennsylvania; there's a fair amount of Paleozoic sedimentary rock, and there are localities that are fun collecting sites.

CI   Was the curiosity spurred by a relative or a teacher, or something you read?

AK   It started when my brother went on a field trip to a local fossil quarry. My dad was a driver, so I went along even though I was three years younger. I picked up a piece of rock, broke it open with a hammer, and there were the remains of something that had lived hundreds of millions of years ago lying in my hand. Even as a kid, that was very exciting, and it still is.

CI   One of the dangers of being good at science is that you don't get to do the things that got you into it originally. Do you still get into the field?

AK   Maybe not as much as I'd like, and when I was younger and fitter, there were summers when I spent two months in places like the high Arctic. Now, because of professional and family responsibilities, I'm less prone to take those huge chunks of time. Nonetheless, it's a bad summer if I don't get into the field at all.

CI   In many areas of science there's the danger of becoming so enamored of simulations and computers that students stop observing nature. Is that happening in your world?

AK   All the students who pass through our lab work with real rocks and get field experience. These days a majority of my time in the field is spent helping my students develop their projects, rather than on my own projects.

CI   As you mentor them, do you find a range in their skills with the hands-on aspect of fieldwork?

AK   Absolutely. All the students we get here are smart, one way or another. That said, there are some people who are really good at modeling, some who are really good at thinking geochemically about rocks, and some who are really good at going out into the rock record and reading a cliff the way you and I might read a book.

CI   Scientists are becoming ever more specialized. Is it important to you to retain breadth in your scholarship?

AK   It is. None of us are good at all disciplines, writ large. There are parts of biology and parts of the earth sciences to which I feel I make more contributions. It is much more important than it was twenty years ago to have a working knowledge of other fields, so that you're able to read *Science* and *Nature*, or talk to people at meetings. If you don't have the basic knowledge that allows you to poach off of other people's disciplines a bit, it's going to be increasingly hard to be a scientist over the next twenty years.

CI   Let me move to the history of life on Earth. It's amazing that we can reconstruct those earlier times based on what's lying around today. How have we learned about the first 2 billion years of life on Earth?

AK   We can make analogies between hunting for dinosaurs and hunting for cyanobacteria. If you're hunting for dinosaurs, you have a search object or image based on knowledge of modern vertebrate animals – their distributions, morphologies, development, all that. You then select rocks that mapping tells you are the right age, and experience tells you are the right types where these bones might accumulate. You collect samples, analyze them, and interpret them in light of what we know about living counterparts.

It's the same thing for cyanobacteria. They're much smaller, but there are places we can go today, ranging from the Bahamas to Western Australia, where there are still ecosystems dominated by cyanobacteria and other microbes. We can learn how those organisms are distributed, their properties, and how they end up being fossilized, how they're going to be preserved in an accumulating rock record. We then go to places that geologic mapping and radiometric dating tell us are a billion years old. We look at sedimentary rocks that provide ancient analogs to modern environments where cyanobacteria are important.

In this case, the individual fossils are small. We bring them back into the laboratory and approach them in one of several ways. We either make paper-thin sections and look at them under the microscope, or dissolve the rock in strong acids and plate out the organic matter that's left, or make chemical measurements of the rock itself and the organic matter it contains. The details of the observations differ between someone working on the Precambrian and someone working on dinosaurs, but the overall philosophy of using the present to inform the past still holds.

CI   There are vigorous arguments and debates over how solid the earliest evidence for life is and the date of that evidence. What are the core issues?

AK   The core issue is that when rocks get metamorphosed, their biological signatures get progressively wiped out. As we go deeper in time, we have fewer examples of sedimentary rocks that represent any one interval of time, and those rocks tend to be much less well preserved. There are only two sets of rocks in the world that are potentially informative about life as it may have existed 3.5 billion years ago. One is in southern Africa; the other is in western Australia. Both of those sets of rocks are moderately metamorphosed. We have done some studies looking at a series of paleobiologically interesting rocks and following them from unmetamorphosed areas to areas of increasingly high metamorphism. It gets tough to tell living from nonliving in ancient terrains. But we have no reason to believe that this problem precludes the possibility of finding evidence of microscopic life where the rocks have been cooked.

CI   Does that tie into the fact that there's been a trend towards greater complexity as life evolved? Is it harder to find evidence of the organisms when they're simpler?

AK   In the following way. If you show me a shark's tooth, I will be absolutely confident that it comes from something that was once living, because we have no reason to believe that physical processes alone can generate that kind of form. Chemically, if you show me cholesterol, I'm willing to believe it was synthesized by an organism, because I have no reason in our collective scientific experience to think that cholesterol can be synthesized under purely natural conditions.

On the other hand, show me a one-micron sphere, and it isn't clear whether that was generated biologically or not. Tell me you found an amino acid in an ancient rock or a meteorite – we have reason to believe it's easy to make amino acids from scratch. Therefore, while either a one-micron sphere or an amino acid *could* be the products of biological activity, they don't have to be. It might be difficult, from first principles, to distinguish the early products of biology from the products of physical processes in a world that could generate biology.

Take, for example, rocks that are a billion years old. We have excellent evidence for microfossils of many different types. We have excellent evidence of organic molecular biomarker molecules. We have carbon and sulfur isotopic records that are really interpreted as the records of biological cycles. We have stromatolites, and stromatolites built by the trapping and binding of sediments by microbial populations have no ready analog in nonbiological processes. That's four or five lines of evidence, and then the question becomes, "What happens further back in time?" We can go back to about 2.5 billion years ago, and all of those things are intact.

As we search back even further, into rocks that are less common and more cooked, what goes first? The microfossil record becomes somewhat unreliable, which is exactly what happens if we take younger rocks and metamorphose them. The biomarker record is erased, which is also what happens when we metamorphose a younger rock. The stromatolite record is potentially resistant to alteration by metamorphic processes, but the complication is that as we go back in Earth's history, more and more of the stromatolites are built by seafloor precipitation processes. And unlike stromatolites built by trapping and binding, those can be mimicked by physical processes.

In the physical, textural record of sedimentary rocks, there is *some* record worth taking seriously back to three and a half billion years. The most potentially robust record is that of carbon isotopes, and the carbon isotopic record three and a half billion years ago looks more or less like the carbon isotopic record for the last 2 billion years.

When we bundle all that, what we see in 3.5-billion-year-old rocks is entirely consistent with the products of early biology being deposited and subjected to alteration by metamorphism. Someone trying to argue that there was literally no life present on Earth should think about what we would predict for carbon isotopic chemistry, or for sulfur isotopic chemistry, if there really was no life, and whether what we see is consistent with those predictions.

CI These ambiguities – the nature of the evidence and how we interpret it – take a different slant if we switch our attention to the Allan Hills meteorite from Mars. That rock was subject to extreme physical processes while getting here. How should we approach these issues when the rock is from a different planet?

AK Since we have no assurance that life arising on another planet would share the same characteristics as terrestrial life, we have to look at the evidence in a Mars meteorite, or any place in the Solar System, and ask whether we see anything that cannot be made by physical processes. That's the same question we ask for the fossil record on Earth. With the Mars meteorite, none of the lines of evidence proposed by David McKay and his colleagues fall outside the envelope of physical and chemical features that might easily be made without biology.

CI Because it was *not* a terrestrial rock, some people insisted on the standard of "extraordinary claims requiring extraordinary evidence," and the team used the conjunct of all the evidence as an argument in itself.

AK Which is not necessarily a bad thing. I don't blame them for making the argument of collective evidence, but the counterpoint is, "Are there reasonable physical processes that could account equally well for what we see?" People at the Johnson Space Center have done experimental work on the shocking of low-temperature materials that may have precipitated in cracks in the rock. It looks like they can simulate, with fair fidelity, what was found in ALH 84001.

There's also no particular reason why all the evidence should come to a single source. Simon Clement makes a good argument that the PAH, polycyclic aromatic hydrocarbons, come from Mars, but even he's willing to agree that they may have originated by meteorite influx. If we then consider the evidence of precipitated carbonates in the cracks in this rock – which may well have a different origin from the organic matter – is there any wisdom in trying to use those as two collective arguments in favor of biology? I don't think so. They probably have different origins. I'll grant that we gain strength if a single process can account for several of the features that have been attributed to life. However, there's a body of evidence suggesting that physical processes, in particular those associated with shocks and shock heating, might well account for several of the lines of evidence that were reported.

CI We take complexity seriously because we're such complex and sophisticated creatures, but we share the planet with simpler, smaller organisms that have been hugely successful in evolutionary terms. Could you put on your biologist hat and then your paleontologist hat and talk about complexity?

AK It's easy for a paleontologist, because complexity is understood in morphological terms. In layman's language that's like saying, "How many adjectives do you need to describe an organism fully?" We can describe most bacteria in very few words. To describe an elephant completely requires a much bigger vocabulary. There is an operational complexity born of development. The remarkable thing about

A distant world shows stratigraphic features that are very Earth-like. This image of Victoria Crater on Mars was taken by the Opportunity Rover in 2006, near the place where it subsequently descended into the crater. The crater is 700 meters wide, and located on the Meridiani Planum near the Martian equator. Scientists speculate that traces of fossilized microbial life might exist below the surface of the red planet (courtesy NASA/JPL).

morphologically complex forms of life – plants, animals, fungi, even certain types of algae – is that they start out with a single fertilized egg, and then through cell division and differentiation end up producing a body like our own, with several trillion cells and several hundred different types of cells allied together in functioning tissues and organs. All these disparate cells have the same genome; the complexity arises largely because of communication between cells, where one cell makes a molecular product that influences the fate of its neighbor cells. It goes back to information theory in a formal sense, but it is information theory tied to development in complex animals.

CI   We could track the history of life on Earth in terms of the number of genes or the informational complexity of DNA. Do those look the same?

AK   It isn't a smooth trend with time. Animals tend to have more genes than bacteria, but not as many as you might think. Something like *E. coli* has four thousand genes. A human only has twenty-five thousand. A little roundworm a millimeter long has almost the same number as a human; rice or maize have almost twice as many genes as we do. While it is broadly true that larger organisms as a class have more genes than smaller ones, gene number is a poor predictor of complexity. What's important are combinatorial effects in our bodies, wherein a lot of the genes are molecular middle-managers. They take information from one gene and use it to help control whether another gene will be expressed or not. They're not making muscle; they're not making nerve cells; they're just saying, "I get this signal. I am now going to influence whether another gene is going to be expressed or not." It's that ability to control the working of genes that makes complexity possible.

CI   What level of evolutionary inevitability is attached to complexity that led to plants, animals, and then us?

AK   We don't know. Everything we know about the history of life on Earth and the phylogenetic relationships on Earth is consistent with the idea that simple bacteria-like organisms are probably more common in the universe than complex organisms like animals. We don't know how likely it is that once life begins, something as complex as an animal will arise. On our planet, animals only came on the stage when there was enough oxygen in the atmosphere to make animal biology work, physiologically. Had Earth never developed an oxygen-rich atmosphere, it would probably still be a slime planet. The probability of forming complex life is nonzero when you get microbial life, because we're an example of it. But where does it lie between *very unlikely* and *very likely*? That's a subject for Tarot cards at this point.

CI   Evolution is under some threat in the public arena. How do we show that this is a legitimate and vital field where our knowledge is growing, but admit there are places where we don't know the story?

AK   I'll start from the back and work forward. There are lacunae in our understanding of most fields of science, and certainly that's true in evolution. What we *don't* know about the history of life, what we *don't* know about evolutionary process and how we get to from one level of life to the next, what we *don't* know about the origin of life, is all *much* smaller than it was a hundred years ago. And it's a good deal smaller than it was ten years ago. If there weren't still things we don't understand, none of us would be in the business. Scientists are most interested in what we don't know, because those are the exciting opportunities.

Some of the things that are used as sticks to beat evolution by those who have particular motives for doing so are misunderstood, often even by the people using the biggest sticks. A good body of knowledge has been built up in the last ten or fifteen years about the molecular controls on development, and it's very useful in explaining the transitions not only between different species of the same type – different bird species or different monkey species – but also between the larger branches on the tree of life, including going from the simplest protozoan relatives of animals to simple animals like sponges, and from there to more complex animals. The advent of molecular developmental biology has given us an explanatory power for those major evolutionary questions that complements what we've learned from comparative biology and the fossil record.

CI   Life developed on Earth as soon as you could imagine it, with extremophiles radiating quickly into all these ecological niches. Does that tell us anything?

AK   It tells us that once there are robust life forms on a planet, and assuming the planet has the environmental capacity to support that life – globally distributed habitable environments and nutrients – life will spread quickly. It's very hard to appreciate geologic timescales. For us it's almost unimaginable to think about the time that separates us from the people who built the Egyptian pyramids, yet that's just a couple of thousand years. A million years is a long time for microorganisms that might divide every twenty minutes. We can have untold generations on a timescale that's very rapid in terms of Earth's history. The fact that life seems to be widespread on Earth as far back as the rock record can take us doesn't mean that life was magic; it just means that over millions of years, microorganisms – which of course can get blown from one place to another in a number of ways – will find any habitable environment.

I quoted Stanley Miller in my book *Life on a Young Planet* a couple of years ago. He was once asked how long he thought it would take for life to begin. He answered, "I think ten years is probably too short, a century may be too short, but ten thousand years sounds pretty good, and if you can't do it in a million years, you probably can't do it." I like the logic of that. If biology is going to begin on a planet, it's probably because the chemistry of that planet allows self-replicating molecules to get going. The chemistry that leads to life isn't some exotic, improbable process that only occurs because there are untold eons of time in which inherently improbable events can occur. It's probably a chemistry that happens, and if the chemistry happens, it's going to happen relatively rapidly.

The Duck Creek Formation in Western Australia preserves a record of carbonate deposition 1.8 billion years ago. The Paleoproterozoic era was the time when the continents first formed and photosynthetic cyanobacteria evolved. Microfossils from this formation have been used to trace the early history of life on Earth (courtesy Andrew Knoll, Harvard University).

CI   Why did it take so long for life to go multicellular?

AK   There are two ways to look at this. You can ask what kind of genetic structures are required to become multicellular, or you can take the environmental approach, "Will complex life only evolve under certain conditions?" Let's take the second one first. David Catling makes a reasonable argument that you can only be big when there is oxygen around – that there really is no metabolic substitute for oxygen when it comes to large, heterotrophic organisms. Macroscopic animal life, as we know it on Earth, is only possible once you get at least ten or twenty percent of present-day oxygen levels in the atmosphere. The geochemical record suggests that oxygen reached that threshold late in the Precambrian. The geochemical record shifts toward a more modern composition of the atmosphere and oceans at about the same time we first start seeing large animals. The historical record is consistent with the strictures of physiology that say you need a certain amount of oxygen to make animal life viable. On this planet, one of the great timekeepers of the evolution of complexity has been the redox history of the planet. You might get to an oxidizing planetary surface much more quickly on another planet, but you might never get there at all.

In terms of the genetics, in order to make something large, and differentiate cells, with coordinated cells and tissues, you do need a fairly sophisticated system of developmental control – elaborate networks in which genes either inhibit another gene or facilitate its expression. Bacteria, by and large, do not have that capability. There are a few bacteria that can make simple multicellular structures or differentiate a small number of cell types, but nobody is arguing that bacteria will win the complexity race. Eukaryotic cells have a different genetic organization, one that facilitates the development of complexity through an inherently different way of controlling gene expression. It's not surprising that complex multicellularity has evolved in eukaryotes at least half a dozen times. The only two cases that have been studied in any detail are plants and animals, and they seem to use different genes in parts of development, but follow similar genetic logic. The logic is similar because of the underlying similarities of the way the eukaryotic genome is put together. Any planet with the kind of control of gene expression we see in eukaryotes has the genetic possibility of developing complexity – which then interacts with environmental conditions that either will or won't support complex life.

CI   The growth of oxygen was a fairly steady process. How does that relate to the sudden or catastrophic changes to life's envelope – geologic changes, snowball Earth, huge impacts?

AK   If I ever make a T-shirt, it's going to say that "evolution is the interaction between genetic possibility and environmental opportunity." That's what the history of life really tells us. We do need the underlying genetic latencies, but we're only going to break through these thresholds in complexity when those latencies are placed in a world that can support the products of genetic innovation. The Ediacaran radiation of the earliest animals is quite consistent with that kind of genetic–environmental interaction, and the Cambrian explosion of animal life is also consistent. In later times we have been nudged along by extinction events, which have done us the evolutionary favor of removing a lot of ecological dominance and allowing the survivors to explore new genetic pathways. Again, it's genetics interacting with environment.

CI   Is environmental stasis not necessarily a good thing for biological development?

AK   That seems to be our experience on Earth. In order for something really new to come in, we need permissive ecology. In a full world, most mutations are going to be deleterious, in that when the world is full of things that do their jobs *well*, most mutations are going to produce a product that doesn't do the job particularly well. The fate of most mutants in nature is simply to die out. The possibility of real innovation, of getting to some morphological area we've never seen before, increases in a world where mutants that might not be immediately better in function are able to survive and produce. That gives the raw material for natural selection to hone something different. What could give rise to permissive ecology, a world that seems empty to the mutants? One is the appearance of new environments – oxygen-rich environments, or things getting out on to land; these represent absolutely new possibilities for organisms. Another possibility is mass extinctions. In the last five or six hundred million years, ecology has been at its most permissive after major extinctions removed a majority of the preexisting species.

CI   The prolific likely number of habitable places in the universe seems at odds with the conclusions drawn about the scarcity of advanced, intelligent organisms by Ward and Brownlee in *Rare Earth* and Stephen Jay Gould in *Wonderful Life*. How do we frame an expectation with only one example to study?

AK   There are two ways two think about it: theoretically and operationally. Ward and Brownlee suggest in *Rare Earth* that intelligent life in the universe is likely rare because of all the particular features that, had they been even slightly different, would have aborted the path toward intelligent life on Earth. I would criticize that argument in two ways. One, it implicitly suggests that we have found the only route to intelligence – that another planet in another solar system must have the same relationship to a Jupiter to have had a comparably fertile biological history. I don't know any reason to make that assumption. There might be many different routes to complexity. That's one problem with the Rare Earth argument.

Let's say, however, that they are statistically correct, which they may be: maybe only one planet in a million is likely to give rise to life, and maybe only one in a million of those will give rise to intelligent life. If those are the right statistics, then the universe must be *swarming* with intelligent civilizations, billions of them. The vastness of the universe compels us to think that even things that are comparatively rare might be absolutely abundant in the universe.

The operational question is, what do we do about it? In our own Solar System, we can go to other planets and explore for microbial or past life. In our cosmic neighborhood we can use things like Terrestrial Planet Finder and its successors to image the atmospheres of extrasolar planets in a way that might give us some bio-geochemical clues to life. But that's only a tiny part of the universe. For the rest of the universe, the only way we will ever be able to approach the question of the distribution of life is SETI. Intelligent life is probably statistically the rarest form of life in the universe, but for most of the universe it is the *only* kind of life for which we can ever probe.

CI   Scientists react strongly to SETI. What's your take?

AK   They're exploring. I guarantee you that if Christopher Columbus had needed to go through the NSF, he never would have discovered America. [Laughs] Rather than lump SETI with experimental science or observational astronomy, I'd label it as a form of human exploration, and judge it on that basis.

CI   I'm thinking of the beginning of this conversation – you looking for fossils as a kid. If some authority figure had told you, "There's no point in exploring that valley or rock bank, you're not going to learn anything, don't bother going," that wouldn't have stopped you doing it. You might learn something they hadn't anticipated.

AK   That's exactly right. I understand that most of our science is and should be driven by hypothesis testing, but we will be poorer for it if we exclude exploration as a legitimate means for studying nature.

14

Simon Conway Morris

Simon Conway Morris survived the hazards of a childhood exploring rock quarries outside London and now hunts fossils as his day job. He received First Class Honors in geology from the University of Bristol and his PhD from the University of Cambridge. He is a Fellow of the Royal Society, and was first a Reader and then ad hominem Chair of Evolutionary Paleobiology at Cambridge. His work on the Burgess Shale shaped a large part of his career, resulting in his explorations of fossils, evolution, and paleobiology, as well as engaging him in a continuing intellectual debate on the roles of contingency and convergence in evolutionary theory. Conway Morris also participates in discussions pertaining to science and religion, two fields he finds no personal difficulty in cohabiting. He was awarded the Walcott Medal of the National Academy of Sciences, the Lyell Medal of the Geological Society of London, and the Nuffield Foundation's Science Research Fellowship. His books include *The Crucible of Creation* and *Life's Solution*.

CI   Were you a fossil hunter as a young boy?

SCM   Yes, but not manic. Usual sort of stuff – working beneath towering cliffs, stuck in mudslides, and going into dangerous quarries, where we certainly had no right to be, and collecting bits of fossil reptile.

CI   Was it just unguided exploration?

SCM   Pretty much. When I was quite young, my mum gave me a children's book about ancient life. That sparked my imagination. In England we're lucky with respect to the range of strata near the coastal. We spent quite a lot of time in the brick pits, not so far from Cambridge. We were "keen amateurs."

CI   The building of the railways and exposure of the strata must have spurred both geology and paleontology.

SCM   That's true, undoubtedly. The initial impetus was the first canal building. William Smith was the founder of geological maps; the French had a professional map before that, but he managed a whole synthesis. He was an engineer involved in canal building. As people were digging around the country, Smith looked at the strata, and saw a hitherto unexpected order.

       People forget that Darwin was a geologist before he was anything else, as well as a beetle collector. That crystallized into a dawning realization of deep time and the nature of evolution, and the possibility of an origin to life, which Darwin had thought, oxymoronically, might be preserved in the fossil record. English science went from strength to strength over a very remarkable fifty years.

CI   Where were you a student?

SCM   At Bristol University. I came to Cambridge in 1972 to do a PhD with Harry Whittington on the Burgess Shale, and then got a research fellowship here. I also taught for four years after that in the Open University; they probably taught me more than I managed to teach them.

CI   When you were a research student working on the Burgess Shale, did you realize that you were working on such a spectacular and pivotal piece of the fossil record?

SCM   I learned that simply by going to the Smithsonian Institution. Harry Whittington, my supervisor, had gone through a good part of the collections in what is now the National Museum of Natural History, on the Mall in Washington, DC. He was particularly fascinated in arthropods. We pulled open drawer after drawer and started looking at samples under the microscope. The quality of preservation and the number of fossils made it clear that something had escaped people's notice in a slightly surprising way. It wasn't just that there were lots of pretty fossils, it was a sense of being admitted into a previously unrecognized world. Then I just *knew*; I felt I had walked into a treasure trove, and it was all there for the taking. It was wonderful.

CI   I forget the German name for that type of entombment. Why are those treasure troves so rare?

SCM   Very good question. The term is Lagerstätten; it's got that implication of "treasure trove," but I believe it originally comes from mining terminology. Nick Butterfield has noted particular times in Earth history when this type of preservation seems to be remarkably prevalent. He's speculated on the ways bacterial action might be stopped or mediated to allow this preservation, but it's still puzzling.

CI   At this point there are other entombments of fossils in that same time frame that rival the Burgess Shale. Is there a lot more evidence on the table?

SCM A great deal more, especially from China and deposits in Yunnan province. I've been fortunate to be involved with collections in North Greenland, and each one tells a particular tale. Overall, we're now in a much stronger position to define a coherent evolutionary story; there are loads of loose ends, but it's falling together in a very interesting way.

CI   An epistemological question: when you're dealing with evidence from a small number of locations, how much do you worry about the uniqueness of the situation or potential isolation of species – the worry that you can't represent what happened on the Earth by limited samples?

SCM It's an important question. To a first approximation, what we find in one place, we find elsewhere. Several years ago we re-described some extremely strange-looking animals that looked like arthropods. In essence, we argued that they are very primitive deuterostomes; that's the super-group, the super-phylum, including both ourselves as chordates, and also the echinoderms. That's an interesting discovery if it's correct. Those peculiar animals turned up subsequently in British Columbia, not so far from the Burgess Shale. So although there are a number of species known only from one locality, when I go to Chengjiang or Greenland or the Burgess Shale, if I'm not seeing old friends, I'm seeing nephews and nieces.

A huge wall separates us from ancient reality. The standard fossil record gives us one or two windows, and typically they're frosted over, so we're peering the whole time. With the Lagerstätten, we get pretty close. It's not a perfect view, but believe me, it's a beautiful view. Now, if we move along that wall to different positions, we see a similar landscape, giving us some confidence that there is a reality there: that what we see may be imperfectly glimpsed, but it's no fiction.

CI   Let me ask about fieldwork. Can you describe how you learn the art of reading rocks?

SCM Scientifically speaking – I sit at a table with specimens that other people have collected. But I also collect them myself. Expeditions to places like Greenland were fairly serious, with fixed-wing aircraft, trudging across the tundra, not to mention guns and all the things we needed to protect ourselves there. But apart from one or two hairy moments, when we got to the locality, we just spent a couple of weeks sitting on a hillside collecting fossils. On a hillside in Greenland we can't dig very deep because everything's locked in place with permafrost, so we've got to glean what's on top. There may be a romance to it from the outside world, but most of it is not drudgery exactly, but grunt work.

When I show people fossils from the Burgess Shale, they're a bit disappointed – "Oh, is that all?" Just seeing them with the naked eye can be quite disappointing; you have to get your eye close in. If you put them under the microscope and get

The Burgess Shale in Canada is one of the most spectacular fossil finds in the world, a place that gives a snapshot of diversity of life in Earth's oceans 500 million years ago. The preservation of soft body parts is what makes the formation so useful. The Burgess Shale captures the diversification of body plans and phyla that appeared in a geologically short period of time, in an event called the "Cambrian explosion." (courtesy Simon Conway Morris).

the illumination right so you can see the various features, some of which can be quite delicate, the overall tendency is to disregard it as if it were road kill, sort of flattened out. But there's a depth to the fossil material. The trick is to imagine what the animal was like alive – which obviously is a recipe for circular reasoning. The goal is to test what we believe are the appropriate structures, and put them into a phylogenetic context.

CI   I can see the layers of inference. You have squashed and possibly incomplete residues. You've got to recreate a third dimension and imagine the full assembly. Then you've got to go from the form to the inferred function.

SCM   It's a huge amount of inference. There are quite a few fossils, even among what are almost certainly animals, which are difficult to understand. Classic examples are the Ediacarans, the interesting assemblage that just predated the Cambrian explosion. Imagine we find a planet with a real fossil record, and it was staffed effectively by Ediacaran equivalents. Only let's say there was a gamma ray burst and the whole biosphere was destroyed, so all we've got is a fossil record of strange-looking organisms. There are some interesting questions regarding how we would identify an alien biosphere in such contexts.

CI   With his puckish sense of humor, Carl Sagan was responsible for the cameras on Viking. He argued, "You don't want to miss a polar bear if it's there." [Laughs] The images engaged the public in a way that a mission with biology experiments alone wouldn't have done at all.

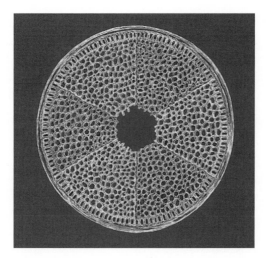

Diatoms are a major component of the web of life in the oceans; most species are single-celled, but some live in colonies. They are generally planktonic. Diatoms reproduce asexually by cell division and they are found in all of Earth's aquatic environments. The exquisite structure is a result of silica biomineralization that forms a pillbox-like shell (or frustule) whose overlapping halves have very intricate and complex markings (courtesy Molecular Expressions).

SCM   I didn't know that story; that's typical Sagan. When I first saw the Viking pictures it was quite sobering, but the Rover pictures are staggering.

CI   In your subject, as in mine, expectations are set by the envelope of accumulated knowledge. Is there tension between your prior expectation and keeping yourself open to something new or unconventional?

SCM   How does one keep an open mind? It's a rather inadequate answer, but you just do. A few years ago, I was invited to talk about biomineralization at a conference in Canada. In the silica biomineralization of diatoms, the molecules involved are a very strange form of protein. Reading about it, I was just gobsmacked. As ever, I didn't know enough about this area, so I went off to the library. What an eye-opener! Diatoms are interesting because they are exquisitely engineered, beautiful structures, made by a bizarre molecular mechanism. It's extraordinary, the molecule has three arms – you'd never guess it could make such exquisite silica skeletons. I never lose my sense of surprise.

Given the competitive nature of science, a lot of my colleagues have a great deal to be proud about, naturally, because they have done extraordinary work. But sometimes I feel they forget that it's what we *don't* know that is more interesting than what we *do*.

CI   Presumably that's a lesson you give your students.

SCM   I'm not very good at it; I'm a hopeless PhD supervisor, just useless. [Laughs] When Derek Briggs and I were lucky enough to work with Harry Whittington back in the seventies on the Burgess Shale, Harry gave us free rein. He gave us all the help he possibly could if we needed him, but we had to get on with it by ourselves. The way education is run makes it more difficult for people to have the confidence to launch themselves and discover things. We were incredibly lucky – partly because we had brilliant material. We couldn't really go wrong, could we?

CI   You learn by doing. I love the quote by Santayana who said, "Art critics talk about theories of art. Artists talk about where to get good turpentine."

SCM   Exactly. I have colleagues who do exquisite work in Evo-Devo or in astronomy. It is a remarkably prosaic, almost dull, fact that we need a piece of machinery to get our data. But the ways we fine-tune these tools to the edge of sensitivity are remarkable.

CI   We only have one biology, one history of life, to study. Is it possible to resolve the issues of contingency and convergence, or chance and necessity, with just one chronology to guide us?

SCM   Indeed, $N = 1$. The failure of imagination means we should be cautious about being too dogmatic about what we might find. Nevertheless, I am convinced that evolutionary convergence is a neglected aspect of the argument. Convergence from the biological and evolutionary viewpoint is well accepted; there's nothing mysterious about it. The way we and insects walk: convergent. The way birds learn to sing, and we talk and sing: convergent. The production of respiratory proteins or certain enzymes, which are absolutely essential for any carbon-based life form: convergent.

Respiratory proteins provide a particularly interesting reverse example; there are three ways to do it. Two are iron-based and one is copper-based. The last one – haemocyanin – is convergent, and there's some evidence that hemoglobin, one of the iron-based ones, is also convergent, though not everybody agrees about that. But crucially, the other iron-based respiratory protein known as hemerythrin has a completely different structure from the more familiar globins. This protein is present in many bacteria, and it's been recruited by animals several times independently.

That doesn't surprise me, since evolution is lazy. We have wonderful examples, for instance, in the crystallins, whereby microbial enzymes have been recruited to make the lens of an eye. Hemerythrin is something that could be used for respiration just as easily; it's there on the evolutionary shelf. But it's hardly used because hemoglobin is better; it is the molecule of choice, not only here on Earth, but I suspect everywhere.

Is there a starting point at which all other things are determined? For instance, would a different genetic code or a different selection of amino acids force evolution into other directions, whereby hemoglobin wouldn't be available? I think that's not going to make a lot of difference; the basic building blocks of life are going to have prebiotic precursors and start off with amino acids like glycine and alanine. We may well have somewhat different amino acids in an extraterrestrial biosphere, but I don't think it's going to be wildly different.

CI   I'm intrigued that you say evolution is "lazy" in terms of basic biochemistry and metabolic processes. Do you mean that not all possible solutions are explored?

SCM   The orthodoxy is that evolution will manage with what it has. There may well be something that's a great deal better, but is effectively impossible to get to – it's on the other side of some invisible chasm. Basic biochemistry, if it's not universal, may only occur with a restricted number of variants – things like the Krebs cycle, and quite possibly photosynthesis.

In terms of more complex structures, superficially there is indeed a vast range of alternatives – but again and again evolution finds similar solutions. For instance, take carbonic anhydrase, which was invented at least three times independently. In each case, the overall structure of the protein differs radically from the other variants. So why are they convergent? Because the active site is identical and revolves around an atom of zinc and three amino acids, usually histidines. At the crucial level of function, carbonic anhydrases are the same.

I'm interested in brushing away what understandably makes us enthusiastic about biology – the natural history and diversity, the bizarre stories that are both entertaining and interesting – and pinning down the fundamental structure of biology. Apart from the Darwinian paradigm, there hasn't been one. My argument for convergence is that if we could identify the common features of convergence at all levels, from molecular to societal, we might be able to move towards a general theory of biology. It would be based on evolution, but we would now have predictability. Say we find an exoplanet, broadly similar to the Earth. I could tell you before we go there that you would find carbonic anhydrase. Like many other examples, it seems to be the molecule of choice. I'll put a million dollars on it, with interest.

CI   Presumably bets get harder to place for more advanced organisms. Exoplanets are going to have a wide range of physical conditions to shape natural selection. Even if there turns out to be a universal biochemistry, does convergence apply to the more advanced stages of evolution?

SCM   We'll see, won't we? I strongly suspect that the envelope may be narrower than expected. Recent work on both cetacean and corvid intelligence shows that they map to an extraordinary extent onto great-ape intelligence. That's convergence. What's particularly intriguing, especially with the crow brain and the parrot brain, is that the macroscopic brain structure is totally different from the mammalian structure. Yet oddly enough the same mentalities are emerging. This also applies between apes and cetaceans; they're both mammals, but the brain structures are rather different. Why be intelligent? Metabolically, it's very expensive. Arguments about social arrangements are probably the most convincing, but there are some intriguing alternatives, such as the ability to manipulate objects. Either way, I believe intelligence is evolutionarily inevitable; it certainly is not a fluke of nature.

But what about planets that are unlike the Earth, even if they are much the same size? If one was dealing with a planet that had a much denser atmosphere, there will be physical and chemical constraints that will dictate the relative Reynolds numbers, for example, of a flying organism. But would that affect the evolution of intelligence itself? I don't see why it should.

We made two programs for National Geographic, about hypothetical alien worlds. One of them was a planet tightly locked to an M-type star, and the other was a moon of a Jupiter-like planet. The Jovian moon, or Blue Moon as it was called, had a very dense atmosphere; we had a lot of fun imagining what the flying organisms and plants might be like in that context. The thinking was driven by convergence. Want a plant that floats in the atmosphere? I'll show you how to make one. Apart from algae-like kelp, it doesn't occur on this planet. But if you want a plant floating in a forest, convergence and functional morphology show how it's going to work, in terms of bladder construction, gas-proofing, hydrogen generation, and so on. You can recombine those into any number of contexts to suit particular planetary conditions.

CI   You didn't mention them, but I presume cephalopods are also interesting?

SCM   The jury is divided. They have been quite neglected. The work by J. Z. Young on the octopus brain was so monumental that until recently people have not quite known where to go next – and of course there are experimental difficulties working with an aquatic animal. One view, I think now fading, says that they are moderately intelligent but are effectively hardwired, more robotic than anything else. Another view, with the evidence accumulating in its favor, says that we've underappreciated the depth of their intelligence. Recent information on octopus personalities and, in my view even more interesting, octopus play, point to a high degree of sentience in these fascinating animals.

Then there have been papers about the way they use their arms, which form quasi-levers. They've got flexible tentacles, but when employed they are broken into subsegments, so that they act as levers. It's quite convergent with the way our arms are built. Correspondingly, there was a wonderful paper in *Science* a couple of years ago; it shows these brilliant examples of octopus-in-mimicry, tiptoeing away from their predators or walking across a lagoon floor on two legs, for goodness sake!

With the TV program on alien life, one of the animals we had living on land was based on asking the question of what would happen if something like a cephalopod really did come onto land – how might it evolve, what might it look like? If we list all the convergences we see in cephalopods – from intelligence to aorta structure, from cartilage to circulatory systems – there's a good overall argument that there will be different combinations which allow the evolution of intelligence, and sooner or later something's going to end up with a human intelligence and make tools. I think that's inevitable, which is why I can't understand why we haven't heard anything.

CI   You're not alone in being provoked by the Fermi question. Brains are expensive metabolically and they're fragile, but do you think the evolutionary advantage of brains is such that they will emerge in other hypothetical biologies, given time?

SCM   Advanced intelligence has evolved independently at least three times on Earth. Crows and parrots are particularly fascinating, but in a different way so are the dolphins; they're mammals like us but in a completely different environment, oceanic, and they draw on a novel sensory mode, echolocation. My argument is that sooner or later you'll make that breakthrough, advanced intelligence, where effectively all bets are off, technology included. I can't imagine what sticking point there is once you start on a biological trajectory.

15

Roger Hanlon

Roger Hanlon's undergraduate years at Florida State University immersed him in oceans and marine biology. When an octopus blew a jet of water on him during a dive off the coast of Panama, Hanlon jumped straight out of the water in fright – but curiosity with these remarkable creatures sparked his career path. He received his Master's and PhD in marine and atmospheric sciences from the University of Miami after serving two years as a Lieutenant in the US Army. Hanlon has been a NATO Postdoctoral Fellow at Cambridge University, and has worked at the Hopkins Marine Station and Friday Harbor Laboratories. He also served as Director of the Marine Resources Center at the Marine Biological Laboratory at Woods Hole, where he is now a Senior Scientist concentrating on cephalopod research. Hanlon is the coauthor with John Messenger of *Cephalopod Behavior*. He has made over six thousand research dives to film and study their behaviors and abilities, and still finds himself as fascinated by their camouflaging and intelligence as he was when he first encountered them.

CI   How does one become an octopus expert?

RH   I grew up in Ohio. The Ohio River didn't do much for me, but my father took me to Florida once and when I saw that blue water, I was hooked. I was not all that serious as an undergrad – more interested in sports and parties – but I eventually got into marine biology. The turning point was swimming on a coral reef in Panama after my junior year. I went over a tide pool and something blew water on me and I rocketed straight out of the water – I've never been so scared in my life. I turned around, and looked in a depression and there was an octopus. And I thought, "How can this little one-pound creature scare the hell out of something as big as me?" I found out these animals had short- and long-term memory and big brains and interesting behaviors. I got into this field by sheer fascination with a bizarre creature. Eventually I went to grad school and I've studied this animal group my whole career.

I'm a marine biologist at the Marine Biological Laboratory at Woods Hole. I'm trained as an ethologist – animal behavior – and I also do behavioral ecology. I've spent a lot of time in the field, underwater, and also testing capabilities of their behavior in the laboratory. Now I'm studying their camouflage capabilities. They have, almost inarguably, the most sophisticated camouflage system on Earth. The principles by which they achieve camouflage are applicable to all animals.

CI   I want to come back to that. First, a general question: where do cephalopods fit in the evolution of life on Earth?

RH   Cephalopods are marine invertebrates – they have no backbone. They are in the phylum Mollusca; the mollusks are shelled oysters and snails. It's a huge, ancient group, and the cephalopods and their predecessors were much more mollusk-like. Imagine a six-foot diameter nautilus shell – that's what they looked like. Ancient cephalopods, which date back to the Cambrian, dominated the seas for the first millions of years. They dominated the oceans until the fishes arrived; fishes dominate today's oceans, both in diversity and biomass.

The cephalopods changed their tactics tremendously in the face of competition from diverse fishes. They got rid of big, heavy, outside shells, characteristic of other mollusks – think of the whelk or the snail – and evolved to be shell-less. They developed big brains and a marvelous suite of sensory organs, especially eyes, and sophisticated skin. My mentor, Martin Wells of Cambridge University, said they learned to live and compete by their wits, not by their external armor.

CI   Some of their attributes tie into vigorous and even heated debates about the role of convergence in evolutionary theory.

RH   You're right. The human eye and the octopus eye are remarkably similar, except that their retina is right-side-out instead of inside-out like ours. Some people use the eye as a case of convergent evolution.

I'm more cautious about cephalopod intelligence, especially compared to human intelligence. I wrote a book with John Messenger, titled *Cephalopod Behaviour*. The driving question in the book is, "Why do cephalopods have such big brains?" We address the intelligence issue, and the book is full of examples of extremely diverse and sophisticated behavior. It takes a big brain to coordinate that. The cephalopods – along with many invertebrates and so-called higher vertebrates – experienced convergent evolution to produce complex behaviors that require memory and learning in a wide range of situations.

These weird, soft-bodied animals have high capabilities, not only through the visual system, but also through the smell system. In the human arena we call it intelligence. I call it extreme adaptivity to a wide range of marine habitats, and protection against the best predators out there. Those predators include marine mammals, twenty thousand species of bony fishes – many of which eat these animals as their main diet – and diving birds. The predators they face are extremely capable. Selective pressures have led to this very odd creature with its head on its foot – "cephalopod." How many animals have a huge brain, and short- and long-term memory, but a life cycle of one year? This is possibly the only animal group that has solved the problems that way. It's truly bizarre – the anatomical and neural cost of such a sophisticated brain system and skin system in these animals over such a short lifespan is hard to explain.

CI   How much of the growth of brains, or the necessity for a capable processing unit, comes from the sheer amount of sensory data?

RH   An octopus has eight arms, each arm has two hundred suckers, and each sucker has ten thousand neurons for taste and receptors for touch – that alone would account for a large part of the neuronal mass of the animal. However, what's clever about the octopus organizationally is that they have separate ganglia of cells just outside the central nervous system that operate the arms. They still have a central nervous system, which is extremely complex, with separate stores for touch learning and a whole separate set of stores for visual learning. Outside of the central nervous system are the neurons to operate the arms: eight times two hundred times ten thousand of them, or fifteen million. They don't clutter up the central nervous system by handling all the musculature and locomotion and sensing. A lot of the sensing is collated outside the brain.

CI   There are interesting issues with the higher functions of animals of any kind – for example, deciding which capabilities are driven by natural selection pressure. Adaptation and survival don't *require* sophisticated strategies or high degrees of information processing.

RH   That's exactly what I'm finding with the camouflage system. This is an incredibly diverse system, but counterintuitively, there are only three camouflage patterns in this animal for all its habitats. In fact, I think there are only three camouflage patterns on Earth. I'm trying to prove that now.

CI   What are they?

RH   One is a uniform pattern, which means uniformly light or dark, with no contrast – all kinds of dogs, bears, fish, insects and octopuses have that. Put a uniform pattern on a uniform background and you'll achieve camouflage. Most of the world is not uniform. On nonuniform backgrounds, animals can put on a mottled pattern, which is relatively small-scale light and dark splotches that

The changing pigment and texture of the octopus skin is controlled by the chromatophores, where pigments are contained in tiny sacs that can expand in surface area by fifty times when activated (courtesy the Tree of Life Project and the University of Arizona).

blend in with the visible background. That's what most people associate with camouflage. The third category is disruptive coloration. Disruptive coloration is counterintuitive: it breaks the animal's visual pattern into large mosaics of light and dark – think panda or zebra. This breaks up the body outline. Sometimes individual bright pieces of that mosaic are shown when there are comparable-sized bright pieces in the visual background, and those parts become a random sample of the others.

CI   A distracting gambit implies higher functioning in the predator, too, because it must actually work as a distraction.

RH   Exactly. It's taking advantage of the sensory capabilities of the predator. The pattern might draw the eye to a light spot that is now represented as one new entity, so the smaller pieces don't add up to the larger shape of the whole animal. In disruptive patterns, animals create false lines and shadows that lead viewers not to see their real edges and lines.

   After analyzing over five thousand images of animals from all phyla, I can fit them all into uniform, mottle, or disruptive categories – with some artistic license – to make the point. If I'm right, or if the animals are right, we essentially have three patterns under two visual mechanisms that seem to fool every visual system ever evolved.

CI   With that fantastic and subtle capability, do they also use coloration socially?

RH   [Laughs] They're color blind! That's probably the most vexing question. How do they achieve color-blind camouflage?

Here's another aspect of this intelligence issue. If I give you an outfit that can change into uniform or mottled or disruptive patterns and say, "Go out in different environments and hide yourself," you have a new problem. You have to decide which of your three patterns is going to work as camouflage for different visual backgrounds. Cephalopods do this brilliantly and uniquely on Earth. And they do it in milliseconds; at the most, in seconds. They have to come to a quick solution or they get eaten and they're out of the gene pool. I have experimental proof that the animals are picking up simple visual cues that tell them which pattern category to use. For disruptive patterning, they only pick up large light areas in the immediate visual background, and those large light areas have to be about the same brightness and total area as the so-called "white square" they put in their skin neuro-physiologically that tells them to use disruption as the main visual trick. It's been bugging me for thirty years, and we've finally got an experimental grip on it.

CI   How do you distinguish between behaviors that are innate or learned?

RH   It's tough. The initial guess is innate. We challenged the animals in the laboratory with unnatural, bizarre, high-sensory overload backgrounds, and it's taken them much longer to make the decision. We did an uncontrolled experiment to see whether they could perform better with a really bizarre pattern if they had some experience with it. We let them stay in it for twelve hours continuously, or forty-eight hours, or one hour, and tested them again. The ones that were in there for two days and had a chance to get used to it were then taken out and tested. They made a choice instantly, so there was learning when they encountered a particularly novel, difficult background. We know they have very good learning and memory, especially for spatial situations. Most of the time they don't use such capabilities for camouflage, but they do if they encounter a new or difficult visual background.

CI   Can they use these same capabilities as predators, to recognize the camouflage of the things they eat?

RH   We don't know. That's a great question that hasn't been studied at all. They're color blind, but they are looking at visual backgrounds and putting on these body patterns. The patterns are beautifully color-matched, as well as pattern-matched, brightness-matched, and texture-matched – in the latter case, three-dimensional rugosity of the skin. But they make the color match without color vision, and they do it with great facility.

CI   What's the difference between what you can learn and infer about intelligence from observing cephalopods in the field versus a controlled situation?

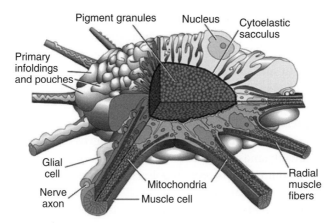

The extraordinary ability of the octopus to blend into its surroundings depends on mimicking texture, as well as color and pattern. The skin of an octopus can change in a fraction of a second, a task helped by distributed brain function. A highly complex layering of skin is involved in the camouflage ability, with different responses to iridescence and polarization, as well as to intensity and color of light (courtesy Roger Hanlon).

RH   I love that question. Fieldwork to me is absolutely critical, because we observe the animal's behavior in the context in which it lives. In the field, as long as we're smart enough about our approach – which means we can't interfere with the animal at all – we let it habituate to us, which we can do quite well. I have had all my main inspirations from fieldwork. I've made over six thousand research dives, so I have spent a lot of time there.

CI   Isn't it hard to be a noninterfering part of a fieldwork experiment?

RH   We learned this by trial and error. We find out where the octopus lives, what den it lives in, mark it, and come back the next morning near dawn. We'll put one dive team after another there, like a rock, at a distance – very discreet. The animal will come out, peer around, and when it's happy that everything out there is okay, including us, it will start foraging naturally. Then we follow it at a distance and videotape its behavior. By doing that we become an acceptable part of the landscape. We're not invisible, but we're not dangerous.

      This brings up a keen question about their intelligence. They have to make an assessment as to whether we're dangerous or not. They gauge us very rapidly and carefully; if we sensitize them by chasing them – which I've done – they'll treat us like very dangerous critters. They assess our behavior continuously, and judge whether we're predators or not. They remember us over successive days.

CI   What happens if you try and get close, and habituate them into getting close?

RH   I've got plenty of video where I'm all over them and they're so acclimated to me that they're doing all their behaviors between my legs or on my arm.

CI   Is that also high-level behavior?

RH   I think so. These animals are watching other fishes and other animals – they're continually assessing who is dangerous and who isn't. It's reflected in their behavioral changes. One of the hardest things to learn doing the diving research was how to approach them and get them completely habituated to us. Imagine being a cephalopod on a remote coral reef. This creature comes up and it's not only big, but it's making a lot of noise, there are bubbles all over it, and it's got shiny equipment on – all of those are really, truly bizarre attributes. But these animals, for the most part, habituate very quickly. That takes brain processing. They use criteria and they take calculated risks.

CI   With the vast amount of fieldwork you've done, do you still observe behaviors that surprise you or that you don't understand?

RH   All the time. That's the beauty of it. When that stops, I'll probably lose interest. But it's not going to stop.
     We've obtained the best data by what we, in behavioral circles, call focal animal sampling. We focus on one animal, and we follow it and film it doing its natural behavior – it's a disciplined form of collecting data on behavior. By doing that, we've made major breakthroughs in the last ten years. We've seen sequences of behavioral dynamics that we would never see any other way. It's hard, because there are other behaviors going on near us, including predators or prey, and we have to hang with the one animal. That's led us to see some quite incredible behaviors.

CI   How much variation in behavior among individuals do you see with that method?

RH   There's a difference between individuals. How much? I don't know. When we're studying cuttlefish or squid or octopus, there are generalities among the animals, but there are also nuances of how they express it and what they do, and how they combine those behaviors. One of my colleagues calls this "octopus personalities" – I abhor that term because it's human and I don't think we have to personify it, but the concept is legitimate. These animals – like all animals of a given species – have a significant amount of individual variation. That's what the natural selection acts on.

CI   What about controlled experiments? Presumably there you can access memory or learned behavior in a cleaner way. What do you learn?

RH   Imagine you're living on a coral reef and you're a bottom-dwelling octopus. These animals have a den that they keep as a home for short periods of time, maybe a few days or a few weeks at the most. They hide there most of the twenty-four-hour cycle, but they'll come out for long forages on average twice a day, for anywhere from two to four hours in the morning and again in the afternoon, during which they're exposed to predators as they're trying to find their food. They traipse around coral reefs, which are spectacularly complex, three-dimensional environments, but they can find their way back *every* time. They can do that kind of maze learning or spatial learning exceptionally well – far better than my diving volunteers, who always get lost. [Laughs]

CI   That's sobering, that the researchers are the ones getting lost.

RH   It's unbelievable. When you see that kind of behavior, you need to know how it's done. We set up a tank system, cut out all the variables, and included only a few visual cues for the animal. We gave it a potential den. We trained the animal, gave it a testing period until it performed the test quickly – cephalopods tend to learn very fast, and in just a few trials. The mistakes early investigators made, me included, is we gave them too many trials. They learn it almost right away; then they get bored and don't pay attention, and the data make them look like they're stupid. But it's just the opposite – we're stupid.

     Once the animals learn the landmarks we've provided, we change them during the night, or when the animal is out of the tank, and then we see how they adjust. Or we take them out and wait a few weeks and test their memory, based on how many exposures they had. We studied memory retention, but also the details of which landmarks they learned. It's too hard to study the animal in the field – we can't control things – but in the lab we can control the variables and give it only one task. These animals are quite good for those kinds of behavioral experiments.

CI   The encephalization ratio is as high as the larger land animals. What's the upper end of the size scale for the cephalopod brain?

RH   If you compare brain weight to body size of an octopus, a bird, a mammal, and a fish, the cephalopods rank right up there with birds and fishes, and very close to mammals. They have a lot of brain per body weight. That's a crude measure; it doesn't tell us much. For example, compare a pygmy octopus weighing twenty grams and a giant pacific octopus weighing two hundred pounds. The brain of the giant pacific octopus is much larger, but it doesn't seem to be more capable. We don't see a relationship between animal size, brain size, and behavior.

CI   That may be true of mammals too, if you consider bird brains. The octopus has great manipulative capabilities, which leads to the obvious question: could they use tools?

RH   [Laughs] Tool use is a slippery concept. We have a project for DARPA for which we're trying to create a new class of robotic arms – in this case, based on the ultimate flexible appendix: the octopus arm. We're using it as "bio-inspiration," to learn exactly what an octopus arm can do. An octopus has a phenomenal range of functions with its arms and suckers. I've called the octopus arm "a device that delivers a sucker somewhere to do some work." I used to think the arm was doing an awful lot. But the arms and the suckers work in concert, and the suckers do nearly all the work. These animals will touch and manipulate things to the nth degree; they're endlessly inquisitive, and most of that is not visually guided. The animals will look at something and get to it, but the operation of the arms and the suckers is done mainly through tactile coordination and sensing. Where does the use of tools fit into that? An octopus will take things near it and build a fortress, a place to hide. Some people would define that as tool use, and some wouldn't.

The octopus has complex behavior and the ability to learn, but as a mostly solitary animal it is difficult to observe learning directly. Recent research has shown that mating rituals are surprisingly complex; here the male octopus' mating arm is inserted into the female's mantle. In mating behaviors, the full versatility of the camouflage comes into play (courtesy Roy Caldwell, UC Berkeley).

CI   My question is probably poorly posed. They have an extraordinary, flexible, projective capability with tactile sensors, all integrated into units. They've already solved some of the problems faced by other animals with appendages of a particular shape and function that might make use of a tool. They bypassed the tool question in favor of a solution that's more broadly flexible.

RH   I love your way of thinking. That's exactly what I think.

CI   Let me return to the question of learned behavior.

RH   Observational learning is a very carefully defined term. I don't think octopuses learn from specifics, because they're asocial and they don't see many octopuses. They sit in their den and look at the world a lot; even when they're out foraging, they'll stop and look for a long time. And they aren't sleeping. They're using their cognition and memory to learn what their prey do and what their predators do, and how they will move around that environment. They have that capability, but no one's proved it. Observationally, we see behavior that would suit that explanation, which is to say they're not motoring out there, making mistakes, doing the wrong things. They're going slowly, looking at what's around them, and doing the right things.

CI   Do they play or recreate?

RH   Another great question. They will investigate novel objects, but I don't know of a serious, good, scientific paper that has addressed or shown play. They are curious animals – they will approach and look at things, and touch them and manipulate them. We need what's called an ethogram of an animal – that is, its total inventory of behavior – before we can sort out play. I've done ethograms of body-patterning behavior, but I haven't done enough to consider play. It's hard to sort out play from investigational curiosity, especially with an octopus. If you put an octopus in a tank, it'll have its eight arms going everywhere, investigating everything.

CI   Astronomers, and SETI folk in particular, are looking for alien intelligence, but maybe we share our planet with alien intelligence and can learn a lot about the possibilities out there from what's right here.

RH   I agree completely, to the extent that I interested the BBC in a natural history film called *Aliens from Inner Space*. It was all about the intelligence and behavior of cephalopods. It's certainly an alien body style, even among marine creatures. They come up with novel ways to solve problems. In my mind, that's intelligence. If we find a planet with water on it, or some equivalent medium that's necessary for all living organisms, we're going to have to look for intelligence in that medium, in that context, however strange it may be.

Intelligence comes in bizarre forms. For that reason alone, the octopus is worth talking about. It looks so alien and stupid, like a piece of Jell-O, but then you see what it actually does! Cuttlefish can perform sexual mimicry, using their body patterning to camouflage their sex. It's an extremely clever cognitive ability if you start thinking about ways people would like to disguise themselves to look like something else; we can relate to that in human behavior and intelligence. Here's an animal as low on the evolutionary scale of complexity as a marine invertebrate, doing something very sophisticated. If we're going to look for life on other planets and decide whether it's intelligent or not, we might think an awful lot about what those organisms are doing to compete well in, or even dominate, certain environments, however small and recently evolved they are.

CI   With SETI, we try to communicate with intelligence of unknown function and form. Could the communication barrier be broken with a creature as different as an octopus?

RH   Cephalopods may have a hidden channel of communication. We know they are highly visual animals – they put on beautiful body patterns that you and I can see, and they can either disappear or be highly conspicuous. But they have another set of cell types that are not pigmented and cause iridescence, a structural reflectance. Those particular cells put out a polarized signal, and we know these animals can see polarization. We think cuttlefish or squid can communicate in very different ways from our conception of communication. They can stay camouflaged in the presence of a normal visual predator, while communicating through the polarized signal, which is hidden to the predator which can't see polarized light.

In communication, sometimes the response *is* no response. Other times, the response is overt behavior. It's very hard to prove that the receiver got the signal and acted on it, because unless we see an overt behavior in return, we don't know if communication occurred. It might be that we'll get to a different planet and want to communicate with something. We may send a signal out, tune it to the animal's sensory system, and try several different methods. The problem is: how are we going to detect a response? That's the hardest question when looking at different environments, and we face that as zoologists all the time. That's why there's so much data on bird song: there's almost always a reply by the other bird. But other animals are communicating, and the response doesn't come right away, or comes through a different channel, and then we don't know what to study next. Communication with alien intelligence, on Earth or far away, is going to be inherently difficult.

CI   It's a fantastic challenge in the fieldwork, to approach that kind of awareness.

RH   If we send a probe to a different planet and put out signals, we're going to have to be able to receive them in as many forms as possible, including strange things like extremely long radio waves or polarized radiation. Cephalopods transmit information from one animal to another by taste, touch, smell, and vision. We ignored a lot of those other sensory modalities for a long time because they're such visual animals. Only recently have we discovered with good experiments that they're using a variety of other sensory systems.

CI   It doesn't sound like you'll ever get bored and move on to another field.

RH   I'm only scratching the surface. It's a world of biological discovery out there, and we've been able to make some fun discoveries. That's the excitement for me. New species in my animal group are still being discovered, for example, on coral reefs in Indonesia. There's a lot happening, and not many people looking, so there's plenty for me to do.

16

Lori Marino

Lori Marino has interests that span the sciences and the humanities, including noninvasive models of science, animal welfare, advocacy, and ethics. She got a BA in psychology, then an MA in experimental psychology, and finally a PhD in biopsychology from the State University of New York at Albany. She is now a senior lecturer in neuroscience and behavioral biology and an adjunct faculty member in the Department of Psychology at Emory University. Marino works on cetacean and primate intelligence and brain evolution, with particular interests in brain–behavior relationships and self-awareness in other species. This has led to numerous occasions of consulting for, or appearing in, documentaries and films. In 2001, she published the first evidence of mirror self-recognition in bottlenose dolphins in the Proceedings of the National Academy of Sciences, and her work on the evolution of brain size in cetaceans has great relevance for the likelihood of convergent evolution of intelligent species on other planets.

CI You have quite an interdisciplinary background. Tell me about that.

LM My background isn't in astrobiology – it's in psychobiology, or neuroscience and behavior. I always wondered what other minds were like. There is a great variety of minds on this planet, and I took up the study of intelligence in other species. I study not only cognitive abilities, but also the physiology and function of the brain, particularly in members of one of the most intelligent classes of animals on the planet: the cetaceans. I study the brains and behavior of dolphins, whales, and porpoises. Then I compare them with primates, including humans.

Any life form out there is most likely going to have a separate origin – an origin that doesn't have to do with us. Primates and cetaceans are both mammals, but they've had 95 million years of separate evolution with different environmental constraints. Dolphins and whales have complex intelligence, just like a number of primates, including us, but they took a different neuro-anatomical route to get there. Through comparison, I'm trying to determine the principles that govern the development of complex intelligence. Astrobiologists will then be able to make more accurate empirical estimates in terms of convergence and intelligence.

CI You bridge the cultural divide between a hard science and a soft science, biology and psychology. How are the methodologies different?

LM A clinical psychologist studies brains and behavior at a very different level. But somebody interested in the biological basis of behavior is going to be studying the brain and chemistry, so the two disciplines aren't as different as you'd think. They both use the scientific method. When you're very interdisciplinary, when you're broad, people don't know what to do with you.

CI You don't fit into the silos at most universities.

LM Right! There are career ramifications. There's much greater difficulty in obtaining positions where what they want and what you do aren't neatly aligned.

CI What about funding?

LM Funding is also more difficult. I've had National Science Foundation funding and some funding from private sources, but I can't apply to NIH and say I'm going to look at something that will eventually cure Alzheimer's disease. Although the astrobiology community has been extremely interested in the kind of work I do, it has taken a while to be accepted. I've had to convince people that what I'm doing is relevant to what they want. It's definitely more difficult. But my colleague Kathryn Denning from York University and I have received support from the NASA Astrobiology Institute to develop a database on intelligence for the astrobiology community. We plan to follow that up with web seminars and conferences. I think astrobiology has finally come to embrace the study of intelligence as part of its mission.

CI There's a perception that our culture is special, that there's something about us that truly places us apart from all other creatures on Earth. It's assumed to be our superior intelligence, which has also led to our technology. In what ways are we and aren't we special?

LM   Special means many different things. All kinds of species have special abilities. For instance, the ability to fly is special. I don't deny that we're sophisticated and we have language, which give us capabilities beyond other species. But I don't think we're qualitatively different. Other species have aspects of tool usage and technology. Though we build 747s and chimpanzees make termite sticks, these activities may not be so different on a qualitative level. Language is a tough issue because we can't understand the communication of other species. We have hints that some other species might have complex communication ability, but we can't crack the code. There's complexity in the sound production of dolphins and whales, but we're not even close to knowing what they're saying to each other.

CI   Perhaps they're the aliens we share the planet with. We can't communicate with primates, which share 99 percent of our DNA. So what are the odds we'll be able to communicate with aliens of unknown function and form?

LM   We've been applying objective, quantitative techniques to comparisons across species. Laurance Doyle at the SETI Institute and Brenda McCowan at UC Davis School of Veterinary Medicine have applied information theory to communication repertoires in different species, and what they've found is astounding. They can't tell us what dolphins or elephants or sea lions are saying, but they can tell us about the structure of their communication system.

CI   When people think of cetaceans, they may remember clever tricks they saw at Sea World. What are the capabilities and levels of intelligence of cetaceans?

LM   There are twenty-seven species of cetaceans and we've only studied a handful to any extent, but cetaceans have many cognitive abilities that we always thought were restricted to humans or apes, including self-awareness – the ability to recognize themselves in mirrors – which is very rare in the animal kingdom. Their capacity for understanding an artificial language, a symbol-based language, is equivalent to what you'd find in chimps.

CI   How do you get a marine mammal to work with symbols and language?

LM   You present them with a series of sounds or a series of visual displays that are paired with objects, and you build up a vocabulary. Then you use that vocabulary to probe and ask questions about what they understand. Lou Herman has done a lot of work over the years showing that bottlenose dolphins can answer questions like, "Is there a green hula hoop in the pool?" or following instructions like, "Take the blue ball and put it inside the red box." It may seem simplistic to us, but those abilities are very rare in the animal kingdom.

CI   I've heard that chimps go further, and use elements of the language and combine them in anticipated ways.

LM   We haven't studied language production in dolphins because that's even more difficult than with a chimpanzee, which can use its hand to press a keyboard or make signs. Another key aspect is the ecology of natural behavior. Humans and many primate species have extremely complex political lives. There is coalition formation, cooperation, flexibility, and interactions you see in a complex society.

We see this in dolphins and whales, so the behavioral ecology is similar in a fully aquatic mammal. Then there's the brain, which is my main focus.

CI   You have studied the historical record of brain size in orcas and dolphins, right?

LM   I studied brain evolution in over a hundred species.

CI   Human or pre-human brains made some fairly dramatic size increments a million or two million years ago. What does your research show?

LM   We recorded the evolutionary pattern of the brain and body size in cetaceans from their terrestrial ancestors right up to modern species. We used computer tomography to look inside the skulls of fossilized whales, going back 40 or 50 million years, to when they were first becoming fully aquatic. We developed a database with sizes of the brain and body using specific measurements of the cranium. Then we applied a statistical technique to pinpoint where and when in their evolutionary history their brains grew the most in size relative to the body. That happened about 35 million years ago. Dolphins have had their large brains a lot longer than humans. Two million years ago, before the appearance of *Homo erectus*, the brainiest animals on the planet were not our ancestors – they were dolphins. That puts things in perspective, because it tells us that things had been different for a very long time, and that this can change in the blink of an eye.

CI   That's intriguing and it leads to lots of question. What kinds of evolutionary pressures lead to brains increasing in size? Why did the cetacean brain size stabilize 35 million years ago? Were they so perfectly adapted that they didn't need to change?

LM   That's the six-million-dollar question. The fossil record shows that the most dramatic increase in brain-to-body-size ratio in dolphins and toothed whales occurred 35 million years ago. After that, brains continued to increase in size in some lineages, but by about 15 million years ago most dolphin brains stabilized in relative size. Some dolphin species did evolve larger brains 15 million years ago, and then stabilized.

    We're looking at patterns between brains, ecology, and cognition, and trying to identify correlations that allow us to evaluate what kinds of environments support and even encourage the evolution of large brains. There are similarities across a number of species in the correlation between brain size and ecology. Brainy species within each group – including birds and insects – tend to be socially complex. They also tend to have evolved in chaotic environments and to have more sophisticated communication systems. There are clusters of characteristics that go with having the largest brain within your taxonomic group. We see signatures of complex intelligence across the board, and if we can apply rigorous methods to looking at those patterns, we might be able to pick out some first principles about what kinds of ecologies encourage large brains.

CI   There are some highly successful species that have done exceptionally well and haven't really changed.

LM   Absolutely! And some species developed large brains despite the drawbacks, one of the main problems being energetics.

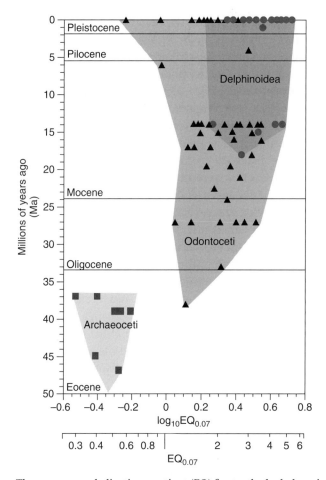

The mean encephalization quotient (EQ) for toothed whales – including dolphins, porpoises and belugas, and their ancestors – as measured from fossilized skull casings. Over nearly 50 million years the brains showed two evolutionary surges, taking the maximum EQ to not far below the level of primates (courtesy Lori Marino, Emory University).

CI   Oh, I thought you were going to say existential angst.

LM   [Laughs] Well, existential angst certainly exists, but nature doesn't care about whether we are happy or not. For us and for many other animals, like dolphins and chimpanzees and parrots – anything with a large brain for its body size – it's a monumental task to keep the brain going.

CI   What do you mean by that?

LM   Brains are energetically expensive. They eat up a lot of the body's metabolism. From the point of view of survival and energy, you need an excellent reason for maintaining this metabolically hungry organ. Our brain weight is only 2 percent of our body, but it uses 20 percent of our metabolism! There has to be a reason why we need this big brain, and why bottlenose dolphins need their big brains, and so on. Despite the energetic costs, some animals are walking or swimming with huge, metabolically expensive brains.

CI   Among all the creatures with large brains, perhaps the fact that we ended up with the ultimate success of controlling our whole environment is a fluke of evolution?

LM   "Fluke" is a strange term because nothing *really* is a fluke, but there's nothing preordained about humans topping the technological ladder either. I'm trying to move our perspective to a more objective view, because it's only in the past 2 million years that humans have been the most encephalized – roughly having the largest brain relative to body weight on this planet. If an alien creature had visited Earth earlier than 2 million years ago and wanted to meet the brainiest species on the planet, they would have passed over our ancestors and visited the dolphins. This scenario tells us that things can change in the blink of an eye evolutionarily, and we shouldn't assume that, just because we're so intelligent *now*, we're always going to be the most intelligent. Nothing in nature suggests that kind of certainty.

CI   What about the argument of evolutionary convergence? In how many lineages of life has intelligence arisen, or substantial-sized brains developed?

LM   Several. We see convergence in a lot of realms; that's what's so interesting. We see convergence across different mammal groups, and across mammals and birds, and to some extent across vertebrates and invertebrates. The brainiest birds, mammals, and even insects have something in common. They're socially complex. They're all tool users. Why should it be that there are few if any brainy vertebrates around that are also extremely solitary? That's interesting. We see functional convergence everywhere, including between dolphins and primates. The neocortex, which is the part of the brain involved in higher-level cognitive processing and memory and so forth, has a very different architecture in dolphins and primates. And yet, the two groups have reached similar cognitive capacities via different neurobiological routes.

CI   In going beyond brain *size* to brain structure, and the implications for behavior and intelligence, what methods do you use?

LM   It's more difficult and more involved than just looking at brain size. There is a number of factors we can look at with brain structure. We can take an in-depth view of the cortex and see its structure: what kinds of cells there are and how they are connected. We're doing that with dolphins. Other people have worked with other species, including a number of primates, so we compare the architecture, how the bricks are laid and how the bricks are connected. We see substantial differences across brains. Two brains may be large, for instance, but they can be built differently. We can look at how the cortex is organizationally mapped. We can also look at scaling factors – such as the connectivity level between the two hemispheres. There are a number of measures of organizational complexity in the brain.

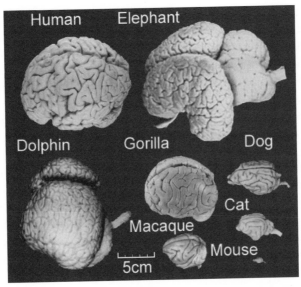

Brain size in a variety of mammal species. Size is a simplistic measure of intelligence, and normally brain mass is divided by body mass to get an encephalization quotient (EQ), which is still an imperfect measure of the superiority of a species. Human EQ is about 7.4, and the next closest is the bottlenose dolphin at 5.4; other examples are chimpanzees at 2.5, and dogs at 1.2. More important than size or ratio to body weight is the degree of folding of the cortex and the neuronal complexity, and that is usually poorly known (courtesy the University of Wisconsin and Michigan State Comparative Mammalian Brain Collections, and the National Museum of Health and Medicine).

CI   There have been human experiments where electrochemical activity can be localized for stimulus and response. Are those experiments done with cetaceans or apes?

LM   For ethical reasons we cannot conduct invasive brain procedures with dolphins, whales or great apes. Those kinds of methods are done with monkeys, but that doesn't condone them on ethical grounds. However, there are ways around that. One of the ways is to use less invasive means. Researchers use brain-imaging techniques to determine function in chimpanzees and dolphins. PET tomography, functional MRI, EEG – all of these technologies are now at the point where we can start applying them to other species.

CI   Has there been research on transmitted knowledge in other species?

LM   If you define culture as learned intergenerational behaviors, then we're not the only animals who have culture. Other big-brained taxa have this capacity as well, including some birds, whales, dolphins, and other primates, to name a few.

CI   It's striking how these intelligent creatures play and have an emotional spectrum in their lives. Who knows what kind of inner lives they have? I visited the orca pods in the Puget Sound, north of Seattle, and learned about their complex social structures and the different dialects of the sub-pods. There was a beach on one of the San Juan Islands where – at significant risk to themselves because of the shallow, curved beach – they rushed in and rubbed their bellies on the pebbles because it felt good.

LM   Intelligence is really about flexibility – coming up with solutions to problems you aren't programmed to solve, new ways of doing things. That's what culture is: manufacturing new ways to interact beneficially with the environment.

CI   It fits perfectly into a standard Darwinian notion, because if you aren't adapting to your environment over many generations, then your behavioral adaptations can accelerate the process.

LM   Exactly. Some ecologies support large brains, and the larger your brain, the more behaviorally flexible you are – the more decoupled you are from the restrictions of the ecology.

CI   We've talked about cetaceans and primates. Many people wonder how intelligent elephants are, and I've read some striking papers on avian intelligence. In terms of the encephalization ratios, what are other interesting species?

LM   The elephant brain is complex. It's highly convoluted as well. In terms of birds, parrots and crows are the most encephalized. Their brains are complex – they're small, but big for their bodies. There is convergence in cognitive and behavioral capacities across birds and mammals. Irene Pepperberg showed that parrots are capable of understanding an artificial language at a similar level to dolphins and chimpanzees. We now know that elephants, and magpies, members of the crow family, can recognize themselves in mirrors. We see that same general pattern with insects. The brains of social insects, such as ants, bees, and termites, are large and complex compared with nonsocial insects.

CI   Social insects seem to have taken steps along the road to a biological neural net, with complex parts and information transmission by pheromones. Should we be contemplating completely different brain architectures?

LM   Absolutely. Even on this planet we see diverse architectures, but they lead to the same behavioral and cognitive patterns. It's a sure bet that any brain that evolves on another planet will be different architecturally than any brain on this planet. But there may be higher-order commonalities. The question is: can we apply analytical techniques to say there's something about being a bee, a crow, and a social carnivore, about being a dolphin, a great ape, and a human, that is similar on a higher level?

CI   Speculations on intelligence elsewhere run into the usual problem: we only have one example to study. There's a spectrum of opinion, ranging from brains and intelligence as a fluke, to those attributes as almost inevitable outcomes of evolution. What do you think?

LM   The contingency versus convergence argument is endless. The argument that we're the only intelligent organism is a total nonstarter; we're throwing all the data out, we're ignoring the evolutionary context of our species. It's as if we popped up *de novo* with our big brains, and nothing else matters. We have an evolutionary history like all the other species, and we're embedded in the fabric of nature. There's nothing about our brains that makes us qualitatively different from other animals. We don't use different principles for processing information. If you allow for all of that, the whole planet is an experiment.

   By opening up a whole planet to study, the argument is based on empirical data. Stephen Jay Gould and others have made a mistake. It's a concrete argument that if you "rewound the tape" of life, you would never get anything like humans. That is almost certainly the case; we may never get another humanoid, another brain like ours, or even similar to ours. But we will get alternatives. I don't think that if you rewound the tape, we'd be here again in exactly the same way. But that's a trivial point. *Something* will likely evolve that has complex intelligence.

CI   Another version of that argument goes that in 4 billion years of evolution, with hundreds of millions of species in the history of the planet, intelligence developed late and rarely.

LM   I appreciate that for a long time on this planet life was microbial – single-celled – and it had to become multicellular in order for something like a brain to appear. However, that didn't take long. Once animals became multicellular, they quickly developed nervous systems and brains. If you get multicellularity, it's inevitable, because there will be a similar problem on every planet, and that's to process information and act accordingly. It's a universal problem, and it's what brains do for a living.

CI   In information terms, the apparatus with which you sense your environment becomes more complex as you evolve, and the data rate increases along with the brain's ability to process it. Eyes are a good example because they've arisen multiple times.

LM   Absolutely! When you get multicellular organisms, you get nervous systems. Single-celled organisms have the rudiments of our own nervous system. Microbes, *E. coli*, paramecium, and amoebas are simple organisms compared to us, but they use the same principle to "process information." They have membranes that are electrochemically excitable; that's how amoebas can identify nutrition, or danger. If there's nutrition, they move towards it, and if there's a dangerous chemical in the environment, they move away from it. The principle they use to decide is the same principle of electrochemical activity used in human neurons. It's never changed. We also use ion exchange – sodium-gated channels and potassium.

CI   Biochemical building blocks seem to be universal. Planets form around stars, and cosmic chemistry produces the same sets of molecules in a similar environment. If there's a clear path from complex biochemical systems to complex pieces of an adapted organism, you can connect the dots to brains.

LM   Yes! There's synergy between the people who study single-celled organisms or simple multicellular organisms within the context of astrobiology, and the people who study mammals. I'd like to see that acknowledged a bit more.

CI   Is there an adaptive advantage to consciousness, as far as anyone can tell?

LM   I've done work on mirror self-recognition in dolphins and consciousness and self-awareness in other animals. Evidence suggests that we're not the only animals with self-awareness. As you can imagine, it's a difficult area of research, but there are ways to make inroads into the problem. The mirror study is one of them. You can also ask how one animal knows what another animal is seeing. You can probe experimentally, if you're in control and you're clever. The work done so far suggests that dolphins, great apes, rhesus monkeys, humans, magpies and elephants all share some capacities related to self-awareness – the ability to know that you have mental states, the ability to have a sense of identity. Some of these animals pass the mirror test while others don't, but they still show the capacity for monitoring their own mental states. So the picture is complicated.

CI   We know most of the higher-order land animals fairly well, but there are several poorly studied creatures that come to mind: for example, the giant octopus. I've heard amazing stories about behaviors in the aquarium and in the wild.

LM   Octopus, cuttlefish, and squid are all highly encephalized for invertebrates. Behavioral studies show that they can process information at a mammalian level. They have behavioral flexibility and can learn in a similar way to some mammals. They're the most encephalized of their taxonomic category. There are taxonomic groups that haven't been studied much, and we have no idea of their capacities.

CI   The ground is shrinking in terms of what marks us as special. But it's unequivocal that we have taken technology to a level – for good and for bad – where we are beyond nature, beyond the normal modes of natural selection, able to mold our global environment.

LM   The fact that humans are in that position right now doesn't tell us whether or not you need a human to get in that position. There's nothing in the biological record that tells us what is going to happen 5 or 10 million years down the road. Another group of animals may be asking the same question then.

We've taken our technology and culture to the highest level, more than any other animal. Our communication system allows us to store and accrue information. Chimpanzees are clever enough to find a new way to hunt for termites, but they can't write a book about it. They can only show the kids how to do it, and pass it on. Cultural evolution is slow if you're a chimpanzee. Humans can write a book, and the next generation doesn't have to slowly relearn information. Accumulated knowledge makes a very big difference, whether you're building 747s or termite mounds.

CI   Writing knowledge, or dealing with the data of your world, is another version of the strategies that have been followed since simpler organisms were sensing their environment. It's not an unnatural progression.

LM   Absolutely not. Language has allowed us to do some amazing things. There's nothing about language per se in humans that is fundamentally different from the communication systems of other animals. Any brain mechanisms that gave rise to the evolution of language are contiguous with the brain mechanisms that occur in all species. Every large brain has, in some sense, the capacity to develop complex abilities like language.

CI   What do you make of the timing argument – the idea that we have developed the technologies to explore the universe, other environments, and other species in a short and recent time span? There was carbon to make planets and organisms 5 or 6 billion years before the Earth formed. If biology in the universe makes brains, it's likely that some will be much more advanced than ours.

LM   Timing is an interesting notion. It tells us that there probably aren't vast numbers of space-faring, telescope-using organisms out there – we would be bombarded. We haven't detected anything up to now, but we've only been *able* to detect in this narrow window of time. We haven't been successful with SETI yet, but I don't think we should be daunted.

   Even if there aren't a lot of spaceship-building, telescope-using cultures out there, there might be something equivalent to a squirrel or a crow, a chimpanzee or a dolphin, on another planet. That would be of great importance. If we went to Europa and found something like a bottlenose dolphin under the ice pack, that would be enormous! Some people would be disappointed because we probably couldn't communicate, but from the point of view of what it says about our place in the universe, it's just as interesting as a species that used a telescope. It would be complex intelligence, but not detectable from a distance.

CI   When people think about astrobiology in popular culture, they would probably not get excited over finding anaerobic bacteria on an extrasolar planet. What they want to know is, "Are we alone or not?"

LM   We want to know if we are alone in terms of our existential situation, and whether there's somebody who can offer an alternative.

CI   Has your close study of intelligence on this planet led you to be optimistic that there's companionship of some form out there?

LM   Yes. But we also need to realize the companionship we have on *this* planet – we aren't as alone as we think. From a scientific point of view, yes, I am optimistic. I don't know if anyone from another planet is ever going to make contact with us, or solve our problems for us, but I certainly understand the emotional need for contact with an organism that could say, "This is how you cheat death." It comes down to mortality. That's what drives people emotionally when they want to see a spaceship land on Earth.

Part III   SOLAR SYSTEM

17

Chris McKay

When the Viking Lander returned evidence that Mars had all the elements necessary to support life, but no life, Christopher McKay wasn't discouraged – he was intrigued. As a PhD student in the University of Colorado's astrogeophysics department, he applied for NASA's Planetary Biology Student Intern Program, which brought him to Ames Research Center. He is a coinvestigator on the Huygens atmospheric structure probe, the Mars Phoenix Lander, and the Mars Science Laboratory. He is currently the Program Scientist for the Robotic Lunar Exploration Program and a director on the Board of the Planetary Society. He was awarded the US Antarctic Service Medal, the Uri Prize of the Division of Planetary Sciences, and the Arthur S. Fleming Award. His research ranges from planetary atmospheres to extremophiles to the origins of life, and his fieldwork reflects that diversity, taking him to Death Valley, the Atacama desert, Siberia, Antarctica, and (in thought if not in person) Mars and Titan. Fieldwork inspires him, and he could easily go on so many excursions that he never came home – but he does return, at least long enough to write up his research.

CI   Tell me about your early history, and your path into astrobiology.

CM   I got interested in astrobiology when Viking landed on Mars and sent back curious results. Here were all the elements needed to support life on a planet, and no evidence of life. I got casually interested, then more engaged. It wasn't called "astrobiology" at the time.

CI   Apart from the ambiguous results of the biological experiments, Viking dampened the frantic speculation from preceding decades because Mars suddenly seemed dry and sterile.

CM   It wasn't what people thought it would be. To me that was part of the puzzle: here was evidence that Mars had water in the past, but why was it so different now? I was intrigued. A lot of people at the time were becoming less enthusiastic about Mars, but for me it was the opposite. If Viking had found life, I probably wouldn't have got interested.

CI   So it was a scientific puzzle. Were you trained as a geologist or a physicist?

CM   I'd been trained as a physicist, period. I did know a microbe from a planet. It was a learning experience to work with people in geology and biology.

CI   What was the career path that took you to Ames?

CM   NASA started a program called the Planetary Biology Student Intern Program, a summer program for graduate students. I applied for it that first year and ended up at Ames, working with Jim Pollack. After the summer they asked me if I'd come back. I was excited to be at Ames because that was where Viking had been put together, the center of NASA astrobiology. That summer was a key event for me. As astrobiology has grown and become more popular, Ames has continued to be heavily involved.

CI   It was a great career choice, because you started as the field was maturing and it grew around you – and now you're at the center!

CM   It's been fun. People used to say, "You're crazy, being interested in this stuff." Now everybody's interested in it, and you've got to work hard to stay on the "crazy" fringe.

CI   Did you experience disincentives early in your career?

CM   I wouldn't say disincentives, but I did get feedback from people who thought it was pointless. In the early eighties, people would talk about what we should do on future Mars missions, and I would push searching for life – evidence of past life early in Martian history. I got a lot of grief about that: "Viking did that, it's over; we're doing other things now and we don't want to hear about life." There were some strong antibodies in the system from the Viking mission. I got arguments against making biology an important part of future Mars missions.

CI   It sounds like you have an iconoclast or contrarian streak that meant you were headed in that direction anyway.

CM   Exactly. I didn't care that they didn't think it was a good idea; I thought it was a good idea and I'd argue back. It's a question of logic. I was sure that eventually this would be what was driving not only human but robotic exploration of Mars.

CI   NASA's a pretty big bureaucracy. Do you ever feel that the entrepreneurial route, or the privatization of space, might lead to more rapid advances?

CM   I'm all for that. I'd be happy if a private rocket company started doing launches for a tenth the cost of NASA. That would be tremendous.

CI   Let's get to Mars. What have we learned from Mars about how terrestrial planets can evolve?

CM   In broad brush, what we've learned from Mars and the fleet of Martian missions is that there *was* water activity, a lot of it, and it extended until surprisingly recently. But the planet as a whole was a dry, cold world, so the water activity was localized. I like to say it's a planet that had rivers and lakes, but no rain. This is what I call the paradox of Mars: the evidence that there was water activity – channels, extensive erosion in localized spots – and at the same time, evidence that on large scale the planet is unweathered basaltic rock, without rain.

That's something we're not familiar with on Earth, but it's not unprecedented – we see it in the Antarctic dry valleys. I've been arguing for some time that what we're learning from these missions is that even when Mars was wet, it was cold. But that's okay – from a biological point of view, we can go to the Antarctic dry valleys and find ice-covered lakes teeming with life; not a problem. The notion that came from Viking and the optimistic interpretation of the models was that Mars had an Earth-like phase. I describe it instead as Mars having an Antarctica-like phase. The one thing it *really* must have had, compared to the present atmosphere, was pressure high enough that water – melting ice or melting snow – could form a stable liquid.

CI   We see evocative pictures suggesting run-off. How recent could that be?

CM   In localized places, some of the so-called gully features, water could have been flowing in the current epoch, the last million years or so – essentially now.

CI   Do we know the census of water on Mars, compared to Earth?

CM   No, we don't. The only direct measurements we have are the water vapor in the atmosphere – which is very small, about a cubic kilometer – the water vapor in the visible polar caps, and the direct detection of ground ice by the Odyssey neutron spectrometer, which was only sensitive to the top meter. Theories based on morphology of craters suggest that there should be massive subsurface ice deeper than one meter. The ground may be ice-saturated kilometers deep. Some theorists suggest that there's even a system of subsurface aquifers, globally connected underneath the frozen ground. But there's no data.

CI   What melts the subsurface ice and bubbles it up to the surface?

CM   There's a lot of debate on that. One school of thought says it's snow melting, not subsurface water. The other school of thought says it's water from subsurface aquifers. Exactly how that water is melting and getting close enough to the surface to come out is unknown. The evidence of these gulley features is clear; the interpretation, the theory as to what causes them, is not so clear.

CI   Does geological activity play any role in water getting to the surface?

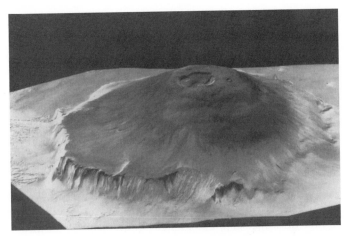

Olympus Mons is the largest volcano on Mars and one that dwarfs any volcano on Earth. It's 370 miles across and towers nearly 17 miles above the Martian surface, three times higher than Mount Everest. The few dozen known Martian meteorites are mostly young, a few hundred million years old, indicating that the planet was geologically active until relatively recently (courtesy NASA).

CM   Mars is geologically quiet compared to Earth, but it's probably not extinct. The best evidence that Mars has activity now is the meteorites. Martian meteorites are volcanic rock, and the age of the youngest is only 150 million years. These rocks, which have landed on Earth, are evidence that there was volcanism on the surface of Mars relatively recently and that volcanism was extensive enough to be a target area for an impact. It couldn't be a tiny fraction of the surface of the planet – the odds of half of the meteorites here being that age are way too small.

CI   Speculate a little on what Mars might have been like 3 to 3.5 billion years ago.

CM   We have evidence that 3.5 billion years ago there was stable liquid water flowing on the surface of Mars. That's the direct conclusion from images from the orbiting Mars Global Surveyor. For that to be the case, Mars must have had a thicker atmosphere to stabilize that liquid. Mars now is close to the pressure at which liquid doesn't even exist thermodynamically, the way CO_2 doesn't exist as a liquid at the surface pressure on Earth. That's about all we can say with confidence: water on the surface and a thicker atmosphere.

I don't think it was necessarily that much warmer than it is today. I don't see evidence that 3.5 billion years ago there was rain, because we see surfaces that old that don't look like they've been eroded. There are some mysteries, such as the northern plains – why are they so smooth? What caused that? Was there really an ocean? There may have been an ice-covered ocean at that time.

CI   What happened to the thick atmosphere?

CM   There are three ways to lose an atmosphere, and there's debate over which one is responsible. One is the combination of a lack of plate tectonics, the formation of carbonates from the CO_2 cycle, and the inability to recycle those carbons. In other words, the carbon gets mineralized. That's why people have been so keen on trying to find carbonates on Mars. I still think that's the best explanation.

The other explanations are more obvious ones: that Mars has lower gravity, thus loses its atmosphere to space. Depending on the model, depending on how you treat the early solar ultraviolet flux, that may or may not be an important factor as well. Another one is the lack of a magnetic field for most of Mars's history, and the resulting impingement of the solar wind on the Martian atmosphere and the loss of CO_2 due to that wind. Depending on how you model the evolution of the Sun, the solar wind can be dominant for the atmosphere. All three of those factors, in relative amounts that we can't gauge, caused the atmosphere to thin, and as the atmosphere thinned the planet got cold. More importantly, as it got cold, the hydrological cycle stopped because water couldn't be a liquid, and it became the cold desert world we see today.

CI   Given what we know about extremophiles on Earth, could Mars and Earth have been equally habitable 3.5 billion years ago?

CM   I think we could say that. In fact, looking back to 4.5 billion years ago, Mars may have been *more* habitable. Earth experienced the catastrophic Moon-forming event; it would not have been a good place. Mars didn't have such a catastrophic early event.

CI   The presumptions about how long it takes to evolve complex life are always based on the only example we know. But I know you've made arguments that it could happen much faster.

CM   The conventional wisdom – not my idea – is that complex life arises in response to oxygen; so the Cambrian explosion is a result of the rise of oxygen. If that's true, it's a powerful handle on this major biological event, the development of complexity. It says that if you can look at the geophysical problem of oxygen rising, then you can deduce information about complexity. There's not a hundred percent agreement among paleontologists that oxygen and complexity have a causal connection, but it's the majority opinion.

CI   Didn't oxygen-producing microbes exist several hundred million years before the oxygen content started rising?

CM   Yes. So the hypothesis hangs together. Photosynthetic algae develop – they make oxygen, so the atmosphere and the ocean system become oxygen-rich. That allows for the development of complex life, because of the energetic efficiency of oxygen. If you accept that hypothesis, you can then ask, "Could there be complex life on Mars?" That question's hard to answer, but you can turn it around and ask, "Could there be oxygen on Mars?" or, "What are the factors that create oxygen?" Well, it's simple: it's just biology. The reason it took so long on Earth was not because biology wasn't making it, but because the Earth was so good at getting rid of it – recycling it, bringing up reducing sediments.

An active planet like the Earth is hard to pollute. If you think of oxygen as pollution, it took life a long time to overwhelm the natural recycling and cleansing mechanisms of Earth. But on a planet like Mars, it wouldn't be as hard. The same biological production rate on Mars as on Earth would produce oxygen in the atmosphere orders of magnitude faster on Mars than it would on Earth. I did a calculation, and concluded that it could be a thousand times faster. In that case, you could speculate that oxygen levels and complexity of life on Mars could have arisen on a timescale of millions of years instead of billions.

CI   And we have such a potential abundance of terrestrial planets that if something like that *could* happen, it probably did happen somewhere.

CM   Exactly. We have to be careful when we take Earth's history as the gospel truth of how life evolved. With complexity, we have a mechanism for timing. We don't have such a mechanism for the origin of life, and we don't have such a mechanism for the origin of intelligence – the other two big events in the history of life on Earth – but we do have a handle on the origin of complexity, and we can extrapolate from that.

CI   I've seen arguments that plate tectonics played a pivotal role in the evolution of atmospheres and the development of life – what's your thought on that?

CM   Peter Ward and Don Brownlee, in their book *Rare Earth*, make the best summary of arguments for this. Earth's habitability over billions of years was maintained by plate tectonics. The history of this idea is interesting. In the sixties, Sagan and Mullen published a paper pointing out the young Sun paradox, which is: "How could the Earth have been habitable 3.5 billion years ago if the Sun was so much different then than it is now?" And they said, "The gases here must have had a different composition, with a thicker, stronger greenhouse gas."

Then Jim Lovelock said, "That's curious: as the Sun changed brightness, the Earth changed its atmospheric composition in just the right way to compensate for that. That's too much of a coincidence." Lovelock argued there must be a feedback mechanism; that there must be a thermostat. He said, "I think the thermostat is biology," and he coined the Gaia hypothesis. But geophysicists, in particular Jim Walker, said, "You're right, there's got to be a thermostat, but I don't think it's the biosphere." He pointed out in an important paper that it was the feedback cycling of plate tectonics, and the carbon cycle, that controlled the atmosphere of the Earth. There's a temperature dependence in the carbon cycle, and particularly in the weathering rate, that tends to stabilize or buffer the Earth at temperatures near the temperatures that allow liquid water to exist, and that weathering requires liquid water. It was an important conceptual breakthrough, and it has made plate tectonics the dominant paradigm for how the Earth has maintained its habitability over 4 billion years.

CI   What would a "dream" NASA mission in the near future do?

CM   If the NASA administrator said, "Here's a couple of billion dollars; do what you think is the best thing to do on Mars," I would send a mission to the south polar region – in fact, to the crash site of the Mars Polar Lander, 76° S, in that ancient ice-ridge-crater terrain with the crustal magnetic features. I would send a sterilized deep drill to go down into that ancient ice and bring back samples of the ancient permafrost material. Then search it, not just for fossils, but for actual preserved, frozen, dead Martian life forms.

CI   Given the uncertainties of subsurface water aquifers, is the door still ajar on continuing microbial life?

CM   It's an open possibility that there's a subsurface ecosystem. The problem is that if that same NASA administrator gave me those few billion dollars, I wouldn't know where to send that mission now. I couldn't point to a place on Mars and say, "Drill here, and we're going to find an aquifer." If we had evidence from ground-penetrating radar that there was indeed an aquifer on Mars, then that would become my number-one choice. But until we have direct evidence of subsurface aquifers, I think our better bet would be to drill in the permafrost, where the water is frozen, because it's holding a record of the early history.

CI   If we got our first tangible evidence of an alternative biology, would it be a pivotal event in the consciousness of the world?

CM   It would be headlined, but it wouldn't be as big a deal as a spaceship landing on the White House lawn, or alien invaders attacking Los Angeles. It would be a big deal in the science community because, for the first time, we would have another example of biology.

CI   And it's identical to ours or it's not. Either way we'd learn something huge.

CM   Exactly. If it's different from ours, then it's really going to be interesting. Pick up any issue of *Science* or *Nature* and you can see that most scientists in the world are biochemists, molecular biologists who work with genes and DNA. Most of those scientists don't care a whit about the space program; they're off doing biology. If we brought back to them another example of life that was a completely different way of doing all the things life on Earth does, they would be fascinated. They might learn something that would help them in their day job, from curing cancer to controlling pests. I think the biggest impact, the revolutionary impact, would be on biological science.

CI   Talking about potential biologies, you've also worked on Titan's atmosphere. Apart from the fact that it pries open the idea of a habitable zone, what can we learn about prebiotic chemistry from Titan?

CM   It's hard to predict what we'll find on Titan. Here is a world with organic molecules and organic energy produced by sunlight, and it has a liquid, but the liquid's not water. Those are interesting ingredients. You could imagine life on Titan that's carbon-based, but it's hard to imagine liquid-methane-based biology because water is such a good solvent. We take for granted the role of water as a prerequisite for life. But we don't know if that's a prerequisite, or if it's just that life on Earth has taken advantage of it.

CI   There's always a tendency to assume that Earth is the best of all possible worlds, but the parameter space of astrobiology may be larger than we imagine.

CM   Right. And the counterpoint is that just because we can't think of how it works, we assume it *can't* work. When I give a seminar and say that we might find an alternative to our biochemistry on Mars, somebody often asks, "What would that alternative biochemistry look like?" And I say, "Well, I don't know."

CI   It's not invalid just because you can't specify it. This is a field where induction is very difficult.

CM   Yes. This is not a question that will be resolved by theory. It will be resolved by observation. It would be like answering the question, "What are the New World organisms like?" If you were a European scientist in pre-Columbian Europe, you couldn't deduce from logic what the New World animals and plants would be. You would have had to go there and look, and I think that's the same for life beyond the Earth.

CI   As an empiricist, I'm sure you'd love to have a ticket to Mars, but you also spend as much time as possible visiting the Mars proxies on Earth. Maybe you can talk about your fieldwork.

CM   I go to places on Earth that are Mars-like, in the microbial ecology sense. The most interesting places are the dry valleys of Antarctica, which are very cold and relatively dry. And the Atacama desert, which as far as we know is the driest place on Earth – incredibly dry, to the point that when I first took my instruments out and recorded two years of data, I thought something must have failed because the signal was so flat in terms of water, rain, or moisture. [Laughs] The Atacama desert is the only place on Earth where Viking could have landed, scooped up soil, and failed to find evidence of life. It would have gotten the same results: no

Taylor Valley in Antarctica is one of the dry valleys used to test for the limits of life on Earth. Despite the barren appearance, there is fairly abundant microbial life in the dry valleys. These high-altitude parts of Antarctica are reasonable proxies for Mars (courtesy Frank Stewart).

organics, no life, but the presence of some kind of chemical reactions in the soil. Yet walk or drive a hundred kilometers south and there are a million bacteria in a gram of soil. This core region of the desert is a little bit of Mars on Earth. In a sense, we're hoping to understand the boundary between them.

CI   I presume our biosensors for the upcoming missions are much more sensitive, so there's no place on Earth where they could land and not find life.

CM   No, they're not more sensitive. One of the problems for Viking was that it didn't heat the samples hot enough to look at refractory organics; it only heated up the samples to 500 °C. In the Atacama, we don't see anything at 500 °C; all the volatile organics are gone – we have to heat it up to 750 °C. Several of us on the Atacama team are trying to push for capabilities that would at least be able to detect what we see in the Atacama. That doesn't guarantee that we'll find something on Mars, but we want to up the capabilities compared to Viking.

CI   I guess it's not just a problem of detectability. As with Viking, it's whether or not the evidence you get is unambiguous.

CM   Yes. There are oxidants in the soil that can mimic biology.

CI   It sounds like fieldwork is a pacing item on preparing for these missions.

CM   That's the way we view it. If you don't know how to do it in the Atacama desert, if you can't identify the oxidants, if you can't detect the organics there, then you're not going to do it on Mars. The converse isn't necessarily true: just because you can do it in the Atacama doesn't guarantee it'll be a success on Mars.

CI   The stakes are pretty high. In the upcoming fleet of Mars missions, which one will have the most sophisticated biogenic experiments?

CM   In the USA, it's the Mars Science Laboratory. It will have a gas chromatic mass spectrometer. That's got the best capability. The Europeans have a mission called ExoMars, which will also have some organic capability. Those instruments will be the next chance we have to analyze samples on Mars for organics. Both the European team and the US team are pushing hard to use the Atacama experience to learn how to do that right.

CI   Do any of these missions have something like Polymerase Chain Reaction so they can search for DNA?

CM   None of them do. I would like eventually to do something like that, even though in my heart of hearts I hope it would fail. I'd hope that if there's life on Mars, it doesn't amplify with PCR. If it does, then it's just the same as us. But PCR is so sensitive that we can't move forward without having done that. We've got to deploy it on Mars in any serious biological search.

CI   You go to Siberia or Mongolia. You seem to like isolated places.

CM   The interesting thing in Siberia, and also in the Canadian Arctic, is the old ice. If I could do a mission to Mars, I would drill into the ancient ice and look for organisms preserved there. The Earth-based lesson for that comes from Siberia and the Canadian Arctic, and also now more recently in the Antarctic, where there's ancient ice. In that ancient ice, we find organisms preserved. Now on Earth, "ancient" means 3 or 8 million years old; on Mars, "ancient" means 3 *billion* years, so it's a lot longer. But we take what we can get on Earth and study the survival of organisms in Earth ice, and then try to extrapolate to Mars.

CI   Have you ever had any difficult or dangerous experiences in the field?

CM   We've had our share of close calls. I have to admit I'm very, very careful – careful to the point of being a real chicken, because the last thing I want to do is fall off a cliff or die in a diving accident in the middle of nowhere. We've never had serious injuries on our field trips because we are so careful. The worst that has happened has been a dive tank bursting open underwater, and some equipment rolling down a ramp towards people but missing them. There's been stuff like that, but we've been lucky and careful.

CI   Do you go out every year?

CM   Several times a year. In fact, if I didn't say no, I'd be gone continuously. Between the summers in Antarctica, Boreal summer in the Arctic, fieldwork in the Atacama, the work we're now doing in Africa, it's like going to conferences – you could easily string together so many trips that you did only that.

CI   Most of us live our professional scientific lives endlessly distracted by e-mails and interruptions. Do you get to think more deeply about your subject when you're out in the wild?

CM   Yes, but not so much because I'm cut off from e-mail. One of the things I really like about fieldwork is that you have all these scientists who come together, so we're all out in the middle of nowhere sitting around the campfire or the dinner table and have excellent discussions. We basically have mini-workshops out there in the field. I find it incredibly stimulating and enjoyable, and that's where we make most of our breakthroughs in understanding, sitting out there in the wilderness, talking about it.

CI   Let's return to Mars. Beyond a sample-return mission and a manned mission, do you think we'll ever have a settlement there?

CM   Yes. I think Mars exploration will follow Antarctic exploration. I don't know when, but I think we will establish a permanent research base on Mars that will be operated somewhat like the permanent research bases in Antarctica. They're small; people don't live there in any real sense; they work there for a certain period of time, a year or less on Antarctica, on Mars maybe two or four years. They will go there on field assignment, and there won't be families; there'll be scientists and engineers doing exploration and staying for a certain amount of time.

That will probably continue for ten or maybe fifty years. The main US station in Antarctica has had scientists continuously since 1955. Nobody *lives* there in any real sense of the word; we haven't colonized Antarctica. I wouldn't even call it a settlement; it's a research outpost, and I think that's what we're going to establish on the Moon, and that's what we're going to establish on Mars. With Antarctica, the motivation for establishing the base was political activity at the height of the Cold War. We've learned about ozone holes, killer whales, and penguins, and science has grown as the base has grown; now that research base is operated essentially as a scientific activity. The political motivations faded long ago.

I think that will happen on Mars, too. Right now there's still a political motivation for human exploration. But as that activity matures and a base is established, science returns will start coming in; people will find interesting things and new results, so the base will become a scientific research outpost. Graduate students will sign up to go there to do their PhD thesis, just as I had two grad students do their PhDs in the Arctic.

CI It's a nice perspective. The visionaries have had a hard time lately. It's nearly forty years since we've been to the Moon, and the Space Shuttle is on its last legs, but do you believe we have a future in space?

CM Absolutely. I don't think it will be soon. It will be when the cost goes down. When a graduate student can do research on Mars as part of his or her thesis, it'll mean that the cost of transportation and support there will have gone down by an order of magnitude. It may be in thirty years, or a hundred. You could say, "What's the rush?" For me, the rush is that I'd like to see it.

18

David Grinspoon

David Grinspoon has long been comfortable exploring the divide between the scientific pursuit of exobiology and the public's perception of it. He earned a PhD in planetary science from the University of Arizona, and then went to work for NASA, where he is now an advisor on space exploration strategy. Currently he is the Curator of Astrobiology at the Denver Museum of Nature and Science, and an Adjunct Professor of Astrophysical and Planetary Science at the University of Colorado. He served as a scientist on the European Space Agency's Venus Express mission. Grinspoon's technical papers and popular writing have been widely published, and he makes frequent appearances on radio and television. His most recent book, *Lonely Planets: The Natural History of Alien Life*, won the PEN literary award for nonfiction. Grinspoon received the Carl Sagan Medal from the American Astronomical Society for excellence in the public communication of planetary science and the search for life. He's an accomplished musician who has played in several rock and jazz bands over the course of his career, but he's happy to keep his day job.

CI   Science was in your blood from your parents; how early was your path decided?

DG   I had an idea by the sixth grade. I remember my first "What do you want to be when you grow up?" answer that went beyond fireman or policeman ...

CI   ... those are always good answers.

DG   Yes, and they're what every little boy wants. In sixth grade, I read a biography of Louis Agassis and decided that I wanted to be an oceanographer. That morphed into space science. I was obsessed with science fiction, starting in fifth grade with Isaac Asimov's juvenile science fiction and going right through junior high and high school. The excitement of space exploration was a formative experience. For my generation, the first Moon landings were real, new, and futuristic. In high school I decided that I was going to be a nuclear physicist.

CI   You must have thought that was a better way to meet girls.

DG   [Laughs] Exactly, I thought nuclear physics would be a *great* way to meet women. But no, it was because I wanted to help perfect nuclear fusion as an energy source and thereby save the world. I went to college thinking I was going to do that, and in my freshman year I was bored by my physics classes, but excited by a couple of classes in planetary science.

      I got a job in my freshman year as an undergraduate research assistant working for the head of the Viking Lander Imaging Team, the camera team, which gave me a chance to work with the first images of the surface of another planet. That was cool, and I was getting paid to do it. By that point I was hooked. When I finished college it was natural to apply to graduate school and continue in planetary science at the University of Arizona. My advisor was John Lewis. Then I did an NRC fellowship at NASA Ames, and Jim Pollock was my advisor. I've had great mentors, including Carl Sagan.

CI   When did people start realizing the complex interplay between the biospheres and geology? When did planetary science connect with astrobiology?

DG   Some people were thinking about it all along. The term "exobiology" was coined by Joshua Liederburg in a *Science* paper around 1960. As soon as we started sending spacecraft to other planets, people worried about planetary protection and the need to explore carefully. There was a fringe of planetary science that was interested in exobiology.

CI   But it was disreputable for quite a while.

DG   That's right. I was lucky to be influenced by some of those disreputable people, including Sagan. I knew him growing up as a family friend and then I worked in his lab as an undergraduate. He urged me to take organic chemistry and biochemistry courses, even though they weren't part of a standard planetary science curriculum. Chris McKay and Jim Pollock were also doing astrobiology, long before it was called "astrobiology." It wasn't renamed astrobiology and made respectable and mainstream until the late nineties.

CI   I sense the iconoclast in you – maybe the disreputability was a draw.

DG   Part of it was being around influential people. Sagan and his associates made it seem respectable, even though they received a certain amount of ridicule from the community. You could get away with doing exobiology as long as you did something else as your "serious science." If you were observing infrared spectra of Mars you could also speculate about microbes on Mars, but you couldn't get funding for the microbe speculation alone the way you can now.

CI   What happened to let the field burst forth in the late nineties?

DG   I attribute it to four things. The most important was the discovery of apparent signs of life in the Martian meteorite ALH 84001, which made huge headlines all around the world, led to a presidential press conference, and led Dan Golden, the NASA administrator at that time, to conclude that people are really excited by this research. It went from something we weren't supposed to talk about to something that tapped into latent public interest.

CI   It was maybe a no-brainer for Golden, because he was also faced with an aging Space Shuttle and the unpopular and hugely expensive International Space Station.

DG   Exactly. That was the watershed event, but three other things were going on. One was the Galileo spacecraft in orbit around Jupiter. The notion that Europa might have an ocean and the other agreed-upon requirements for habitability went from exotic speculation to almost a sure thing. At the same time, extrasolar planets were discovered and that field ramped up rapidly. This term in the Drake equation that had been taken on faith – or at least faith supported by *theory* but not by evidence – suddenly became concrete. There are lots of planets out there. The fourth thing was the discovery of terrestrial extremophiles, which was going on in the eighties but accelerated in the nineties. People started to connect the dots between that and possible extreme environments on other planets.

CI   This work rewrote the book on habitable zones, as well.

DG   Absolutely. The surprises of the Jovian system made us realize that if we actually go out and explore, rather than just relying on theory, our assumptions might be wrong. There could be a lot of activity in realms of the Solar System – and therefore the Galaxy and universe at large – that previously weren't considered remotely habitable.

CI   What would you judge the likelihood of Mars being alive?

DG   I think it's mostly dead. I would love to be proven wrong, but my view is that if a world is mostly dead, it's probably all dead. Looking at Earth as the only example we have, I see life as a phenomenon that thoroughly infests a planet and becomes inculcated in every pore and realm of that planet – in a sense, life transforms it.

CI   It's a paradigm shift – instead of thinking of life as painted on a surface, it forms a biosphere. Life is integrated into the geology and the atmospheric physics.

DG   It's related to the Gaia hypothesis, which was misunderstood and caught a bad rap – people said some silly things, like "The Earth is alive."

CI   New Agers picked it up, too.

DG   Obviously the Earth is not a living organism. It hasn't evolved like a living organ-ism and it can't reproduce. Nonetheless, it has interesting properties – life has fundamentally altered the conditions of the planet. The planet and life have affected each other and coevolved in a way that's very integral to the physical functioning of the Earth, including the realms of the Earth that are not obviously "alive," like the atmosphere itself. Even the crust and plate tectonics have been modulated by life over time. In this view, life is not something that happens on an otherwise dead world – it's something that a planet takes on.

CI   Some of these connections and cycles involve negative feedback and are self-reg-ulating, while others have positive feedback and run rampant. That must make it hard to predict conditions on any particular planet.

DG   That's absolutely true. If this whole relationship is true, planets will be either fla-grantly alive or dead; therefore, because Mars is not fragrantly alive, excuse me, *flagrantly* alive...

CI   [Laughs] It might be fragrantly alive.

DG   Yes, with its methane. But I think life on other planets will mostly be obvious. There are some exceptions to this. When life started on Earth, it was probably very fragile. But it quickly reaches a state where it takes over a planet – as it has the Earth – and becomes a stable entity that can last for billions of years, through catastrophes and planetary changes like those the Earth has suffered. However, it can't hang on in a barely existing state, in isolated pockets on a planet that otherwise dies out – as has been postulated for Mars – because you don't have the reinforcing structure of these global cycles.

CI   The public may focus on whether Mars is alive now, but for astrobiologists the history of biospheres is just as interesting as their existence.

DG   Absolutely, and that opens up a lot of Solar System real-estate exploration. It's not just Mars; there are many other places that could have supported life in the past. Astrobiology is largely a historical science, like geology – we are interested in reconstructing the past to understand the present. Mars isn't boring at all if it's dead; it just creates a different set of scientific problems. And in some ways it solves problems. If we find that Mars has an extant biosphere, extant life, then it raises some ethical issues about what humans ought and ought not to do on Mars that just don't come into play if I'm right and Mars is dead.

CI   You stressed that the dynamism of the system could be the thing that correlates with the likelihood and abundance of life. That's an interesting concept. Maybe you can talk a bit more about dynamic planets.

DG   I'm bothered by the fact that we have this whole science built upon assumptions that come down to one data point that supports them, as everything in astrobi-ology does to some extent. People have written some eloquent papers on why carbon is *probably* the best way to make life, or why water is *probably* the best solvent. They may be right that only on water worlds can you have life, but I'm still bothered by the fact that everything we bring to bear has been learned by us, on a world that has been wholly shaped by what I would call "carb-aqueous" life. We know about the potential for organic life by reverse-engineering life on Earth. I'm not sure we could have learned it if we had to start from first principles and invent carbon-based life through physics and chemistry.

CI   Perhaps we tell "just so" stories of how Earth got to be this way.

DG   That's the fallacy of the Rare Earth hypothesis. I would like to ask if there are other criteria we can apply to habitability, rather than looking for conditions similar to our own and a natural history that mirrors ours. That leads me to think of planetary properties as a whole. What is unusual about the Earth, other than the fact that it has this narrow range of temperatures and pressures and chemical constituents that make it friendly for our kind of life? If you compare Earth to all the other bodies in the Solar System, with the possible exception of Io, it's by far the most geologically active world. I don't think that's a coincidence.

CI   That is one of the Rare Earth arguments: tectonics-as-driver.

DG   Right. Stuart Kauffman talks about life in an abstract sense as a system that uses energy and builds complexity out of flows and gradients of energy and matter, resulting in something that self-replicates, so Darwinian evolution can take over. If you look at that as an abstract idea of what you need – constant flows of energy and nutrients to provide templating building blocks – then you ask, "What kind of environment provides those sources of energy that facilitate complexity?" A planet with continuous geologic activity: it provides not only a source of energy, but also a constant renewal of chemical materials. It provides the ultimate physical basis for cyclic geochemical behavior.

   Those ideas – complexity theory, abstracting life from a thermodynamic point of view, and looking at the Earth's uniqueness as a planetary body with the eternal evolution that facilitates constant cycling of energy and matter – mesh to create this idea of living worlds, where the geologic activity is going to be the most important criterion, maybe even more than liquid water.

CI   Given the range of extremophiles on Earth and the interesting and varied chemical environments in the Solar System, how weird could life be?

DG   I don't think we have a good handle on that from the point of view of theory. Steve Benner has written about the possibility of life in nonpolar solvents and the possibility of using silane, which is made of silicon–carbon chains. I think we're naïve about those possibilities. We're relatively smart about what carbon can do in water because we've had a lot of incentive to study that. Planets are complex, and that's why planetary science is not a reliably predictive science. Life is even more complex than planets, so it's much harder to predict.

CI   Let me roam through the Solar System, asking about habitability. Venus is an old favorite of yours. Do we take it off the list?

DG   I don't think so. Venus was the first planet we explored in the Space Age with actual spacecraft, and it was a big disappointment to know the surface was so hostile to our kind of life. The first successful experiment on any spacecraft that went to another planet was a microwave radiometer on Mariner 2 that proved

The surface of Venus, as seen by Russia's Venera 13 spacecraft, which landed in
1982. Conditions of extreme high pressure and temperatures hot enough to melt
lead are the result of greenhouse warming caused by the thick carbon-dioxide
atmosphere. Venus may have been almost as habitable as the Earth 3 billion years
ago, but if life exists there now it will be unlike any form of life we are familiar
with (courtesy NASA).

Venus was hell. It was a big disappointment because some people thought the
clouds might be made of water, and perhaps it could be clement on the surface.

But we overreacted. There can't be life as we know it on the surface of Venus,
but there is the possibility of life in the clouds of Venus – they're within the right
temperature range for life as we know it, and they are in a continuous dynamic
environment, one with a lot of interesting energy sources and a certain amount
of chemical equilibrium in the atmosphere that has not yet been well explained.
It is an aqueous environment, albeit one that is suffused with concentrated sul-
furic acid.

We keep finding more and more acid-loving life on Earth. We haven't yet
found something that lives in conditions as acidic as the clouds we find on Venus,
but I think when we restrict our imagination to places on other planets where
terrestrial extremophiles could live, we are being extremely conservative. Life
on another planet is not going to be a terrestrial extremophile – it's going to be
something that has adapted to conditions on its own planet. I don't rule out life
in the clouds of Venus.

CI That's a slight echo of Sagan's idea that Jupiter could have buoyant creatures
floating in its clouds.

DG Sagan and Salpeter talked about floaters and sinkers; they made up this whole
imaginary ecology of Jovian life, and there is definitely an echo of that in this
concept. There's a second possibility that's much more far out – maybe there is
some kind of life on the surface of Venus. It couldn't be organic-based life that
needs water, but it's a fertile environment with supercritical CO_2 at the surface,

where it's not really a liquid or a gas. I prefer not to completely rule out something exotic going on there that could be called life, with a completely different chemical basis than our life.

CI   In terms of temperature and pressure, Venus is not too dissimilar to a hot vent at the seafloor on Earth.

DG   That's absolutely true, except that life on Venus could not be made of polymeric carbon compounds. It would have to be a wholly different thing.

CI   You've also speculated about Titan and an acetylene cycle.

DG   It's fun to think about life on Titan, to help us break out of our normal set of assumptions, which are very Earth-centric and might be wrong. Astrobiologists broadly agree on the basic requirements for life. You need an energy source, a liquid medium, and organic chemistry, or at least polymeric chemistry.

We've been interested in Titan for a long time because it's so organically rich and it has some potential for teaching us about prebiotic chemistry and early Earth. People haven't talked much about the potential for life there today, mainly because it's so darn cold and chemistry proceeds very slowly. Then Cassini and Huygens told us that Titan is an active world with a young surface and very few craters. It has apparently recent and ongoing endogenic geology, producing cryovolcanism and flows of various kinds on the surface; it also has active meteorology, forming rivers and other fluvial forms. Titan is a dynamic place, and as a planetary quality that's encouraging.

There are available energy sources; you can't have cryovolcanism without something melting material and gushing it onto the surface. There are reservoirs of liquid hydrocarbons – we've seen evidence for them flowing on the surface, and we've seen the active clouds. There's a methane cycle analogous to Earth's hydrological cycle. There are also liquids – whether you can have life in liquid hydrocarbons is an interesting, unanswered question. Cryovolcanism creates reservoirs of liquid water near the surface. It's probably liquid water–ammonia because ammonia is such good antifreeze, but there is no reason water and ammonia can't make a good basis for life.

Is there any life? Are there nutrients to go along with the energy sources? That's where acetylene comes in. Methane is broken up in the upper atmosphere and is reformed into more complex organics, many of which are dense and would then rain down on the surface. One of these is acetylene, which is energy-rich and apparently there's a lot of it on Titan. We did some simple calculations – reacting acetylene with gaseous hydrogen back into the methane – and it's an exothermic reaction, it releases a lot of energy. Acetylene is on the surface and presumably is mixed into the subsurface, too, because of all the activity turning over the surface. There have to be places – in hotspots underneath the surface – where acetylene is in contact with liquid water. This reaction of acetylene back into methane, which releases a lot of energy, could be some basis for metabolism.

CI   So you don't have to depend on solar radiation – you can get chemical networks to harness that energy.

The pockmarked surface of Jupiter's small moon Io reflects the activity of volcanoes across its surface, one of which can be seen in silhouette on the limb of the moon. Volcanism on Io is caused by tidal flexing of the moon by the much larger, nearby planet. If activity and a source of energy are prime requirements for life, and if sulfur chemistry has the possibility of coding information and function, then Io is potentially habitable (courtesy NASA).

DG   That's right. The sunlight is being harvested in the upper atmosphere by powerful UV photons that split up the methane, and then it's falling to the surface. So it's indirectly solar-powered by upper atmosphere photochemistry. This neatly solves the problem of why the methane hasn't gone away: methane in Titan's upper atmosphere has a lifetime of 10 million years against photochemical destruction.

CI   You make Titan sound more alive than Mars. To wrap up the potential habitable places, would the iconoclast in you like to make a pitch for Io?

DG   Definitely. Io has a lot going against it if you're attached to water- and carbon-based compounds, but if you just like continuous energy sources and are willing to consider other liquids, there are levels within Io with liquid reservoirs of sulfur and perhaps sulfur dioxide. Sulfur is underrated as a basis for life. It has many different phases and allotropes and a lot of strange chemistry that hasn't been completely characterized. It makes long-chain polymers in some conditions. I'm not willing to rule Io out yet. On a scale ranging from worlds that are pretty darn dead – like our Moon – to worlds like Earth that are obviously alive, I would put Io somewhere in between; where you put it depends on how much weight you give these different qualities.

CI   You've made a side study of belief systems. Scientists know that intelligent life beyond Earth will be rarer than microbial life, but they also know that we'll never know without evidence, while a substantial percentage of the public already think there is intelligent life and we've made contact. Where does this strange parallel universe of belief systems come from?

DG   There is a fascination with the idea of aliens – and a certain percentage of the population take for granted that we have already made contact. [Laughs] Maybe nothing would change for them if SETI succeeds, because they already believe it. Nonetheless, if the scientific authorities say, "This is real, we've actually heard from them, they are out there," then belief might go from whatever it is now – fifty percent of the public – to eighty-five percent. But the thirty-five percent that would be swayed would be the highly educated, scientifically literate public, who are overrepresented among those making decisions and running things.

CI   Many astronomers, astrobiologists, and planetary scientists grew up on science fiction and they used that fascination to fuel a scientific pursuit of answers to those questions, whereas part of the general public – who may or may not even read science fiction, but see it represented in pop culture – have bought the whole premise, hook, line, and sinker.

DG   If you poll scientists who grew up on science fiction and ask them, "Do you think they are out there?" most will say, "Yes, we simply don't have evidence yet." The difference is whether there is definitive evidence or not. But, in an odd way, the astrobiology community and the SETI community are in agreement with the UFO community regarding the ultimate question of "Are we alone?" It's just that some of us want to believe.

CI   Like Agent Mulder with his "I Want To Believe" poster.

DG   We are almost all like Agent Mulder. It's hard to find scientists who say there is *no* intelligent life out there. They're so rare that I think most of them enjoy being contrarian and aren't expressing a deeply held, logically derived belief.

CI   Would you care to make a "dime store" Freudian speculation as to why we don't want to be alone?

DG   I can't answer that question without venturing a little bit into the realm of the spiritual, which maybe I should be reluctant to do – "I'm a scientist, dammit, Jim, not a theologian!" [Laughs] I think it arises from a basic, deeply held desire for connection with the wider cosmos and other sentient creatures. What's neat about astrobiology and SETI is that it's simultaneously a scientific quest and a spiritual quest. We are using scientific methods to go about this search that has implications for our place in the universe. Even for scientists who wouldn't really think of themselves in these terms, I think that widespread belief and desire is a spiritual drive we all share.

CI   Let me finish by asking about your other passion: music.

DG   I was in a band called Liquid Earth through high school, college, grad school, postdoc, professorship, and research science. It's always been something I did on the side. What kind of music? It varies. It's always been rock-based, but I also played in a lot of reggae bands and went through an African-music phase in grad school and as a postdoc. Now I'm influenced by hip-hop, African, funk.

CI   Do your more – let's say, rigid – colleagues look at you squint-eyed because you still do that and obviously enjoy it?

DG   I don't know. I was aware of that when I was a postdoc and a young professor. You go through phases when you're more insecure and worry what people think about you. I feel that most of my colleagues appreciate that side of me, as I appreciate the things they do that are beyond strictly science.

CI   Were you ever tempted by that life, which is so different from your day job?

DG   At various times I was *cursed* with being a good enough musician that I knew I would have what it took to be successful if I did *only* that – but I also knew that without their level of dedication, I would get enjoyment but not the success. In the past I've been tempted to pursue that. I know some professional musicians quite well, including some I really admire, and I'm sometimes jealous of their ability to focus on the music and take it to the higher level that I'll never be able to reach without that intensity of focus or time commitment. But I'm not jealous of that lifestyle; in fact, I don't know how they do it. Successful musicians have to be on the road all the time, and that's not something I want to do. I'm satisfied to be a research scientist.

19

Jonathan Lunine

Jonathan Lunine got a taste of Titan and the moons of the Saturn system with the Cassini mission and he hopes to be involved in the next wave of missions to the habitable parts of the outer Solar System. He got his BSc in physics and astronomy from the University of Rochester, and his Master's and PhD degrees from Caltech. Lunine's research interests are broad, including the formation and evolution of giant planets and dwarf stars, the evolution of Titan's atmosphere, and organic chemistry leading to the origin of life. He synthesized his knowledge into two books, *Earth: Evolution of a Habitable World* for Cambridge University Press, and *Astrobiology: A Multidisciplinary Approach*. He is an interdisciplinary scientist for the Cassini mission and the James Webb Space Telescope, and he is Distinguished Visiting Scientist at NASA's Jet Propulsion Lab, where he is also on the Director's Advisory Council. Lunine is a Fellow of the American Geophysical Union and a Fellow of the American Association for the Advancement of Science.

CI   How does a top-flight planetary scientist get sucked into astrobiology?

JL   I got sucked in out of necessity. This field had a lot of traction with the Mars rock and the discovery of extrasolar planets. People who pretended it wasn't there were ignored. I realized that I needed to embrace astrobiology at some level. It's a different name for something we've been doing all along. Now we're doing it consciously.

CI   Writing your book must have been a brutal learning curve. It required a breadth of scholarship that penetrated different disciplines at more than an introductory level.

JL   It was fairly brutal, but I eased into it with a course for nonscience majors, which I started developing in the late eighties. The class became more consciously biological in the mid nineties, when I started writing *Earth: Evolution of a Habitable World*. Talking with colleagues was helpful, as was getting honest reviews. It was definitely not a one-step process; it was a multistep process over several years.

CI   You synthesize as well as contribute to knowledge in your own field. Who have been the great synthesizers in astrobiology in terms of straddling fields and bringing together communities of scientists who may not have been initially receptive?

JL   Long before our generation, before astrobiology was called astrobiology, there was Harold Urey. He bridged several fields that later became planetary science. He felt that the origin of life was an inevitable planetary phenomenon. Then there was Carl Sagan. We think of him as a popularizer, but he also had a deep sense of *why* we explore the universe, and what we're looking for, which is an understanding of life and our origins. He was an astrobiologist before the term was coined.

CI   He was incredibly broad. *Dragons of Eden*, one of his first books, is about human evolution and origins.

JL   Before *Dragons of Eden*, *The Cosmic Connection* transfixed me when I read it as a teenager. Without that book, I would not have made the connections about planetary exploration until much later, if ever. It's hard to say who's leading the field now, because there are a lot of good astrobiologists. Chris McKay is bridging planetary exploration and biology. Jim Kasting's work on habitable zones connects to the search for extrasolar planets. Mario Livio does a good job of making connections between cosmology and why life works; Freeman Dyson was a pioneer in that regard as well. There are also some budding bridge-builders, people like Steve Benner. He started out in genomics, but now he's begun to think about planetary environments.

CI   Writing a book about Earth must have forced you to think about the detailed relationships of the physical and biological properties, but your training as a planetary scientist leads you to look inductively for the general properties of planets. How special is the Earth?

JL   When I started writing *Earth: Evolution of a Habitable World*, the original motivation was a bit pedantic. I wanted to find a way to teach a natural science course that had a theme and was not just a collection of lectures about the universe or the planets. How the Earth became habitable is a core theme of planetary science. In many ways, it's what we're really after.

I had a rather simplistic view of why the Earth was habitable when I started teaching the course and writing the book. The issue of to what extent Gaia-type mechanisms operated was present, but there were also areas we think of as "boring old geology," like plate tectonics. Plate tectonics is probably largely responsible for the persistence of water on the Earth for billions of years. Water has played a profound role in shaping the properties of the Earth's crust and the different kinds of rock. There are lots of subtleties associated with that, which are difficult to transfer to other planets. It's easy to say that Venus didn't have plate tectonics and lost all its water. But how did that happen, and when? What kind of tectonics might have been there instead? We still don't know.

The role life plays becomes an additional layer of subtlety. I don't have a full handle on all of this yet. Nobody does. I was inspired in part by Wally Broker at Columbia University, who wrote a privately printed textbook that looked at the Earth from a geochemical point of view. My perspective, from writing the Earth book and *Astrobiology: A Multidisciplinary Approach*, is that it will be difficult to predict the particular properties of any "terrestrial," or rocky, planet. If you change its size, composition, position around the star, or the nature of the star, you are likely to get different evolutionary outcomes. Even though you can identify similar processes, like the greenhouse effect, plate tectonics, and biology, trajectories of those planets will be dissimilar. We see examples of this unpredictability in the satellite systems of the giant planets. It's hard to predict the properties of any natural satellite system. The Galilean moons of Jupiter look nothing like the system around Saturn.

This doesn't mean other terrestrial planets will be uninhabitable or that habitable planets will be exceedingly rare. The nature of that habitability, the biospheres, the diversity, and the complexity of life on other planets may be quite variable. It's dangerous to say the Earth is the epitome of habitability, because there are some serendipitous properties of the Earth's evolution that caused our planet to be very habitable, as opposed to marginally habitable. On the other hand, I don't buy Ward and Brownlee's argument that we're exceedingly lucky, and everything was tipped in exactly the right way. There may be a lot of potentially habitable planets out there.

CI   Let's talk about habitable worlds. In the increasing census of extrasolar planets, there are some strange solar systems, even admitting that the detection limit is still hovering around Neptune-mass or a bit less. From a theoretical standpoint, do we expect terrestrial planets on nearly circular orbits, or is that an unusual outcome?

JL   My prejudice is that it's not unusual. It's an informed prejudice, because I've been working with Sean Raymond and Tom Quinn on modeling the formation of terrestrial planets. The extrasolar planetary systems we know about are the ones that are easiest to see. Our own Solar System would be tough to detect. Jupiter would be a challenging object to detect by the Doppler spectroscopic technique if it were at 5 AU around a Sun-like star, tens of parsecs away.

CI   We now have a lot of super-Jupiters, and a growing number that are closer to Uranus and Neptune in mass. You've been doing this for about fifteen years. Do you expect to begin finding more varied solar system arrangements?

JL   We're about twenty years into this, if you say that by 1990 we had the ability to measure Doppler effects accurately. Right now there are about 450 giant planets, most of them single objects around stars, but many in multiple systems. They range from abundant super-Jupiters to a steadily growing number of Uranus- and Neptune-sized objects, and even a dozen or so that are just a few times the mass of Earth.

      Ten or fifteen percent of Sun-like stars have giant planets around them. Half of those are systems where the giant planets are close-in, so maybe six percent of Sun-like stars have giant planets that are not so close as to preclude habitable terrestrial worlds. That says two things to me. One is that giant-planet formation is not an uncommon process. The second is that there's plenty of real estate, plenty of empty lots for terrestrial planets, even in systems with giant planets. Giant planets aren't extremely rare, so we shouldn't worry about planet formation in general, and it's not as though most Sun-like stars have giant planets in positions that would rule out terrestrial planets. We're in the middle, which is a good place to be.

      In our own Solar System, the presence of Jupiter probably played a role in accelerating the growth of the Earth to its present habitable form. It might have played a role in supplying water and organics to the Earth by tossing colder bodies inward to the Earth's orbit. It also did us a favor by quickly clearing the Solar System of smaller debris. The initial impact rates and velocities on the Earth were severe, but everything settled down quickly, rather than dragging out for billions of years.

CI   In simulations of planet formation, does the vastness of the parameter space prevent you from determining the percentage of systems with giant planets where the terrestrial planets could coexist in stable orbits?

JL   Yes. It's difficult to make forward predictions of the percentage of Sun-like stars that will end up with systems anything like our own. There are many other uncertainties. Why do giant planets migrate inward? If they migrate inward, do they ruin the system for terrestrial planets, or pave the way for a terrestrial planet system later on? Some dynamicists have done elegant calculations showing how a system can be cleaned out, and giant planets can spiral in and park with enough debris left over to make terrestrial planets. I suspect that these calculations are a better reflection of the skill of the modelers than of reality. We need to get observational evidence. Kepler, which is a Discovery-class mission was launched in 2009 and will discover Earths around nearby Sun-like stars.

CI   Let's talk about "habitable" as an adjective.

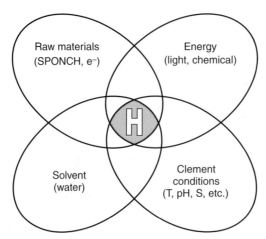

Habitability is difficult to define precisely since we don't yet know the full range of conditions that can permit biology. In this schematic view, it requires a combination of available energy, raw chemical ingredients (SPONCH is sulfur, phosphorus, oxygen, nitrogen, carbon, and hydrogen), a solvent, and suitable physical conditions (courtesy Tori Hoehler, NASA/Ames).

JL   I wish we had a better word for that.

CI   We no longer believe that habitability is defined by a naïve calculation of the distance from a Sun-like star, where water is liquid on the surface of a terrestrial planet. Tidal heating, geothermal energy, and greenhouse warming on moons of giant planets could provide habitability as well as terrestrial planets. How wide would you be willing to open the envelope of that word?

JL   Given our ignorance, we can go in one of two directions. You can go with Rare Earth, the Ward and Brownlee route, and say that the conditions and parameter adjustments needed to make something like the Earth are so narrow and precise that we shouldn't expect to see anything like our own planet anywhere else. Or you can say that planetary processes leave the door wide open to life in many different environments.

Let me focus on the latter for a moment. In our own Solar System, Venus is clearly uninhabitable, but we're getting strong hints that Mars was once habitable. If we could drill into the deep crust, we might find Martian microbes that live on hydrogen. Europa is tidally heated and has a liquid-water ocean. We don't know how close to the surface that ocean is, but there could be life there.

Titan, the large moon of Saturn, is my favorite target. The landscape photographed and sampled by the Huygens probe is weirdly Earth-like. It's shaped by winds and methane rain. It may also be shaped by cryogenic volcanism of melted water and ammonia. When you look at that landscape, you see a nightmarish version of the Earth, something Dante might have described, with organic mud flats, icy ridges, and methane rivers. It doesn't seem like a *great* place for

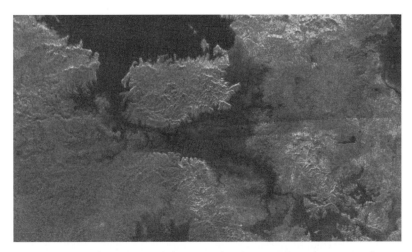

This radar image from the Cassini spacecraft, taken on February 22, 2007, shows hydrocarbon lakes in the north polar region of Saturn's moon, Titan. The island in the middle of the large lake is 60 by 90 miles, the size of the big island of Hawaii (courtesy NASA/JPL).

terrestrial-type life, but some organic chemists have raised the idea that rather than being a good place for the early steps of "life as we know it," Titan could be home to a weird biota that works on hydrocarbon liquids. It depends on how far we stretch our minds.

Enceladus, much smaller than Titan, spouts plumes of water ice mixed with organics. If the source region contains liquid water, there's another potentially habitable world. The variety of different planetary outcomes in our own system is huge. When we look at planets around other stars, we will see other outcomes – some closer to that of the Earth, and some more distinct – that may well allow for life. They might even allow for the evolution of life more rapidly, or more profoundly, than life on Earth.

CI   What's the prognosis for planetary missions in the next decade? When are we going back to Titan? When are we going to Europa to melt through the ice pack?

JL   A few years ago, I would have said we're on the fast track for Terrestrial Planet Finder and going back to Europa, and Titan would have been dependent on what Cassini found, which turned out to be intriguing. Today, I'm more pessimistic about the program overall. We are still considering putting humans on the Moon and Mars, which is going to compete for resources with science programs. NASA's strategic priorities are still very unclear. We could end up high and dry for a while with science programs. It's a shame, because we know what we need to do on Europa and Titan. We also are close to knowing how to build space telescopes to detect and study terrestrial-type planets around other stars.

CI   What's your perspective on our ability to connect the dots between simple chemicals and the first life forms? Is it just a matter of time before we determine all the mechanisms by which the first replicating molecules and cells derived? Or do you think that there will always be gaps in our knowledge?

JL   I was pessimistic when I wrote the Earth book, but I was optimistic when I wrote the astrobiology book. What changed was the fact that biosciences conquered the genomics problem. There had been a lot of pessimistic predictions that we'd never get even a fraction of the human genome sequenced. When that project was accomplished, it demonstrated that biology could "do the impossible" much faster than even many of the optimists had predicted. Proteomics is coming along in many ways as well, and the techniques are becoming more powerful. We will soon have a deep understanding of how life functions, what proteins do, and what sequences build the proteins.

It will then be possible to put enough of the steps together to have a reasonably good story of what aspects of life came first. Was it metabolism with no genome, or were genomic molecules like RNA built outside of the cell and then combined with naturally forming vesicles? The precursor molecules to DNA and RNA and the mechanisms by which RNA form will be understood. The ultimate problem is that the formation of life was undoubtedly messy, and happened innumerable times in innumerable environments. We will not be able to point to a particular time and place and say, "This is where it happened." It's an emergent phenomenon, an epiphenomenon at some level. If you had snapshots of the Earth at different times, you couldn't say, "There's no life now, there's no life now. Wait! There's life."

CI   That sounds like a paradigm shift. What about the tree of life and last common ancestor?

JL   Lateral gene transfer is now thought to be fundamentally important at the most basic level for most primitive life forms. People talk about the earliest forms of life in an environment of "rivers of genes" that move back and forth between different types of organisms. The root of the tree of life may be hopelessly jumbled. The nomenclature also changes. It's no longer the universal ancestor or the last common ancestor, it's now the "last universal common ancestor," with lots of caveats attached. That still only takes us to a point in evolution where there had already been countless generations of living organisms. Identifying a transitional form between nonliving organic chemistry and life itself will be tough. We can enumerate the steps in a general way, but we can't trace the specific set of events leading to any organism we see today.

CI   What's a greater limitation: the fact that the molecular or atomic tracers are hard to find, or that the evidence is not strongly and clearly present in the archaeology of an existing primitive organism?

JL   The latter. For what we think of as truly primitive organisms, there's no good evidence. The roles that RNA plays now are different from the role it might have played in a precursor cell, so there's almost nothing that reflects that earlier function. Another approach is to construct pseudo-primitive organisms that work only on RNA, either in the laboratory or in the computer. You'll never know if your model shows how it really happened at the dawn of life, but you will get a perspective on whether it's a viable approach.

CI   It might be as hard as making planets. Two things strike me about the history of life on Earth. Early on, life formed almost as quickly as imaginable and radiated into outrageously diverse evolutionary niches. Then nothing happened for a long time. Can we explain the gap separating single cells from multicellularity?

JL   Astrobiology hasn't yet found a complete fusion between geologists, who worry about the effects of planetary evolution and climate on life, and biologists, who worry about how evolution happened. The first gap is between the first cells and eukaryotic cells, which might not have been such a huge leap. Eukaryotic-like or symbiotic cells might have been around for most of Earth's history, and we just happen to be the descendents of the most successful forms that took advantage of the changing oxygen content of the atmosphere.

Bacterial colonies, where cells are dependent on each other in a communal arrangement, go back at least several billion years. That's one of the big frontiers of astrobiology. It's crucial in creating and informing our expectations when we look at exoplanets. Should we expect that billions of years is the time required for complex, multicellular life to transform the environment? Photosynthesis came before multicellularity and transformed the atmosphere, but do we have to wait billions of years for macroscopic forms to transform the land? Was it a fluke on the Earth, or were we unusually slow? We don't have a good perspective.

CI   Let me ask you an out-of-the-box question about life...

JL   Forty-two!

CI   Exactly! Does your training in physical science make you susceptible to ideas of information content and complexity, where life could deviate radically from our known example of biology?

JL   Yes, certainly there are possibilities. Talented organic chemists like Steve Benner are always there to remind us that some of the properties of life that we assume to be fundamental are actually workarounds. Although water is good as a polar liquid, it creates lots of problems for biochemistry. Alternative organic biochemistries are conceivable, for example, in liquid hydrocarbons. That would make the ethane–methane seas on Titan habitable environments! In terms of alternative systems that aren't organic, but have the same information content and low entropy, there's no violation of thermodynamics. We were born in a universe that began with reasonably low entropy, and there is absolutely nothing wrong with the idea of self-organizing systems arising spontaneously where there's a large flow of free energy.

Another area that intrigues me is whether we'll get to the point where silicon-based computational systems will have equivalent levels of information content to sophisticated life. We're a long way off with current computers, but it's within the realm of imagination. Science-fiction authors have thought about this for decades and decades. Will these then be "ghosts in the machines?"

CI   What is one major question you would not only like to see the answer to, but think will be answered in five to ten years?

JL   In five years, the breakthroughs are going to be in evolutionary biology and geo-biology, in understanding ancient life on Earth. Five years is too short for real breakthroughs on the astronomical front. In ten years, the big breakthrough is going to be the discovery of Earth-sized planets around other stars. In twenty years, maybe the discovery of life on Titan, or in Europa, or even Enceladus.

CI   If we get a Mars sample-return mission, which would be more interesting or excit-ing: finding microbes with a biological basis identical to ours, or different from ours?

JL   I doubt we're going to be able to find organisms on Mars in the next fifteen years. If they're there, they will be deep and hard to find. If we do find something similar to us, but can't do sequencing on the genomes, we won't know if it was blasted from Earth to Mars from an impact, or whether we share a common origin with something that formed originally on Mars. Interplanetary transfer between Earth and Mars is easy enough that determining a separate origin is a real problem. That experiment is going to be difficult to do on Mars. I'm astounded that Spirit and Opportunity were so easily able to show chemical evidence of water in two places. That tells us that Mars had pervasive wetness in the past. But suppose we returned to Titan and found life in the hydrocarbon seas – something really dif-ferent. It would be astounding. It would open up a whole new set of questions.

20

Carolyn Porco

Carolyn Porco has likened leading an instrument on a planetary space mission to raising a child; nearly twenty years of challenges and satisfaction, frustration and thrills. She got her PhD in planetary sciences from Caltech, working on data from the Voyager mission. In 1990, she was named the leader of the imaging team for the Cassini mission to Saturn and its moons. She's also an imaging scientist on the New Horizons mission to Pluto and the Kuiper Belt. She is a senior research scientist at the Space Science Institute in Boulder, and an adjunct professor at the University of Colorado and the University of Arizona. Porco has been a central figure in the NASA advisory and planning process for planetary exploration, and conveys the excitement of her field as a guest analyst for CNN and in frequent appearances on radio and television shows. She also combines science and art as the CEO of Diamond Sky Productions and has had Asteroid 7231 named after her.

CI   How did you get into science?

CP   I was attracted to science for as long as I could remember – even as a child I found it very appealing. The lack of subjectivity appealed to me especially. Truth suddenly wasn't the opinion of some great authority. Mother Nature was, and is, the final arbiter. To me, science means that there *is* an absolute right and wrong, and it's not determined by someone's opinion. I found that very compelling, and right from the beginning, I gravitated towards it. I was also always very curious, though I can't say that I was a terrifically good student when I was a youngster.

   As I became a young teenager, I got interested in philosophy and religion. When I was about thirteen, I became very interested in Eastern religion, which is what eventually got me into astronomy. I didn't grind telescopes and get into it that way. Philosophy, religion, and thinking about the big questions got me interested in astronomy.

CI   Your interest in big questions could have taken you in a different direction. Why did you choose astronomy rather than physics or biology?

CP   I was interested in understanding our cosmic situation. What are humans doing here? What's out there? Because here we are on this little planet, and it's so big out there. Physics is the basis of all science and life, so I was very drawn to physics.

   I think that religious people get out of religion what I get out of astronomy and participating in the exploration of our Solar System. It means involvement and engagement in something so much bigger than I am; something so much more important and meaningful. People want a connection with something bigger and more eternal than they are. Being a scientist lets me put my mark on the future, and sign my name to the great declaration of human thought. You can't buy that.

CI   You've described the juxtaposition that makes research or science intoxicating: the ability to ask and potentially answer very big questions. It's quite profound.

CP   There's been nothing more intoxicating to me than those moments – and I've had a few of them – when I thought that I had discovered something that nobody else on the planet knew! Eventually, of course, you tell your colleagues, and everyone scrutinizes your idea to make sure it's worthy. But there's nothing more giddy and intoxicating than that brief moment when you know that you've found something about nature that nobody else knows. That's the rush – the "eureka moment."

CI   Tell me about one of those moments.

CP   I got that rush while I was working on Voyager; that started when I was a graduate student. Voyager had discovered the spokes in the rings of Saturn, but no one knew how they behaved. There were so many things discovered by Voyager and there were not enough scientists on the science teams to analyze it all, so some very interesting topics fell into my hands as a graduate student. One of them was the spokes. This research wasn't even promoted or encouraged by Peter Goldreich, my thesis advisor at Caltech. It wasn't that he discouraged it; he just didn't tell me to study it. I looked into the business about the spokes all on my own. They were new and no one had seen them before, so I had to start with something simple. I did a time-series analysis of their appearance in the rings and found that they came and

In the late seventeenth century, Giovanni Cassini first saw structure in the rings of Saturn. It was studied up close by the Voyager probe and then in exquisite detail in the nineteen nineties by the probe named after Cassini. The complex structure of the ring system is caused by the gravitational interplay of ring particles and Saturn's many moons. The Cassini probe bounced radio waves off the rings to show that most of the particles in the rings are a few millimeters up to a few meters across (courtesy Cassini Imaging Team, NASA/JPL, and ESA).

went in time with the magnetic-field period. I was so excited that I forgot to eat and couldn't sleep.

CI   When Isaac Newton really got into a calculation, he would forget to eat and even bathe. His servant would have to come and force him to take care of himself!

CP   There are times when we share that with the greats. Another time was when I was working on the Neptune ring arcs and figured out how they worked. Those things are very exciting. You know you're going to make big discoveries with a mission like Voyager, because you're going to an alien place with equipment that has never been carried there and is a quantum leap over anything that's ever been there before. That's the rush, and that's the fix.

CI   The experience with Voyager early on must have convinced you that you wanted to be associated with space missions in your career.

CP   I went to Caltech because I knew I could work in the planetary program. When I chose my graduate school, I knew that I wanted to be in the American space program. I consider myself very fortunate that I ended up doing what I wanted.

CI   What was it like working on those big teams – as a very young scientist?

CP   It was frightening in the sense that the politics of it was very hard. I was not even a team member when I first started out. I had to hold my own and work in a very competitive environment – the kind where everybody tries to outdo each other and come up with the most brilliant ideas. This happens in real time, and with data coming in and people standing around computer monitors; I did find it rather difficult. Everybody was male in those days; my team today is still mostly male, except for one or two female junior associates.

Women scientists have their work cut out for them. Women in my generation generally feel that it's not acceptable for a female to behave the way an individual has to behave to be successful in a scientific environment. For example, there is a lot of mental arm-wrestling that goes on when scientists get together. Much of a person's reputation depends on how well she handles these circumstances and many women are ill-prepared and don't fare well under such conditions.

CI   It seems ironic that it still happens in space mission teams – they have a very tight, collaborative framework compared to other science projects.

CP   I think it's getting better. For example, I think our female team associates have an easier time of it on Cassini than I did on Voyager, and that's good.

Some of us have already fought those battles. It was difficult breaking in, but I'm from New York, and I grew up with brothers and no sisters. I knew about arguing, fighting, and holding my own because of them. I'm talking real brothers, really aggressive boys. They used to chase me around the house with dead bugs, lock me in closets, and I was forced to play football and baseball at an early age. As a result, I'm not uncomfortable to be the only woman sitting in a roomful of men, because I got used to that as a kid. But you can imagine how difficult it would be for a young woman who's not used to that.

CI   Let me ask about how planetary science is perceived by the public. People know that Earth is special in the Solar System, but they might generically think that the rest of the planets and moons are just rocky or gassy and not very special or interesting. However, when you see them up close, they're worlds with their own personalities and interesting features. Is that a new awareness?

CP   Yes. The field of planetary exploration is barely fifty years old. And we've seen that these bodies *are* worlds. We find that when we look at all of the bodies in the Solar System, almost all of them that have solid surfaces. I'm not talking about comet-sized things, but even looking at the pictures of Comet Tempel-1 from Deep Impact, I'm amazed by its geography. If you look at the larger bodies in our Solar System with solid surfaces, many of them don't have matching halves; one hemisphere doesn't look much like the other. That indicates that planetary-scale processes alter their surfaces. They're not bland surfaces that got pockmarked with impacts for 4 billion years. Many have undergone internal processes that have evolved their surfaces.

CI   It's also changed our thinking about life in the universe. When we look for life in a solar system, we have to consider moons in addition to terrestrial planets, right?

CP   Yes. We also have to consider the interiors of these bodies as well, and not just the surfaces. It seems that given the right ingredients, life can crop up anywhere. Even in places where there isn't any sunlight; some life forms can feed off chemical energy. And where there's abundant chemical energy and water, you might expect to find life.

We can find life at the mid-ocean vents, where there's no sunlight – just lots of chemical energy and lots of heat. But we can also find life in cold places, like the Antarctic ice. That opens up a much broader range of environments throughout the Solar System where we might be able to find life. Organic materials are abundant. There are probably no organic materials on the surface of Mars now because they've been oxidized and torn apart by UV radiation, but if you go into the subsurface where it's warmer there may be liquid water and organic material. It's almost a foregone conclusion that there will be living organisms on other planets. But we don't know for sure, and that's why we explore.

CI   We used to define a habitable zone as a region having a terrestrial planet with a nearly circular orbit where water could exist in its liquid phase on the surface. What would be the definition now?

CP   We can discuss habitable zones for water-based life. The issue is whether we could even recognize life that isn't water-based. I think one of the main goals in exploring the Solar System is to understand what the cosmic or planetary context is for life. I think a cardinal goal is just characterizing the surface and subsurface environments in our Solar System, and then attempting to search for evidence of life or at least prebiotic chemistry in some of these environments.

CI   Processes like tidal heating or subsurface processes that could generate enough energy to support life may be universal. In the Solar System, we've focused our efforts on Mars, Titan, and Europa. Are there any other moons that are intriguing enough to send a probe to if we had the resources?

CP   Callisto and Ganymede. They're also believed to have subsurface oceans.

CI   What's the evidence for that?

CP   Magnetometer evidence. It can be explained by a salty subsurface ocean.

CI   Are these moons big enough to have some sort of internal heating?

CP   Ganymede and Callisto are big – about Titan-sized. Io, Europa, and Ganymede are in a three-body resonance together, so there's probably enough flexure for heating, too. It would be good to study those three nonvolcanic Galilean satellites as a trio. If there's subsurface water, it's probably not as close to the surface as it is on Europa.

CI   People think of Earth as the water planet. Europa was a surprise, and now we seem to have more water worlds. Should we be surprised?

CP   Water is abundant in the outer Solar System – the moons out there are mostly water ice. You also need either enough rocky material to have a heat source – rocky material is mildly radioactive and gives off heat – or you need a process like tidal heating, something to flex to produce heat. Currently, we think liquid water is a prerequisite for life. Enceladus is very peculiar; it's tiny but it certainly looks like at one point in time it was heated, and it now contains water.

CI   What has Cassini been like for you?

CP   It's not so much a mission as a way of life. I knew this was going to happen when we were selected. I was gleeful for about fifteen minutes when I was told I had been made the Cassini Imaging Team leader, but I sobered up really quickly when

One of the biggest surprises of the Cassini mission was the discovery of ice geysers on Saturn's tiny moon Enceladus, only 500 km across. This backlit view of the moon's southern rim was taken during a close flyby in November, 2005. The geysers erupt from eight locations that are associated with surface fractures; the water just under the surface is just above freezing and much hotter than the –200 °C temperature of the surface of Enceladus. Cryovolcanism so far from the Sun's warmth expands the idea of the habitable zone to include the active moons of giant planets (courtesy Cassini Imaging Team, NASA/JPL, and ESA).

I fully realized what I had bought into, and how much work it would take to pull it off. It turned out to be far more than I expected.

CI   When did it all start?

CP   November 13, 1990.

CI   I imagine it would be like raising a child. The payoff would come fifteen or twenty years later.

CP   It's similar in terms of the degree of commitment. There are joyous moments, but also outrageous frustrations, and lots of hand-wringing. This mission required inordinately long periods of time when I had to be obsessively devoted to it. To pull it off, I had to clear the decks of everything else, including any semblance of a normal life, for years on end.

CI   The timescales are dictated by factors out of your control. If there's a crisis in the mission, you have to drop everything, right?

CP   Probably every project could say this, but it would have been much better if we had been better funded. We were woefully underfunded. I think it's a result of this push to have more and more missions. NASA wants to look very productive, and we as a community want to *be* productive. We want to go to comets, aster-oids, planets *and* moons. We want to conduct fanciful missions like smashing things into comets, which is outrageously great. But the downside of all that is that we're so overcommitted that any one mission is not given enough money to do what it needs, and people are asked to work excessive hours. It's like being asked to sprint for the duration of a marathon!

In the days of Voyager, there were years between encounters. For those several years, you could have a reasonably normal life. You looked at the data from the last encounter and published papers. You had several years to plan the next flyby. Then as you got closer to the flyby date, it became more intense. But there was only a period of several weeks around each encounter when you got into an obsessive mode. Here we've had to be in an obsessive mode for years on end; it's become a way of life. I'm sure the Deep Impact people, for example, had to do the same thing, but their whole mission was only about six years long, and ours has been eighteen so far. It really requires a lot to commit oneself to one of these missions.

CI   Was the mission ever in jeopardy?

CP   There were tough times in 1992. Just a couple of years after we were selected, we had to descope, which is a word coined by NASA, incidentally. It means that we had a budget cut and had to remove capabilities from the instruments and the spacecraft. Then a couple of years later, the administration started looking closely at the project. They had a budget crisis and Cassini looked like an attractive option to cut because it was a big, expensive mission. At that point in time I was asked to go and speak to the Office of Science and Technology Policy to try and explain the value of the mission. I also spoke to the House Authorization Committee. So yes, there were times it was politically in trouble.

CI   Descoping sounds like an insidious concept. Scientists are usually so ingenious that they save most of their science despite being descoped. NASA can then turn around and say that they could do what they needed to do with less money. What was it like when Cassini finally got to Saturn?

CP   It was enormously gratifying to see the fruits of our labor.

CI   Where were you when that was happening?

CP   I was at JPL; my team had to be the entertainers. We stayed up all night to get all the pictures ready for the press conferences the next day. We had the world's attention, so it was wonderful. Of course, it's also exhausting, but you're really happy because all your work has paid off. On top of all of that, you have the privilege of seeing things nobody in the history of the universe has seen before – or at least in the history of Earth. That's the rush. It's like winning the gold medal at the Olympics.

CI   I imagine that it would be quite emotional, too.

CP   Yes, very emotional. The subsequent months were completely heady. There was picture after picture full of surprises. Our instruments continue to work beautifully and the spacecraft is working flawlessly as well. There could have been a disaster, or we could have had something fail. But so far nothing has. We were very fortunate.

Another superbly thrilling moment was when the Huygens probe landed. I didn't have to be the one doing the entertaining for that event; my team didn't have to stay up all night to get everything ready. That almost had a bigger impact on me than the Saturn orbit insertion, maybe because I didn't have any responsibilities, was more relaxed, and had the freedom to absorb the significance of the whole thing as it was unfolding. When we landed on Titan, I felt like a different person. I was stunned; it was just such an amazing, amazing achievement. To have those pictures tell us right off the bat, unambiguously, that stuff flowed on the surface of Titan was the thrill of thrills.

CI   Right! I think it took a while for it to settle in with people. One of the problems is that it's been a generation since we set foot on the Moon. Young people think that not much happened after that. But a mission where you're doing real-time exploring of new worlds hundreds of millions of miles away vaults us back to the frontier again. I hope that will generate a new wave of interest and encourage young people to reengage in this pursuit.

CP   I think it will. Many people – not billions of people or even millions of people – but many, visit our CICLOPS website. They're looking over our shoulders. But I do think we should work harder to get science into the public arena more. If we make use of other avenues, we can connect to people directly and get their financial support. We need to use television in the way that other disciplines and enterprises use television.

CI   What have been the most scientifically interesting or exciting things from Cassini?

CP   Well, we were expecting to discover things that would puzzle us. The stuff that I found most intriguing is the complexity in the rings. We had a very simple model to explain what we thought we knew about the rings, but now that we're looking more closely, we've found this simple model inadequate. I'm intrigued by what we'll learn about the rings once we understand why the model is inadequate. It is going to be a real thrill to work on how this gigantic expanse of debris behaves. It's also going to be a touchstone for other disciplines in astronomy that deal with accretion – subjects like protoplanetary and protostellar discs.

I am also intrigued by the morphology on the surfaces of the satellites, because they're like autonomous worlds. I look at them and think, "I could be walking on that." That's how I think about it. I say, "This picture is so many miles across, and it would take me five days to walk across it." It's like hiking. I think about how long it would take to hike across one of our pictures, and what it would feel like to be there. That is the physical adventure by proxy. It's not a real physical adventure, but our pictures allow us a means of at least imagining what it would be like to be there. That's the explorer in me, and I think it's the explorer in all of us who get involved in this: new terrain, new territory, new horizons.

CI   The problem with a success, especially with what we saw in Titan, is that you have a whole new set of questions that you want answered right away, but there is no mission coming up. What's next?

CP   There are various plans. No one has settled on any one thing, but they call for aerobots, balloons, and airplanes in the atmosphere. It would be good to have something that could go up and down to sample both the upper atmosphere, as well as closer to the surface. Most Solar System bodies have a very diverse geography. In other words, we shouldn't just touch down in one place, take a few rocks and go back to Earth, because that would be like landing in one place on Earth. The Antarctic is nothing like French Polynesia. The polar region on Titan is completely different from the equatorial regions. The same thing applies to other bodies. There's so much variety on the surfaces that you could probably spend a lifetime studying any one body.

CI   Where do those follow-up Titan missions sit with respect to a Europa mission?

CP   This goes back to the work we did on the Solar System decadal survey, and what we ought to be doing in the next ten years. The priorities were: visiting Pluto, orbiting around Jupiter, landing on a comet and returning a sample, going down to the surface of Venus, and returning samples from the lunar south-polar region. Those were the missions we considered most important to address the breadth of unanswered questions after the first forty years of planetary exploration. Basically, those are the reasonably middle-sized missions that we thought could be accomplished in the next ten years. We also hoped for one big mission. We chose Europa.

CI   How much will these upcoming missions cost?

CP   Medium-sized missions are something like six or seven hundred million dollars and a big mission would be a billion plus.

CI   Cassini was a few billion dollars in the end, right?

CP   Cassini was about $3.2 billion, but that $3.2 billion was spread over eighteen years.

CI   These are hard choices, because as budgets and politics in NASA change, you don't know how many new starts you'll get.

CP   Yes, that's very frustrating because I'm such a purist. I still have this child-like idea that science is the most pure, beneficial endeavor that could be undertaken by humans. It should be supported in a very special way and have a high priority in our social consciousness, and yet it's always such a struggle. I don't understand that. It's nothing but good, yet it's always a struggle.

CI   Space exploration really is a young enterprise. We've only had our civilizations for a couple of thousand years and technology for a century or so, but we've only been doing space travel for half a century. Do we have a future in space?

CP   If we don't destroy ourselves, I believe that we do have a future in space. But it's not going to be steady progress. I think it's going to be two steps forward and one step back. Even in my lifetime, I've seen this process: we commit, we retreat, we commit, and we retreat. I think it's going to go like that. But I don't think it's stoppable, because I think it's a drive that's very innate, and part of what we need to do to survive. When you live long enough to have watched and observed people, you come to realize that there will always be a small group of people who will be willing to do crazy things. It won't be hard to find people to sign up for a one-way trip to the nearest star. It might take twenty-five years or so until we figure out how to go faster, but people will want to do it.

CI   So despite the struggles of funding the big missions, you're still an optimist.

CP   If I could come back in 500 years, I would expect to see communities on Mars and possibly people taking extreme vacations or excursions to the Saturn system. We might be mining water on comets – I think that's going to happen.

In the sixties, I knew that I was interested in studying astronomy – this was encouraged by movies like *2001: A Space Odyssey*. I thought that by the time I was fifty we'd have telescopes on the Moon, and I'd be going there to observe. That hasn't happened. In a sense, it's disappointing that we haven't come as far as we thought we would in the sixties. The initial stepping off the planet didn't lead to the results we were expecting by the time we all got to middle age.

CI   I agree with you, but I think the curiosity and momentum are unstoppable. I have one other question. You have an asteroid named after you. I'm jealous of that because it's so cool. What's it like? How big is your asteroid?

CP   I'd have to go look it up. It's either 25 kilometers in radius or diameter, but I never remember which. It was thrilling – an honor. It's not something I dwell on, but when I do, I think it's as close to being immortal as I'll ever get.

CI   Is it observable? How bright is it?

CP   It's in the Main Belt. I tell people that at the moment it's in a stable orbit, but "Don't piss me off."

CI   The best thing would be having an Earth-crosser named after you. You could go down in history for some future extinction.

CP   People would say, "We expected no less from her."

CI   "She went out with a real bang."

21

Laurie Leshin

NASA's strategy in looking for life in the Solar System is to "follow the water," and Laurie Leshin has been following that strategy her entire career. She got a BSc from Arizona State University and a PhD from Caltech, then returned to ASU to become a professor and curator of the largest meteorite collection in any university. As a cosmochemist, her specialty is deciphering the record of water in objects in the Solar System. Leshin has fifteen years of experience working in Antarctica, and has used her lab skills to assess the history of water in meteoritic material from Mars, and so assess the potential for life on the red planet. She is a member of two science teams on the Mars Science Laboratory. In 2005, she joined NASA as Deputy Director for Science and Technology at Goddard Space Flight Center. Prior to joining NASA, she was awarded their highest honor, the Distinguished Public Service Medal, and she was the first recipient of the National Meteoritical Society's Nier Prize. Asteroid 4922 is named after her.

CI   Most Arizonans aren't *from* Arizona, but you're a native. What path led you to where you are now?

LL   I'm too young to remember the Moon landing; my generation got its excitement about space from Viking. The images of the surface of Mars brought back by the Viking Landers were in *Time Magazine*. That landscape felt familiar. I wanted to reach out and touch those rocks. I was mesmerized.

In my memory, there was a picture of the surface of Mars on the cover of *Time Magazine* back in the mid seventies. I remember standing in the kitchen, looking at this picture. I told this story to a reporter from the *Chronicle of Higher Education*, and he called me back two days later, and said, "I talked to *Time*, and they didn't ever run a picture from the Viking Landers on the cover!" I was devastated. I had been sure it wasn't on the inside, because what ten-year-old opens *Time* to read the articles? I went on a quest to prove this guy wrong. But he was right; it was on page 23, this beautiful panorama of the surface. Nadia Comaneci was on the cover! [Laughs] That was her perfect performance and the US gymnastics gold medal in 1976. That's the reason I opened it.

CI   You could have been a gymnast instead of a planetary scientist. A lot of people were disappointed by those images because popular culture had conditioned them to expect a lush world.

LL   I didn't know any of that. A lot of it had to do with being from the desert and identifying with the landscape. I was a bit of a rock hound. I wouldn't say I was overly scientific, and I wouldn't say that "from that moment on, I knew I wanted to study Mars." I was a chemist as an undergraduate. When I was nineteen, I did a summer internship at the Lunar and Planetary Institute in Houston and got to work on Viking data every day, which was amazing. That was the lightning strike.

CI   A lot of your research is on meteorites. What can we learn from space rocks?

LL   I got into meteorites because they combine space science and chemistry. I'm studying space, but I can do it in incredible detail in the laboratory. It combines my two loves.

Broadly, there are two different kinds of meteorites. I can start with the primitive meteorites or chondrites, which are the remnants of the formation of our Solar System. They're the leftovers that didn't make it into planets. They survived because Jupiter's gravitational pull prevented them from assembling into a planet. We're lucky they didn't, because they're a window in time to 4.5 billion years ago, when the Solar System was forming. In fact, they are the first solids that formed in our Solar System. They give us an opportunity to witness that event close up and understand the environment. I call myself an extraterrestrial environmental geochemist – I'm trying to apply the tools of environmental geochemistry to understanding rocks from space.

CI   Is it the case that chondrites haven't been altered by heat or pressure? Are they primitive material?

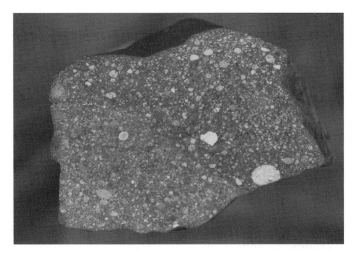

Meteorites typically come from the Asteroid Belt and they are usually unaltered since the formation of the Solar System. Most meteorites are rocky, like this chondrite that landed in Romania. Chondrites formed at the same time as the planets. The bright patches are nearly spherical inclusions called chondrules, rich in calcium and aluminum (courtesy Laurie Leshin, Arizona State University).

LL   Some are, but not all. To a certain degree, all of them have some overprinting of what we call a parent-body process, an event on the asteroid after they had conglomerated. You can hold most of them in your hand and see little flecks, each of which was a free-floating object in the solar nebula before the planets formed. If you swept up an armful of solar nebular dust and patted it together like a snowball, that describes the formation of chondrites. It's like cosmic sediment. Each one of those little dust grains or chondrals has a story to tell about its history and its time and place in the solar nebula.

    We spend a lot of time studying these rocks in incredible detail, more detail than on any Earth rocks, honestly. People are starting to study Earth rocks like this in the origins-of-life field. They're getting a lot of techniques from us because we're incredibly limited in terms of our samples, which are complex at microscopic scales. We've become experts at tearing them apart, atom by atom.

CI   Are these the materials that give the most accurate age of the Solar System?

LL   Yes, absolutely.

CI   When we say Earth is old, the public thinks we mean *roughly* 4.5 billion years, but it's more precise than that, isn't it?

LL   The best age is 4.566 billion years, at the moment.

CI   What determines the precision of that dating?

LL   It's determined by the abundance of the different radioactive isotopes and how well we can measure them, and how well we know the decay rate. We found a subset of objects within these meteorites that are slightly older than any of the others, and when you look at them chemically, they are the most refractory. If you take a cloud of solar composition and heat it up so everything's in the gaseous state, then start cooling it down to see which things become solids first, the first thing you get should be sapphire or aluminum oxide. Then you get a little magnesium, and the most high temperature materials. That's exactly what we see in the flecks within these meteorites. Chemically, they *look* like the first things to have condensed, and when we date them, they're a bit older than everything else. They really were the first solids in the Solar System. That's fascinating – in these meteorites, we can watch the solar cloud condensing.

CI   What's the time frame for the whole process?

LL   The nebular lifetime is about 10 million years. There isn't an *end*, exactly – we're still adding material today. The Earth accretes 40 000 tons of extraterrestrial material per year.

CI   In terms of the processes of planet and debris formation, did anything special happen in our Solar System?

LL   Absolutely not. It's typical from what we know. The Spitzer Space Telescope allows us to see zodiacal dust in other solar systems, as well as disks and their rocky material. These smaller objects, from pebbles up to planetesimals, are difficult to detect because of their size, but we're starting to be able to see them. Again, it looks pretty standard.

CI   Are there any chemical composition anomalies in meteorites?

LL   Absolutely, all kinds of them.

CI   Are those curiosities interpretable, or are they mysterious?

LL   There are lots of possibilities. From an astrobiology perspective, the organic matter in the meteorites is fascinating. In our research, we study mostly the inorganic, rocky parts of the meteorites, but we use organics to understand the environments of these asteroids and to construe what happened with the organic material. In some cases the meteorites are up to 7 percent organic matter, and it's quite complex. They often contain amino acids, the building blocks of proteins. The precursors to life are found in these meteorites. They're probably easy enough to make chemically, so it's not that meteorites supplied the Earth with amino acids; rather, as long as you've got carbon and some energy, you're going to make organic matter.

What interests me is that these are our only natural example of purely prebiotic chemistry, because on Earth everything is contaminated by life. We cannot go back in the rock record to a time when life didn't exist. We'd like to be able to walk down the Grand Canyon and see the layers of rock and the dividing line between chemistry and biology. We don't have that, because of plate tectonics and impacts. But frozen in time in the meteorite record are chemical reactions that led to the development of prebiotic molecules, a huge gold mine of processes we need to identify and understand.

Was it very wet, or not so wet? How long does it take to build up? What were the timescales involved? That's an active area of research with primitive meteorites. The jury's still out. We don't understand it nearly enough; we don't really even understand the full diversity and complexity of the organic matter in these things. They even show mixtures of left-handed and right-handed amino acids.

CI   When there are two classes, the components are always equal, right?

LL   Right. There are equal component mixtures, called a racemic mixture. When you have as many right-handed as left-handed it's not a biological process, because we use only left-handed amino acids. There are excesses of the left-handed amino acids in some of the meteorites, but they're slight, only a few percent. These amino acids are clearly not contaminants, because some of them aren't used by terrestrial life, and they show these excesses. How did that happen? Where did those come from, and is that process – offering up more of the left-handed stuff – important in the origin of life?

CI   Most of the amino acids contain 15 to 25 atoms, right?

LL   Yes, they're small, and they come in decreasing abundances in carbon-chain number. It looks like a nice chemical sequence.

CI   A nonscientist might not know whether to be surprised that there are amino acids in meteorites. Apart from the fact that they're basic components of life, is the existence of that type of molecule in a meteorite surprising?

LL   Cosmochemists tend to think of our Solar System forming as follows. Things got really hot, especially near the forming Sun. We tend to think most things formed in the vapor phase, in which case they're mostly either single atoms or very simple molecules.

Finding extremely complex molecules – amino acids 20 atoms in size – in meteorites was a surprise to many scientists. But when we stop and think about it, not everything was heated up and vaporized. Some of these organics are refractory and heat-resistant, so they might not have heated up, and they could have survived. During the formation of many of these asteroids, especially the carbonaceous and volatile-rich ones, they accrete ice and are heated up by the decay of short-lived radioactive isotopes in the early Solar System. The ice melts and flows – real geological processes happen on small rocky objects in combination with a lot of chemistry. *Where* these amino acids and other organic materials formed is a major debate. Are they purely remnants, the chemical memory of things that happened before the Solar System, or irradiation on ice grains in the interstellar medium? What fraction of them formed in the Solar System on the asteroids themselves?

CI   Either possibility is interesting. Given billions of years, you wonder what might happen with interstellar material in general. We stereotype life as needing a planet with a lot of liquid, but interesting things could be happening on the surfaces of smaller rocks with a modest amount of ice and water.

LL   Absolutely. There was a liquid medium on a lot of these asteroids, but it probably didn't last very long. Maybe that's what kept the process from going further. These short-lived isotopes have half-lives of a million years or less. They heated up and cooled down quickly, which is why meteorites have been floating around unaltered for more than 4.5 billion years – all the action took place in the first couple of million years, and it's been cold and dead ever since.

CI The sum of all amino acids that have been found in meteorites includes variants that are not present in our life. How many have there been?

LL A few dozen. A couple of them are used as indicators that fit into a sequence and show that what we're seeing is not contamination after the meteorite landed.

CI Is there anything more complex than an amino acid?

LL There's a complex macro-molecular material. It's basically tar. That's insoluble material, so we don't know to what degree it's involved in these prebiotic reactions. Amino acids are soluble in water, so if there was water flowing around, they were involved in the chemistry, *somehow*.

CI Let's look at it the other way around. With the Mars rock, people started to ask whether you could find life in a meteorite, and how well life might be preserved in space. Extremophiles here exist in environments not so different from the interior of a rock in the solar nebula. What do you think of those possibilities?

LL I've tried to separate the possibility of life on Mars from the hypothesis that there are fossils in Allan Hills 84001, the famous Martian meteorite. I'm optimistic about the possibility of life on Mars, but I'm not so optimistic about the possibility of fossils in Allan Hills. We stand here with our catcher's mitts, waiting for whatever Mother Nature sends us from Mars. We're unable to go there and interrogate the rocks in sufficient detail, or to find the right rocks and bring them back to Earth so we can throw everything we've got at them analytically. We know *how* to answer this question, if we could get the right material. We can't do it yet. It's exciting because we can see where we need to go, but it's frustrating at the same time.

CI It's nice to get a free rock from Mars, but you don't know where it came from and it's been altered by the process that got it here.

LL Yes, although not nearly as much as you might think.

CI From a geochemist's perspective, why was it so hard to draw a firm conclusion about primitive or extinct life from ALH 84001?

LL For two reasons. The rock itself is extremely ancient, over 4 billion years old. Ninety-nine percent of the stuff in the rock has nothing whatsoever to do with life. It's the most boring gray rock you could ever hope to lay your eyes on. Even among igneous rocks, it's boring. It's monomineralic orthoperoxine igneous rock.

CI How many other Martian meteorites do we have?

LL About three dozen, which is wild. When I started this business, there were eight. [Laughs]

CI They must be expensive to buy, even a chip.

LL Yes. ASU has the largest university-based meteorite collection in the world, so we can trade for almost anything we need. It's money in the bank for us. Some of the interesting ones come from Antarctica.

CI   As a long-time desert dweller, do you have a hankering to go down to the cold places and find them yourself?

LL   I've gone! I went down in November–December of 1996, right after the Allan Hills announcement. There was a lot of media attention because that was an Antarctic meteorite, so reporters visited us in the field, and there was a big controversy swirling around. It was the first time in my life I ever shoveled snow.

CI   That must make a nice change from lab work.

LL   It was awesome. We'd drive around on the snow all day looking for black rocks on white ice. I'm a lab rat, but even I can find a black rock on white ice. [Laughs]

CI   You've been involved in a series of Martian missions. I gather you had a slightly painful early experience with the Polar Lander?

LL   I was live on CNN when it crashed!

CI   You had to be the commentator?

LL   I was team spokesperson. There I was on CNN, all ready to jump up and down, and there it wasn't. That was tough, although honestly, getting so close just gives you the taste for going back. You hide for six months, and then you come out determined to get there. Being a member of a mission team is an incredible experience. The scientists and engineers work together towards a common goal that's so challenging. It's extraordinary.

CI   Let's talk about upcoming missions. You're involved in Mars Science Lab, right?

The Mars Science Laboratory, scheduled to launch in the fall of 2011, will be the most advanced set of instruments ever sent to Mars. Closer in size to an SUV than the kid's go-kart Mars Exploration Rovers Spirit and Opportunity, MSL will descend on a tether as the spacecraft that carried it hovers above the surface. Its suite of a dozen instruments will carry out complex analyses of surface and subsurface rocks, and the atmosphere. It will be able to look for subtle signs of biological activity (courtesy NASA/JPL).

LL   Yes. I'm on two of the instrument teams for MSL.

CI   What's the status of that project?

LL   It's moving at light speed. The instruments have been selected. It's an ambitious payload. I haven't seen the latest design for the Rover, but I've jokingly referred to the MERs as golf carts and MSL as the Hummer. It's going to have a much greater capability in terms of driving distance. MER has already driven five or six times its spec capability. The frightening thing is that MSL nominally has a one-year mission lifetime; MER nominally had 90 days, and they're still going. This could take the rest of our lives, driving this thing around Mars!

CI   A dry version of the Ancient Mariner. Funding might become an issue.

LL   True. MSL will have precision landing capabilities. Its landing ellipse will be significantly smaller than the MERs, which opens up the planet. With MER, you had to be able to plop down in a 100-kilometer ellipse, where everything was safe. That fundamentally limits the number of places you can go – maybe a few dozen sites. MSL has a 10-kilometer ellipse; there are many more places to go. It's terrifying, but it's also exciting in that we can optimize our landing.

One of the main instruments is SAM, Sample Analysis and Mars. It is the most complex analytical laboratory ever sent to Mars. It's got the capability to analyze organics, which MER does not. It's so frustrating, we see beautiful aqueous sediments, and we don't know if they've got even carbon in them! With MSL, we'll be able to identify different kinds of organic materials and analyze isotopes, and get detailed and sensitive chemical information. That's extremely exciting.

CI   Can it dig, or is it just for work on the surface?

LL   It can dig. It has an arm, and probably it'll have a little drill. It will definitely have a scoop.

CI   So it's not just for surface chemistry, it'll pulverize things, too?

LL   Yes. It has to pulverize things, because one of the complementary instruments is an X-ray diffraction instrument, which needs powder samples. This mission will give us extremely good chemistry and mineralogy for the first time.

CI   Do we still need to bring rocks back? Are there still types of analysis that aren't possible remotely?

LL   One single instrument in my lab is as big as this entire Rover. There are fifty different kinds of experiments we could do, and we'll never effectively send all of those tools to Mars. The amount of money we spend on analytic instruments sounds enormous, but a state-of-the-art spectrometer on Earth costs $75 000. To build that thing and send it to Mars costs roughly $7 million. That's huge, a factor of a hundred. The analytic lab in my lab costs $2 million off the shelf, so you could spend $200 million to send that one piece of equipment to Mars. If you have fifty others, it quickly becomes unreasonable to try to miniaturize all of these instruments to send them to another planet.

CI   Given the uncertainties in NASA, is there a real time frame on sample-return?

LL   I would say it's not even really on the books. I'm a relatively young woman, and I've been involved in replanning Mars exploration for about ten years. Sample return has been at least ten years away for those ten years.

CI   It's like a mirage.

LL   It is! We just have to keep making the case, and hope the technology will come along to reduce the risk. It's hard to imagine, though. We can think about more palatable, alternative approaches to making the discoveries, because the discoveries are the important thing, not the implementation.

CI   Dropping back to the big picture, would you be surprised if we found either extinct or possibly continuing living subsurface organisms on Mars?

LL   I would not be in the least surprised. I would be ecstatic. The work we've done on other Martian meteorites shows that there is water in the crust of Mars. It flows like water in the crust of the Earth. There are a lot of similar environments on Mars to what we see on Earth in places where life is perfectly happy in the subsurface in the rocks. I would not be surprised at all, and yet I wouldn't be particularly upset if we didn't find it, either, because Mars preserves the first half a billion years of Solar System and planetary history, unlike the Earth. There, we have a chance to observe prebiotic chemistry and understand what was different. Why *didn't* life arise, since it should have had all of the ingredients and energetic environments? What does it mean when it doesn't happen? To me, that's an equally fascinating question.

CI   Mars is worth obsessing about, but what are your other favorite sites in the Solar System? Based on the geology and chemistry, where are you optimistic of finding simple organisms or interesting prebiotic processes?

LL   I'm on the Europa and Titan bandwagon, although I'm not nearly as optimistic about those places because the energetics are much less favorable; they're so cold and dark and far away. I'm prepared to be surprised by nature, because it's phenomenal at doing whatever it likes. [Laughs]

CI   You were on the "Moon, Mars, and Beyond" presidential commission. What was that process like?

LL   It all started with a strange message from the White House on my phone. It was amazing to watch how NASA operates at a very high level. If we have goals for revitalizing the human exploration side of NASA, what's going to be the best way to go about it? How should science and human spaceflight work together? It was an extraordinary chance to look across the agency broadly. I'm proud that we were able to argue effectively that a broad science program is in the best interests of our future exploration, because it's virtually impossible to predict where the next life-changing discovery is going to occur, so you have to plant a lot of seeds and let them flourish.

CI   Was there any tension between the manned and the unmanned vision?

LL   Everybody asks about that, but there was none. Zero. Carly Fiorino, the former CEO of Hewlett Packard, was on the committee. I can't imagine a more different person than me for this discussion, but when we focused on goals, it was easy to agree on the major issues and how we wanted to address them. That was a sign that we were headed in a good direction.

CI   In an even longer time frame, that of your children or grandchildren, we're on the edge of privatizing the space program. Scientists worry about advertising lights in space and people buying and selling astronomical turf and minerals, but it might unleash a lot of discovery, even if it's not driven by pure science. How do you feel about that prospect?

LL   I tend to be optimistic – or maybe naïve. Opening up the space frontier is likely to lead to more discoveries than downsides. We have to be vigilant about making sure that things are thought through and done in a reasonable way, such that we preserve the great resources out there, both scientifically and otherwise. I held a big public space forum during the work of the commission, and 1500 members of the public showed up. The parents were asking complex political questions, and the kids were asking, "How long is it going to take for us to get to Mars? Let's go!" They understand that this is the direction that we've got to go – what are we all bickering about? There's something inherent in kids about exploration. If we could have a little more of that attitude, we'd be in great shape.

CI   There's a lot to be said for the impatience of youth.

Guy Consolmagno

Guy Consolmagno seamlessly lives in worlds that might seem to be in conflict. This self-proclaimed MIT "nerd" is the curator of the one of the world's best meteorite collections, but he also wears the collar of a Jesuit priest and spends half of his time in the papal summer palace on the edge of a crater lake, half an hour outside Rome. Science and religion comfortably coexist in Consolmagno, who is known for a regular spot on BBC's Radio Four as "Brother Guy." He keeps the vows of his order and follows the rigor of his scientific training, having served in the Peace Corps and been employed as an assistant professor of Physics at Lafayette College before joining the Jesuit Order. He has been a visiting scientist at Goddard Space Flight Center and visiting professor at Loyola College and Loyola University. He is the author or coauthor of four popular astronomy books, and was among the first to speculate about the possibility of life on Europa.

CI   Your background is really diverse. How did you get interested in science?

GC   I was a Sputnik kid like everybody else in my generation. My dad was a naviga-
tor in WWII in the Air Corps, and he taught me astronomy. I wound up at the
Jesuit high school in Detroit and studied classics, so I had an odd split interest
between classics and astronomy which wasn't resolved until my sophomore year
of college.

CI   How did you decide between astronomy and classics?

GC   The decision happened in a funny way. At the time I was at Boston College, lean-
ing towards classics. But my best friend went to MIT, and MIT had weekend movies, pinball machines, and the world's biggest science-fiction collection. It just
seemed like a whole lot more fun of a place.

CI   Like nerd heaven.

GC   Nerd heaven, exactly. And Boston College was not a place for nerds – it was a place
for guys who wanted to have a good time. My idea of a good time means being
a nerd. So I transferred to MIT and chose astronomy. Except they didn't have an
astronomy department – they had an Earth and Planetary Sciences department.
I signed up for that, thinking it was astronomy. In fact, I discovered that it was
geology, with the geology of the planets as a subtopic. So basically it was by accident that I got into this corner of the field.

CI   But it hooked you anyway, right from the beginning?

GC   It hooked me completely. The professor who hooked me was John Lewis, who is
now at Arizona. He is a fabulous character. I didn't know it at the time, but the
man is barking mad – in the best possible sense. He's just so enthusiastic! In many
ways he's been a role model for what it means to jump feet first into anything
you do. He's also a good scientist, in that even though he had his pet theories,
he was willing to discard them given the right data. The ability to be objective
was something we learned early on by watching him. Which, I'm afraid, a lot of
scientists never learn.

CI   How did you pick your research specialty?

GC   The original research that I did with John Lewis was planetary chemistry. I got
my Master's degree by modeling the icy moons of Jupiter. In those days, the only
things we knew about those moons were roughly their size and mass, out of
which we could get a very crude density measurement. We had also observed
water ice on their surface.

When I wrote my thesis on the icy moons, I was a science-fiction fan, so I
threw an offhand comment into the appendix pointing out that water in the
oceans under the crust of Europa – which I predicted for all the wrong reasons –
should have chemical reactions leading to salts or even organic complexes. I
said, "I stop short of predicting life in the oceans of these moons and I'll leave
that for others to do." That was sort of a snide way to claim it. As far as I know,
that's the first place any scientist has ever talked about any life in the oceans
of a place like Europa. My little teenage dreams were prescient, but for all the
wrong reasons.

CI   For most people, the notion that the moons of giant planets are real worlds with atmospheres, geologies, water and oceans is a novelty. It resets our notion of what a habitable zone means.

GC   Exactly. If we look at the giant Jupiters that are orbiting close to other stars, we see they're much too hot. But they could have moons with thick ice crusts which harbor life underneath them. My gut feeling is that life is ubiquitous; it's hard to stop it. But I'd still like to have some data.

CI   Most people think that the rocky bodies of our Solar System are just that – rocky bodies. But there's a lot of ice in the outer Solar System, and the Earth had to get its water from somewhere. What is the current thinking of where the oceans came from?

GC   There are two competing theories. One is that a late "veneer" of comet-like material added water. The other is that the water was incorporated early on – maybe in the form of hydrous minerals – in the stuff that made the Earth. Each theory has its strengths and weaknesses. The real trouble with the late-veneer idea is that the isotopic ratios that you see in the water in comets are distinctly different than what you seen in the water on Earth.

CI   When you talk about late veneer, how late is late?

GC   Still close to 4.5 billion years ago.

CI   So it doesn't affect the timing of the emergence of life in any direct way.

The outer part of the solar nebula contained rocky and icy material that probably seeded Earth with its water and oceans within the first 50–100 million years after the Solar System formed. Chondrites, the oldest and most primitive form of Solar System material, could have incorporated water and organic material in the cool outer regions and then delivered them to Earth. The alternative explanation is a "late veneer" of water and organics from comets (courtesy NASA).

GC   No, it'll be indistinguishable in terms of when the water gets there. But, of course, that water would probably have organics with it because there are organics in common with water in certain asteroids.

CI   Can we make any speculations about the ability of other solar systems to create a watery terrestrial planet?

GC   Well, we thought we could, until we actually saw what other solar systems looked like! Our Solar System has rocky planets close to the Sun and icy planets farther out. The Jupiter moon system has rocky moons close to Jupiter and icier moons further out. Having seen two examples of this, we used to think that this was the rule. But the first extrasolar systems discovered have giant planets very close in. So this nice picture that we had of warmer inside, cooler outside, may not be a general rule. All bets are off at this point.

CI   I assume there are mechanisms within solar systems to transport icy material to the inner regions?

GC   Absolutely. That's one of the strong points of the late-veneer model – you would expect it to happen. Convection in the solar nebula could dry out material on the inside. Then, as the water-rich nebula got convected to colder regions, it would freeze and build up the ice in the outer solar system. Then a late veneer would bring the ice back in and add the water at the end.

There's another problem with the late-veneer model, though: the Moon is bone dry. If the Earth got blasted with water, why didn't the Moon? It's conceivable that the veneer occurred before the impact that caused the formation of the Moon. But all of this is pretty much wild speculation.

CI   What about the other critical ingredient for life – carbon? Where did the organic compounds come from?

GC   Presumably, with late veneer, you'd get the organics and the water together. But if the water was built into the rocks that formed the planet, then the carbon would have to be the late veneer. They would be separated.

Then there's a third element most people don't think about, which is where all the nitrogen came from. We have a substantial nitrogen atmosphere in Venus and Earth. You don't notice it in Venus because the carbon dioxide overwhelms it, but if you took away the carbon dioxide, the amount of nitrogen on Venus is comparable to that on Earth. And nitrogen doesn't exist in rocks. It's just not something that's normally formed in minerals – you have to find a way to bring it in. It could be either a late veneer, or dissolved in the iron that makes the iron cores of the planets.

CI   Phosphorous is another important ingredient that gets deposited. It's obviously essential in terms of the molecules that store energy and form life's backbone.

GC   Right. We see phosphates in iron meteorites, so it may be ironically enough – no pun intended – that some of the hotter stuff that you normally think of as being the center of the planet could be carrying two of the essential ingredients to make life on the surface.

CI   You work on asteroids and meteorites. What do these small objects tell us about the history of our Solar System?

GC   Meteorites tell you what chemical elements were around and what their general abundance was. The astonishing thing is that all meteorites are homogeneous at the tenth of a percent level, including isotopes with most extreme chemical abundances. Whatever differences there are, are easily explained in terms of how much heat they were subjected to. The chemical clues are very uniform.

    But when you go to below a tenth of a percent, you see many hard-to-explain differences; that's really exciting. Those differences are in the isotopes present; for example, when you see an excess of magnesium-26, it tells you that there was live aluminum-26 when the rock was formed, and that the atoms were put together in the rock crystals. It must have been formed in our Solar System, yet aluminum-26 is generally thought to be made in supernovas elsewhere.

CI   Does that mean the formation process was very rapid?

GC   Either it was very rapid or that it happened in a neighborhood with lots of other stars. That's consistent with what you see in places like the Orion nebula. It's not a stretch. We've got two different ways of making the isotopes. There are some isotopes that can only be made by supernovas, some isotopes that can be made by an energetic sun, and a whole bunch that can be made both ways.

CI   Does age dating of the most primitive meteorites set the timescale of our Solar System?

GC   Yes, and it's extremely accurate. There are white inclusions in a particular variety of meteorite – the CV3's – the most famous of which is the Allende meteorite. You can date these white inclusions with potassium, rubidium, and lead, which gives agreement to within a million years of a time 4.56 billion years ago. This is consistent between the meteorites. We have more than one of these tracers and they give a consistent answer, so that number is very reliable.

CI   You are the guardian and curator of a truly impressive set of meteorites at the Vatican Observatory. Tell me a bit about that.

GC   I'd been a scientist doing geology and geophysics for fifteen years when I entered the Jesuit Order, mostly as a way to teach. But instead they ordered me to Rome. You take this vow of obedience that goes with all the other vows, so I had to go to Rome and live in a palace and eat terrible Italian food. Tough life!

CI   *All* the other vows? I thought there were only two other ones.

GC   Oh, it feels like a lot.

CI   Maybe they've been adding to the list. So when was this?

GC   That was 1993. I'd only been a Jesuit for four years at that time, but I'd been a scientist about twenty years. When I arrived there, I discovered why they sent for me. I knew about the meteorite collection, but I didn't really know what state it was in. In the thirties, the widow of a French gentleman scientist had donated his rock and mineral collection to the Vatican. The Vatican didn't know what to do with it so they dumped it on the Observatory. The Observatory gave away the minerals but kept the meteorites. The minerals are in Vienna now.

CI   I'm glad somebody knew how to tell the difference.

GC   Yes, well that is questionable at times. The meteorites were in a complete jumble when I got there. It took months just to sort through them, organize them, and do an inventory.

CI   But that must have felt like being a child at Christmas.

GC   It was exactly like that! Only it was more exciting because I had done research on meteorites for ten years without ever really seeing one. It really changed my perspective because I suddenly understood a lot just by handling the rocks and looking at their textures and colors. We had twelve hundred samples.

CI   Wow!

GC   Five hundred different meteorites in different pieces – and no equipment. When I got there, the only microscope I had was one that was designed to help solder wires to electric circuits. The only scale I had was one that was donated to the Pope from a scale maker in Germany in 1929. It was a beautiful antique, but very, very difficult to use. A lot of what I did in the first two years was simply assemble equipment. One of my colleagues in Arizona was great at scrounging equipment and, between us, we put together a very good lab for measuring meteorite density at just the time that people were beginning to measure asteroid density. We compared densities and were able to show that asteroids are grossly under-dense compared to the meteorites that we thought came from them.

CI   Does that mean we don't know where the meteorites come from?

GC   No, it just means that we didn't understand what asteroids have been through. Apparently the asteroids have been broken up and reassembled, and in some cases never really completely compacted. That means there are places within the asteroids, perhaps in their centers, which could be cores of ice and organic material that no one's recognized before. The astrobiology implication is a possible way of transporting organics and water in material that, at least on the surface, looks very dry.

CI   What about the implications for impacts on Earth?

GC   It makes it a little trickier because the things that are hitting us probably aren't as massive as we thought, by a factor of two. But they're so big that a factor of two doesn't particularly matter. It's the speed that they're coming in. It does mean, for instance, that stopping a porous asteroid is a lot harder than we thought. Blowing them up like Bruce Willis does in the movies won't work. With something that porous, it absorbs explosions without cracking or shattering, or doing anything.

CI   You've got this incredible meteorite collection and you've wrestled it into shape – what are the most exotic or fascinating samples? I know there's at least one Martian meteorite.

GC   There are more, actually. We've got a bit of each of the three main classes of Martian meteorite, which is unusual – especially the Chassignite class, since it's very rare. The Nakhlite that we have was actually donated by John Ball, who was the acting head of the Egyptian Geological Survey back in 1912. It's now one of the most valuable meteorites around because it's a beautiful piece of Mars. In

The largest meteorite found intact is located on a farm in Namibia. It impacted 80 000 years ago, weighs more than 50 tons and is about 300 million years old. Falls of meteorites this large are rare and only occur somewhere on Earth once every few years. Contamination of the exterior occurs immediately and gradually penetrates to the interior, so meteorites are most useful when they're recovered swiftly (courtesy Simon Collins).

addition to that, we've got some fascinating and unusual iron meteorites, which are typically very poorly understood. There are also a whole variety of one-of-a-kinds. Some have odd chemistry, like molten metal surrounding pieces of rock that look like they've never been melted or even hot. That's hard to explain.

Then we have a few meteorites that are interesting for historical reasons. Like L' Aigle – the one that was found in 1803 by the French scientist Biot. He was one of the first scientists to convince people that rocks really do fall out of the sky and are not just pieces of a volcano or something.

CI   That's a classic episode in the history of science, because the mythological idea of things falling from the sky had been dismissed. Biot really nailed it with the scientific method.

GC   Absolutely. One of the other meteorites that had led to this being thought a myth was Ensisheime that fell in 1492. The townspeople stored it in their church. By the time of the Enlightenment, people said, "These superstitious people – what do they know?" Well, it's a real meteorite. We have it in the collection.

CI   What's the nature of the meteorite at the Kabbah, in Mecca?

GC   No one has ever gotten a piece of it, so no one knows for sure. The best theory is that it may not be a meteorite but a piece of glass formed by the impact of a meteorite. And it's black simply because there have been a thousand years of people rubbing it with their hands.

CI   You've done a little meteorite hunting yourself, right?

GC   For the past thirty years, the Americans have sent teams to the blue-ice regions of Antarctica. Blue-ice regions are places where the ice has been flowing down from the South Pole towards the ocean and runs into an obstruction – a mountain range or something that makes it stop. The ice has little specks of dust or gas bubbles mixed into it, which causes it to look blue because those points scatter the blue light, just like the sky. Meteorites that might have fallen during the thousands of years that it took to flow to this position are also trapped in this ice. They're kept frozen so they don't rust, which is what happens to meteorites anywhere else. A typical meteorite that falls in North America, unless it falls in the desert, is going to turn into dust within twenty years. That's why you don't notice them. But in Antarctica, they can last for thousands of years. It's very easy to find them once they're on the surface: the meteorites are black, the ice is blue.

CI   If you could recognize all the meteorites that fell in a particular area, how much material would you be looking at?

GC   Something like 10 000 meteorites hit Earth a year. Three-quarters land in the ocean. Given how long they last, and how fast they come down, you might expect to see a meteorite roughly every two square kilometers. But in fact, you don't. They're completely lost.

CI   Do you expect to see them lying on the surface, or in the top few inches of ice?

GC   Embedded in the top few inches of ice. The ice evaporates, leaving the meteorite behind. Stony meteorites, when they hit Earth's atmosphere, slow down and break apart into small pieces. The pieces that survive to land don't make much depth. If they hit houses – which they do every now and then – they'll poke a hole through the roof, and maybe crash all the way through to the basement of the house. But they don't destroy the house. They've already slowed down and are traveling at terminal velocity. Iron meteorites are much stronger and they stay together. So they carry quite a wallop.

CI   Were you ever part of a crew in Antarctica?

GC   I was part of a crew in 1996. We lived in a tent out on the plateau, hundreds of kilometers from anybody.

CI   Wasn't there a nice base with a movie theatre and cozy canteen?

GC   No. There are six people in three tents. You do your toilet business outdoors, you know, in the snow bank.

CI   It's cold enough even in the summer so there must be some hazard associated with doing your toilet business.

GC   There is indeed.

CI   How long are you out there in a stretch?

GC   The American group is out there for six to eight weeks at a time every year. The Japanese build a camp and stay there for eighteen months: a winter, a summer, and then another winter.

CI   That's brutal.

GC   Yeah. And they go down by boat. We go down by airplane, which is ten hours of misery. They have two weeks of misery. Both groups do the same thing: simply

Hunting for meteorites in Antarctica, where the extraterrestrial rocks stand out against the fields of blue ice. Thousands of meteorites have been discovered in this frozen wasteland, including a number of the rare specimens from the Moon and Mars. In most terrains, meteorites are not recognized and they degrade quickly (courtesy Ralph Harvey and NASA/NSF).

traversing the blue ice in mechanized skidoos, and sweeping back and forth and picking up anything that's not ice. With meteorites you have to be very careful. You don't touch them with your hands. You put them into sterilized Teflon bags using forceps. Actually, what we used a lot was sterilized scissors. For some reason, scissors get a nice, tight grip on them. But the whole point is to try to keep them as free from contamination as possible.

CI   How did you do? Did you find any?

GC   We found four hundred meteorites, which is a pretty good haul. There have been places that yielded over a thousand, but eight hundred of those thousand are often tiny pieces of the same fall, so it's kind of boring. We got four hundred meteorites, most of them unique, and one of them was a piece of a lunar meteorite. So we brought back a piece of the Moon.

CI   It sounds like quite an adventure.

GC   I'm reminded of my sister's description of childbirth. In retrospect, it's wonderful. But at the time you're in it you're thinking – why the heck am I doing this? It was very rough. Although I loved it there and thought it was beautiful, I was also really happy to get back to civilization. You know, hot showers are a wonderful idea.

CI   Definitely. How do you see astrobiology going in the next decade or so?

GC   I think the real push – the fastest way we could find life elsewhere – would be if we found any evidence that it had been on Mars. That's simply because Mars is the easiest place to get. But the more we look, the more we realize how hard it is to know if you've found life. All of the chemical tracers that you would think of as signs for life can be mimicked by exotic, inorganic chemistry. And because it's Mars, you don't know what's exotic and what's not.

CI   We're planning sample return missions, but they're very expensive and they're still going to take awhile. With Martian meteorites to study, do we need them?

GC   Yes. Especially because the kinds of rocks that we're convinced come from Mars are not typical of what we see over most of the surface of Mars. What we see over most of the surface of Mars, and the kinds of rocks we think might have life, are apparently too small or too fragile to survive being launched off Mars by impacts. We will have to get samples. I worry about sending astronauts to Mars because human beings are leaky. The worst possible result would be to find life on Mars and say what we found is identical to life on Earth. Because then we'll never know if we brought it there ourselves.

CI   Will we have the answer within a decade?

GC   No. Fifty years, I'd say, for Mars, and a hundred years for Europa. I'm pessimistic about how fast we can advance our technology to explore these places. I think we know where to go, but we sure don't know how to get there yet. Just as an example, there is a fascinating place in Antarctica – Lake Vostok – which we know from seismic measurements is a liquid-water region. It's like a lens of liquid water underneath the Antarctic ice that's been isolated for thousands of years. The Russians have drilled close to it, but no one has been able to figure out a way to drill down and sample the water without contaminating it.

CI   It's our Europa analog.

GC   Exactly. And if we can't do it here on Earth, what gives us the confidence to deal with it when we get to Europa? Someday we'll be able to, but not yet.

CI   I wanted to finish by asking you a little bit about how the pieces of your life are integrated. You're a brother, you work on science, and you're in the heart of the Vatican with meteorites. For some people, that would seem like an unusual life.

GC   It isn't unusual at all. Most of science is collecting data, sorting data, and filing data. It's clerical work. The reason that work is called clerical work is that until the nineteenth century, it was clerics that did it. Only clerics had the free time and the education to do that kind of science.

CI   Also to preserve knowledge during dark times in history.

GC   Not just the Dark Ages and Middle Ages, but also the French Revolution. It's a great luxury to be able to dedicate your life to the stuff that won't make you rich and won't put food on the table, but does make life all the more interesting to live. I think there's a great religious motivation for doing the science. The joy that I feel when I make a scientific discovery is an awful lot like the joy I feel in a really wonderful moment of prayer. I feel it's that same connection to what is out there. Some scientists would call it a connection to the universe; I'd go further and say that it's not just a connection to the universe, but a connection to the Creator of the universe.

CI   You're lucky to be part of a religious tradition that really encourages intellectual thinking and questioning.

GC   One of the great things about being a Jesuit and a scientist is that a lot of other scientists have been willing to tell me about their religions. There are scientists of every religious tradition you could imagine – including Evangelicals – doing good, solid science. A number of people working on the Allan Hills 84001 meteorite in fact attended the same Presbyterian church, which is actually the same church one of the astronauts belonged to – the one who took a chalice of communion wine to the Moon.

There is, I think, a great religious motivation for what we do. The conflict comes when people are afraid of what science is going to teach them. I think a person who's afraid of science is really a person who's afraid for their faith because their faith isn't very strong. Likewise, I think a scientist who feels threatened by religious people tends to be afraid of not being taken seriously as a scientist if it were known that they were religious. It's a myth that you have to choose between one and the other. It means lots of good religious people miss out on all the fun that science can be. And vice versa.

CI   That's a nice perspective. My last question is about the big picture of astrobiology. We're faced with the real possibility that our biology is not unique. Many religious traditions emphasize the specialness of man or the specialness of life. How do you view the prospect of not only life, but intelligent life elsewhere?

GC   It reminds me of the medieval theologians who worked out the phrase that we are made "in the image and likeness of God." I believe it's Thomas Aquinas who formulated it. What we're really talking about is two elements that characterize the soul: intellect and free will. You have knowledge of yourself and knowledge of the other person. You've got self-awareness; and you're free to do something about that, you're free to love that other person. You're free to make decisions for good or for evil.

If we're going to have any interaction with any intelligent being that we discover, it must have those two attributes. If we're going to call them intelligent, we should be able to interact with them, and they should be free to choose or not choose what they're going to tell us. That's more than just talking to a computer. In that case, they are also in possession of what we would call the essentials of the soul, so they are no different from us and no different for the moral challenges they would have to face. Just as I would expect their bodies to obey the same laws of chemistry and physics, I think they'd be faced with the same moral laws. It would be really interesting if we could communicate. I'm not sure in our lifetime or in the next millennium that we'll be able to.

CI   You're excited by the prospect. It will force us to look deeper at ourselves, as well as force us to understand our place in the universe better.

GC   Yes. It will force us to look at all the other assumptions that we've made. There was an announcement a few years ago of the discovery of an object bigger than Pluto. I'm on a committee that decided it's not a planet. It's a fascinating issue, but now that I know that there are more things than Pluto out there, it's changed what I think about Pluto. It's put the entire classification of objects in the Solar System into totally new categories. Finding life elsewhere will do that for our understanding of what life is – probably in ways that we can't even guess at yet.

23

Peter Smith

Peter Smith has tasted both sweet and bitter fruit as a planetary scientist and leader of space missions. He got an undergraduate degree in physics from UC Berkeley then a Master's in optical sciences at the University of Arizona in 1977. It was a long gap until his PhD was awarded in 2009, on the heels of his very successful Phoenix mission to Mars; not many people have a thesis that explored an alien world and brought in $400 million to his university. He has been working at the University of Arizona's Lunar and Planetary Lab since 1978, where he is a Senior Research Scientist. He worked on the Pioneer missions to both Venus and Saturn and he was a Project Scientist for the camera on the Huygens probe that landed on Titan in 2005. In terms of his Mars involvement, his cameras have worked flawlessly but the same cannot be said of the spacecraft. The Mars Polar Lander crashed and Beagle 2 was lost without trace, but the Mars Pathfinder and Mars Reconnaissance Observer missions were great successes, setting the stage for Phoenix and its recent exploration of the Martian polar soil and ice. Despite the ups and downs, Smith says he still has more missions left in him.

CI How did you get involved in so many space missions?

PS I graduated in physics from Berkeley in 1969. I had no idea what I was going to do. I failed miserably on my interviews with the big aerospace companies that usually hired out of Berkeley.

I was working for Sumner Davis. He built a huge spectrometer into the rock of Berkeley Hills, a 50-meter tunnel with great mirrors at the end, 4 feet in diameter. The first time I looked down this dark tunnel, all painted black, I thought I saw two big eyes staring at me. [Laughs] It was the eeriest feeling. I was always a little uncomfortable in that room. I was taking spectra of titanium dioxide, which had implications for cool stars. When I graduated, I went to the University of Hawaii, and worked on sounding rockets. We flew from White Sands, New Mexico, and instrumented them with a spectrograph we built at the Institute for Astronomy in Hawaii. Because we were understaffed, I learned every part of this procedure – design, machining, optical alignment, vacuum systems, and the environmental testing.

CI It was a full apprenticeship.

PS Basically. At the time I didn't know what direction I was headed in. I kept telling myself, "I'm only here for a short time," which was a bad approach to life. [Laughs] I only had a Bachelor's, and after five years I went back to school. I was fascinated with optics, so I went to the University of Arizona and got a Master's in optics. Then I went across the street to work with Marty Tomasko on the Pioneer Venus mission. I started at the calibration phase. Marty was a great teacher. We launched two months later, and after four more months it entered the atmosphere of Venus.

CI In the space game, that's instant gratification.

PS I thought, "What a great field!" Six months from calibration to publishing papers.

CI You probably thought it would always be like that.

PS Indeed. The next summer I was in the control room at NASA Ames as Pioneer 11 flew past Saturn. It was a low-cost mission, and someone like me, fresh out of school with lots of ideas, could do anything. We designed the observations, so took a lot of images of Titan, and later wrote some papers.

CI The seventies was a glorious time for planetary science. Was there a difference in the climate for space missions or the funding?

PS Sure. We approached the Viking Mars missions with a no-holds-barred attitude – let's do the best job we possibly can as a nation, and use orbiters to understand every aspect of that landing site. The Viking team had nearly 200 scientists, all working on every conceivable aspect of what they might find on Mars. When they worried that the landing thrusters were going to disturb the surface, they went out of their way to design a special system to direct the thrust away from the digging areas. Setting aside cost was a great way to do things. It was a matter of national pride, because this was the bicentennial year, 1976: "Look how great America is after 200 years!"

CI   Was it easy to get the science community aligned behind a single mission?

PS   Absolutely – specially one doing the first real exploration of the surface of Mars. We didn't know what to expect. There were lots of wild hypotheses and ideas, especially about life.

CI   Pioneer gave you an early look at Titan. Did that glimpse put Titan on everyone's map as a potentially interesting place?

PS   It certainly did. It was only a year or two before the two Voyagers went by, and then we found out all kinds of interesting things about Titan. Pioneer brought back the absolutely stunning news that Titan is super-polarized. It's the most polarizing atmosphere in the Solar System. It's like a bunch of dipoles lined up.

CI   Interesting.

PS   We had trouble understanding how that could be. Whatever is in that atmosphere is polarizing it nearly 100 percent. There's no way to explain that except with small particles. We had it all worked out and we sent it to a publication. Then Voyager flew by, and found out the satellite was forward-scattering, which meant there were actually *large* particles in the atmosphere. That led to ten years of head-scratching as to how there could possibly be atmospheric particles that are both forward-scattering and highly polarizing.

CI   Are these tholins?

PS   Yes, they're tholins, or tiny particles that stick together in long chains, also known as fractal particles. They're made from the UV radiation breaking up methane in the upper atmosphere, and they stick together very quickly to make nitrogen-rich organic molecules.

CI   Your name is associated with a very successful Mars mission. But it's not always smooth sailing, right? There are ups and downs in the game – long timescales, huge amounts of money at stake, fierce competition.

PS   Ten years ago, I was working with a group at LPL that was trying to find planets around other stars, and I designed a fancy spectrometer based on principles that had been put together by Kris Serkowski. He had ideas about how you could find and measure velocities to an accuracy of three meters per second and have a stable instrument over ten years. He had the ideas, but he couldn't put together the optical system. So I put it together and integrated it with the software, and we started observing twenty stars with the one-meter telescope up on Kitt Peak.

CI   When was this?

PS   Around 1982 or 1983. We had the accuracy needed to see planets around stars; however, because we had a small telescope, we could only observe a small set of bright stars.

CI   They were the wrong ones!

PS   Unfortunately! [Laughs] If we'd looked at the stars that were later observed by Butler and Marcy, we might have seen a four-day periodicity.

CI   It would have jumped out at you.

PS   It would have been huge! Our accuracy was down to ten meters per second or less. But you've got to have a star that's moving. All we saw that was moving were the oscillating red giants. We could track those. But they weren't regular, so it was hard to understand what the heck was going on. That was my first taste of a huge effort going to ultimate failure, because we never did find a planet.

Then I helped Marty Tomasko design the instrument that was going to land on Titan, and did land successfully. After three years as his project manager, there was a chance to build a camera for Mars Pathfinder. We had developed a flight-qualified CCD system for his descent imager. I repackaged that into a surface camera for Mars. It had a high likelihood of success, because we had already space-qualified the parts. That worked out really well, and Pathfinder was a huge success.

We built a similar camera for Mars Polar Lander. We were a little more ambitious. We had two cameras, one of which was on a robotic arm. We had everything all worked out, and it failed miserably. Mars Polar Lander crashed on the surface and never returned any data.

CI   Was the problem ever traced?

PS   Nobody knows; they didn't program communications to be part of the landing process. They prepared for entry into the atmosphere, said good-bye to it, expecting to hear from it from the surface, and never did. There was only speculation as to what happened. But a flaw was found in the entry-descent-landing procedures: when the legs were deployed, they came out on large springs with a lot of vibration and shock. There were touchdown sensors at the end of the feet, so that it'd turn off the thrusters when it landed; those were probably *activated* by the deployment of the landing gear. The software wasn't smart enough to know the difference between the deployment of the legs and landing on the surface. It said, "We've landed on the surface, time to turn off the thrusters," but it was still a hundred meters above the ground.

CI   Ouch. It's brutal to travel so many millions of miles, and then fail within a hundred meters!

PS   Our cameras were working great, the last we heard from them.

CI   I suppose you just have to turn the page. It's a lot of work, a lot of time, and a lot of investment of energy.

PS   Worse than that: I had thirty employees. I had to fire them all, slowly. It was very painful. I had cameras on two other missions, both of which were cancelled after this landing. I thought I was set for a decade with all the projects I had going. A year after the failure, I was left with one employee and no future. [Laughs] I was starting to wonder what else I could do with my life. As a last-ditch effort, I wrote proposals like crazy and got more projects than I know what to do with.

CI   Let's rewind to Pathfinder, because that was pivotal in the public consciousness and in building momentum on Mars. The pictures were so compelling, and it put Mars right back in the public eye.

PS   Some wonderful coincidences helped. We landed intentionally on July 4th, which was a Friday in that year and an American holiday. People were settling down to watch the news just as we got our first pictures back. There hadn't been any

Mars Pathfinder was a technology demonstration project that exceeded expectations, the lander outliving the design lifetime by three times and the rover by twelve times. Pathfinder used a parachute to slow its descent and then airbags to bounce to a halt on the surface. The rover, called Sojourner, returned over 500 images, including evidence supporting a warmer, wetter Mars in the past (courtesy Peter Smith and NASA/JPL).

public awareness of our mission until the day it happened. The press had never paid any attention to us until three days before, when everybody got interviewed. Then we landed, and it was on national TV, and the pictures were revealed. The first day, the Rover had a little problem getting off the platform. It was this tiny little guy that was going to rove around on the surface out in this alien terrain – it had a personality. We had not seen the surface of Mars for twenty years, and there was no real understanding of the surface we were going to land on, compared to the other Vikings. It turned out not to be all that different, but it was thrilling to see those pictures come down, and to watch the little Rover roam around.

CI   With such uncertainty about the nature of the surface, how was the Rover tested?

PS   Arroyo-Seco is right next to JPL in Pasadena.

CI   Where the Rose Bowl is held?

PS   It's just upstream of the Rose Bowl. That's where rocketry was developed in the thirties. In that arroyo they can fire off rocket engines without bothering any-body. It's also a good place to test rovers. There's a lot of rough terrain, and you can drive things around and over rocks. Then they built a "Mars yard" for driving the Rover, and they had interior sand boxes for trying out various ideas. It was difficult putting that Rover through all its paces to do its mission. A lot of major advances had to be made in robotics. It was quite a team. They had $25 million, a strict mass limit, and orders not to fail! [Laughs] There was a lot of pressure.

CI   Pathfinder was a great hit. Then you had dark days with the Polar Lander. But a hook had been set with the public, and NASA responded. Has that spurred your current wave of activity?

PS   Once you get the taste of Mars exploration, you don't want anything else. Mars is complex. There's a lot going on there, and in your lifetime you can pursue inquiry into various aspects of Mars and hope to get the answers. With Titan, we got the Cassini mission, but there's not going to be any follow-up in my scientific career. I may be 95 years old when they get back there.

CI   That's probably true of Europa, too. Let's get to Phoenix. From concept to landing and science delivery, what's the full timeline on a mission like that?

PS   Phoenix became a project in January of 2002. We'll complete our mission in the fall of 2008, six years from concept in the brain to completion. That's as fast as we can do it right now. For example, the Huygens probe was conceived in 1987. We didn't get our data until January of 2005 – eighteen years from start to finish.

      When the Scout program was first envisioned as a cost-capped activity focused on Mars science, I thought the best way to get involved was to build instruments for all the proposals. I tried to get on as many proposals as I could as a camera guy. I figured the odds were in my favor. Not everybody wanted a camera, and not everybody wanted *me* to build a camera. I was on seven out of maybe twenty serious proposals. I had a 30 percent chance of being selected. [Laughs] I worked hard to make those good proposals.

      Then I got a call from Chris McKay and Carol Stoker at NASA Ames who thought there was a good concept that nobody was doing. They wanted me to lead the effort because I had been involved in the 2001 mission, which was four months into integration and testing when it was mothballed. I would take the actual instruments that were developed for it or for Polar Lander and build a mission around them. That way, you get high reliability, already-selected science, and it's already been reviewed and is low cost. Just find a science goal and it'll be great.

      I said, "But what *is* the science goal?" [Laughs] You can't go to the south pole anymore, which is what Polar Lander did, looking at layered terrain. I didn't want to lead a losing proposal; that's too much work. While I was thinking about it, Bill Boynton announced that he'd discovered ice in the southern circumpolar region. He was pretty sure it was in the north, too.

CI   Ice, that's your hook.

PS   Our excitement is that we are following up, as fast as possible, an important Odyssey discovery. You can think of the Mars Exploration Rover mission as following up on Mars Global Surveyor discoveries from the region with shafts of hematite. Programmatically, this makes great sense. We've had Mars Global Surveyor, the Mars Exploration Rovers, Odyssey, and now Phoenix.

CI   Before launch, what did you hope to get out of the mission?

The first view of the frigid Martian north-polar region from the Phoenix lander, the first successful rocket landing on Mars since 1976. Phoenix was a part of NASA's strategy for Mars missions: "follow the water." The lander confirmed the presence of frozen water ice just under the surface and renewed hopes that microbial life, existing or fossilized, might be found deeper under the Martian surface (courtesy Peter Smith and NAS/JPL).

PS Nobody in our group believes there's liquid water, at least, under normal conditions. This is a cold part of the Martian cycle. But we are familiar with the latest results about the obliquity variations on Mars, and how climate is affected by that, and there are some real possibilities that during the obliquity changes, you can actually get a warmer climate, warm enough to melt the upper layers of ice, until you get bottom layers of wetting.

CI As far as subsurface water goes, these erupting gully features happen in the coldest, most unlikely places – is that understood yet?

PS The most likely theory is that snow falls in these areas and collects in these basins, these pole-facing slopes where it tends to stay the longest as the planet heats up in summer. The ice layers form a protective cap under which you can get melting, because the sunlight penetrates through a couple of meters of ice, right down to the absorbing soils, and warms them up from the base. You can have a microclimate down there and water can trickle down the hill. Water's not stable on Mars, but it's not stable on the Earth, either. A glass of water evaporates. It will evaporate on Mars pretty quickly, but not so quickly that it can't run down a hill.

CI I've heard that the water mass below the surface could be as high as 10 percent of Earth oceans.

PS Right. The gamma-ray spectrometer measured only the top meter. They see 80 percent ice in that upper meter, where we landed. It may go down five kilometers.

CI   Let me ask about unmanned versus manned spaceflight, and sample return. You know the amazing things a robotic mission can do. What is it that pushes us towards sample return? It's so expensive.

PS   Everybody who has tried to get an instrument to Mars has been frustrated by the lack of power, by the fact that the thing has to fit into a small and particular space, and that it can't weigh much. It also can't produce so much data that you have trouble sending it back to the Earth. We've designed some sophisticated, wonderful instruments, but nothing like those you have in a laboratory. Sample return is the desire of scientists to get rocks and soil in their laboratories, where they can look at them molecule by molecule, find out what the planet's made out of, and do radioactive dating, which we can't do anywhere on Mars – we don't know the age of anything we're looking at.

CI   Another ambiguity is the degree to which Mars was warm and wet in the past.

PS   There are two camps. A geologist who studies morphology looks at the images and sees evidence for water everywhere: flow features, floods, ripples, deltas, and meanders. But a spectroscopist looks at the surface and sees unweathered rocks like olivine, and various lava rocks. We don't see any altered minerals. Where are the clays, the quartz? We just see basic primary rock, like it was laid down last week with no weathering at all. So spectroscopists say there couldn't have been a lot of water on Mars – look at all the minerals; these aren't the right ones. Geologists who study morphology say, "What are you talking about? It's in the middle of a flow pattern!"

CI   If you could bring back a rock from one place, where would it be?

PS   One obvious place is Meridiani Planum, where this MER team found sedimentary rocks. I don't think bringing back a basaltic rock will be all that fascinating. One of the sedimentary rocks will be a lot more interesting, because you can get into the details of how they formed. There's a slim chance of finding fossils at Meridiani, but I don't think there's any life there now because it's been desiccated for 3 billion years.

   If you're looking for life signatures, the place we sent Phoenix is a good bet. It's the lowest part of Mars, so the air pressure's the highest, and liquid water's the most stable. Biologists studying polar regions on the Earth, the permafrost, are finding viable life 3 to 5 million years old. They can melt that ice in their laboratory and bring those organisms back to life. If they can last that long on the Earth in the permafrost, why can't they do the same on Mars? They're protected from radiation under the surface.

CI   Back to Phoenix. I watched the landing in an auditorium at the university and everyone was riveted to the big screen. I remember everyone held their breath, and it was almost like a collective gasp, the descent was so fast before the retros kicked in, it seemed like it was going to crash. What were the last few minutes like?

PS   We'd been rehearsing this for weeks, if not years, and the tension at the Jet Propulsion Lab was palpable. Phoenix is the sister ship of the polar lander that crashed in 1999, and we'd found and corrected twenty-five failure modes.

CI   Were you worried you'd missed a couple?

PS   I wasn't worried because of the diligence of the search and the quality of people doing it. But upper management at JPL was wondering, "Maybe there's a 26th failure mode that we haven't found." It was scary because the last one was found not long before launch. [Laughs] After the launch there isn't anything you can do anymore because there's no more testing. Mars Science Lab was the next mission; it might have been canceled if Phoenix hadn't landed safely. JPL was ready to tell the public what would happen if the thing failed, if somehow we didn't recover from the entry mode and it spun out of control, or the cruise stage didn't separate or the parachute didn't launch. Each disaster scenario had a press release …

CI   … you had them already written?

PS   Already written!

CI   Like obituaries for famous people that newspapers keep on file.

PS   We had obituaries for every death mode imaginable. It's like you wrote twenty eulogies written for yourself if you died of one disease or another, or fell off a cliff, or had a car wreck. As we were coming down, the press agent was tearing up these things and throwing them in the air.

CI   Wow.

PS   They were coming down like snow or confetti. We were all tense, but on the other hand I was happy to see that confetti in the other room. When we were done, the floor was covered in torn-up documents. Then the communications engineer started counting down the altitude and you could tell the speed by the rate he was calling out the numbers, "1000 meters, 800 meters, 600 meters." I thought it's too fast, we're going to crash. Then the thrusters cut in, and he gets to "200, 150, 100, 80, 75," and aaahhh …

CI   … everyone breathes. So you could relax and celebrate.

PS   Well, you still have a little bit of doubt. It's like running a race and you're three lengths ahead; you can always trip. [Laughs] There was great joy on landing, but the mood didn't lighten up at JPL, because without the solar panels there's no mission. We wouldn't be secure until an hour and a half later, when the orbiter told us the solar panels had opened. The mood was still somber.

CI   These people are killjoys.

PS   Gloom and doom! More than you can imagine.

CI   You must have been euphoric, but I guess then the hard work started.

PS   Right, there was no day or two off. I was in Los Angeles and was trying desperately to get home. When I got to the operations center in Tucson the front entrance was crowded with reporters, and the mayor was there. I gave a little speech and glad-handed everybody. My science team was out there, and we were all hugging and high-fiving. Somebody walked up and said, "I heard you're the rock star of science!" So I was doing the air guitar and singing, "Come on baby light my fire" [laughs] and they put it on the front page of the papers: "Rock star of science."

CI   You deserved to cut loose.

PS   I was all pumped up. I could have done anything at that moment.

CI   What were the main scientific results of the mission?

PS   I'll give you my top four. The first was learning about the ice and the soil above it. Not that it is a huge surprise that there is ice, but now we can be absolutely certain how it interacts with the surface layers. Number two was finding calcium carbonate and alkaline soil of pH 7.7. That's not very different from ocean water; it's the kind of pH and soil that's very familiar on the Earth.

CI   Why is that surprising? What was conventional wisdom about the soil?

PS   We were led by the rovers to expect a very acidic soil, with a pH of 3 or 4, plus sulfates and a sulfur-rich environment. There are many places on Earth like that, sulfur-rich and acidic and associated with volcanoes and volcanic islands, as opposed to coral islands. If you were a farmer, you would know how to deal with either type of soil, and you would put in the additives you need to grow what you want to grow. It was a surprise to us when we found out there were carbonates and an alkaline environment. It certainly speaks to Mars being a very diverse planet, and you can't just characterize it from the two rover sites and say this is the way Mars is. But people had done that because global dust storms have distributed the surface layers around the entire planet.

CI   That raises the stakes for future landers like Mars Science Lab. You only get to land in one place and have to do some inductive extrapolation.

PS   Right. I don't think it's going to be the same as the rovers or Phoenix. The third and most astounding result was finding perchlorate. I'm not sure if you're familiar with perchlorate; I had to look it up.

CI   No, go ahead and give me the primer.

PS   I thought it was bleach, but it's not. Bleach is ClO (chlorine oxygen); this is ClO_4. So it's not a powerful bleaching oxidant; it's a very soluble salt. Once it goes underwater it's there forever. Because of that, it's only found in the very driest environments. If it rains a lot, it goes into the solution and then into the ground water, and washes down the rivers and back to the ocean. Apparently it forms from salts in the ocean, with the chlorides moving into the upper atmosphere and getting oxidized through the action of ozone, which as you know is a powerful oxidant. Through complex chemical steps it becomes a chlorate, and then it rains down on the surface as particles. In the Atacama desert, where it hasn't rained in god knows how many years, they mine the stuff; it's very common there.

CI   But Mars has no analogous oxidizing agent in the atmosphere, so how do you get this?

PS   There is ozone on Mars, just not as much as Earth, and it's only found in the polar regions. Is there perchlorate on the rest of the planet? I'd guess not. So the speculation is that when volcanoes were active they spewed out HCl, and then the chloride pushed into the upper atmosphere, where it reacted with powerful oxidants, particularly ozone, perhaps others. It may be coming down to the surface elsewhere.

CI   So even the atmospheric chemistry of Mars is not homogenized.

PS   It's very surprising. We find one or two percent perchlorate in the soil. That's a lot; on Earth, it's a contaminant in our drinking water at 10 ppm. If you concentrate this material – which might have happened when there was a warmer period, when this stuff got wet and there was a flow – you could lower the freezing point of water down to –70 °C just with these salts.

CI   Interesting.

PS   We saw blobs on one of the struts of our lander. So we asked, "Why are these things growing?" They were getting bigger; two of them coalesced. They act like a liquid, so it might be a concentration of perchlorate.

CI   This is tantalizing, because it's been assumed that aquifers are going to be 100 meters down. But with perchlorate at –70 °C, you can have liquids localized, not very far down. And number four on your list?

PS   Number four is seeing snow fall later in the mission. Snow had never been observed, at least on the surface of Mars. That was exciting, and it really tells us a lot about the height of the turbulent layers in the lower atmosphere and how they change during the season. They get shorter and shorter. Water binds to the surface at night and gets released during the day. It's a very active local water cycle.

CI   That subverts the classic archetype that Mars is dead, dry and boring.

PS   You have dry soil on top, with ice within two inches of the surface, and by the end of our mission we had layers of frost on top. So you have ice on both sides of this soil, and the one on top is evaporating or subliming and depositing on a very regular basis.

CI   What does all this imply for the habitability of Mars?

PS   In weak solutions of perchlorate like in our drinking water, people use bioremediation: they get microbes to eat it. There is a guy named Coates at Berkeley who has spent the past ten years studying perchlorate, and he has a whole zoo of microbes that live on it. So it's a perfectly reasonable basis for metabolism, for life not far under the surface.

We have to consider Phoenix as part of the suite of Mars exploration. If you put this in the context of the ice discoveries at low latitudes, with a lot more ice on Mars than we thought, some of it near the surface, and new discoveries like these vents of methane, I see a real progression. I really believe within the next ten years we are going to see solid evidence of life on Mars. Maybe not proof, but it won't be the Viking era all over again.

Part IV   EXOPLANETS

24

Alan Boss

Alan Boss knows that theorists have a fairly dismal record in predicting the properties of exoplanets, so he has never been afraid to champion interesting but unpopular ideas. He got his MA and PhD degrees in physics from the University of California, Santa Barbara, and has been a staff member of the Carnegie Institution's Department of Terrestrial Magnetism since 1983. Boss has developed the disk-instability model of planet formation, which can make planets quickly and efficiently, as required by the observations. He has a long history of NASA funding, and has served on numerous high-level committees to consider future space missions, including a committee of the International Astronomical Union that set the definition of a planet in 2005. The author of over 200 articles, he is a Fellow of the American Geophysical Union, the American Academy for the Advancement of Science, the American Academy of Arts and Sciences, and the Meteoritical Society. Minor planet 29139 is named after him.

CI   Did you always want to be an astronomer?

AB   Oh no, I was born in Ohio, and my parents moved to Florida when I was in first grade. I grew up on an island in the Gulf of Mexico. I was planning on becoming an oceanographer so I went to a college in Florida to get into a program in ocean engineering. But I didn't want to suffer through engineering; I wanted to take classes like contemporary literature and German literature, instead of sticking to shop and mechanical drawing. [Laughs] I fled to physics.

I went to UC Santa Barbara for grad school, largely because I wanted to live in a beautiful area. The only graduate schools I applied to were Santa Barbara, Santa Cruz, and San Diego. I didn't apply to Berkeley or UCLA because I didn't want to live in a city. I was planning on going into high-energy physics. This was the early seventies when high-energy physics was on a roll – physicists dominated everything, and they were still planning on building the superconducting super-collider. In my mind, that was the place to be.

My first year of classes, I took a dynamics class from Stan Peel, and he hired me to work with him on the origins of the Solar System. It was 1974, and a book by the Russian Victor Safronov had just been published, called *The Evolution of the Protoplanetary Cloud and the Formation of the Earth and the Planets*. That book summarized several decades of work by the Russian school that was founded by Otto Schmidt during World War II. Safronov was one of Schmidt's best students; the book covered analytic techniques for making planets out of planetesimals.

CI   Why were the Russians so advanced in this area?

AB   The Russians had a head start of several decades on the West and they've always been strong in mathematics. Also, in the first few decades after the war Russian computers were not comparable to those in the USA, so Russians were forced to work with analytical solutions, which gave them better physical intuition.

CI   To an outsider, it's amazing that you can use equations to understand something as complicated as a solar system.

AB   When you go into the gory details, you quickly enter a territory where the simple methods of Safronov wouldn't work, but you can make progress by realizing that Newton's laws are pretty straightforward. In the phase of evolution Safronov was talking about, you don't worry about gas drag, or the interactions of the gaseous nebula with itself. He let a swarm of objects stay in Keplerian orbits until they hit each other and grew. I came to the Department of Terrestrial Magnetism in 1981 because George Wetherill was going a step beyond Safronov and putting it all on computers. Wetherill had got hold of a Monte Carlo orbital-evolution code used to study the evolution of asteroids, to determine whether asteroids could scatter into the inner Solar System and strike the surface of the Earth as meteorites.

By the time I arrived, he was well advanced in such problems. You bang together increasingly larger solids and build planetary embryos. First you build comets, then you make one- to ten-kilometer-size bodies, then planetary embryos, which are lunar-sized bodies, and finally, over periods of tens of millions of years, you build up the terrestrial planets.

CI   It's amazing you can go from snowflakes to Earths in a few tenths of a percent of the age of the Solar System. How does it happen?

AB   The slow part is the final phase, growing from lunar-mass bodies to Earth-mass bodies. Accretion is so efficient that there are roughly 500 lunar-mass bodies in orbit around the Sun, but they're well separated and small, so it's hard for them to collide and grow. It takes tens of millions of years and many orbits for these lunar bodies to excite themselves gravitationally, so that through mutual pulls and tugs they end up on increasingly eccentric orbits and have a chance of hitting each other. If they stayed on circular orbits, nothing would happen.

At Earth's distance, it takes tens of millions of orbits. After going around a million times, they get a *little* more eccentric. After 10 million orbits, they finally smash into another one, and now they're the size of two lunar masses. They need eighty of those collisions to go from one lunar mass up to the mass of Earth. These bodies get larger and their gravity increases, but the space between them also increases.

CI   The Solar System seems empty, yet I've heard that dynamically it's full. How can solar systems be full with just a handful of planets?

AB   Looking into space, you have the sense that the Solar System is a tremendous vacuum, based on the size of the bodies compared to the distances between them. But their gravity is a long-range force; when you have a lot of mass and a lot of time, even a small gravitational tug will add up and give you a significant integrated force. If you tried to put another planet on a stable orbit between Earth and Venus, it would not last long. Near-Earth objects, typically 50 to 100 meters in size, orbit the Sun between Earth and Mars, or between Earth and Venus, and those objects can't have a stable orbit. They get pushed and pulled by Earth and Venus and Mars, even by Jupiter, and eventually those random pushes put them on chaotic orbits. They end up either hitting the Sun, or hitting one of the planets.

CI   That's one of the powers of computers – you can simulate one of these situations to show what would happen if you added another object.

AB   Exactly. You can use a Keplerian orbit and put in some small variations and see how things change. Meanwhile, Jack Wisdom and Matt Holman at MIT came up with symplectic integrators nearly twenty years ago, algorithms that allow you to speed up the calculations a thousand times. Take that factor of a thousand, plus the fact that computers are about ten thousand times faster than they used to be, and you can do some serious calculations rather quickly.

CI   The astrophysical situations still challenge the best computers. Is that progress?

AB   Absolutely. When people first started doing these calculations, they did 100 to 1000 bodies at a time; with modern computers, people do N-body calculations of hundreds of millions of particles. But think of how many comets, kilometer-sized bodies, it would take to make the Earth – you need on the order of 10^{12} of those. We're not quite up to running million-million particles yet.

CI   Does Bode's law and the roughly geometric spacing of the planets naturally fall out of our modern understanding of solar system formation?

AB   In a general sense, the tenuous Bode's law spacings come out. But to be quite frank, most people don't put too much stock into trying to reproduce Bode's law

In computer simulations of planet formations, gravitational instability leads to very rapid formation of planet cores. However, the complex and nonlinear processes of planet formation occur over many scales, so are challenging to represent in a simulation.

exactly because it's more of an exercise in numerology, playing with numbers to see if things match up. While making the planetary system, the number of solids per unit area is much higher in the inner disk than in the outer disk. You have a high density of building materials, so you can build houses closer together than you can in the outer reaches of the neighborhood, where you have to go far to find the next brick. You do tend to space them closer together in the inner region, but there's no special neatness to it. *Icarus*, the official journal of the Division of Planetary Sciences of the American Astronomical Society, has put a notice in its cover page that they will no longer accept any papers which purport to explain Bode's law. [Laughs]

CI   Before 1995, you were obsessed with explaining the one solar system you know. How has your work changed with over 450 other systems to understand?

AB   The discovery of the first planet around a solar-type star in 1995 was an incredible milestone for the field. Before that, we would put together a theory and make sure we could reassemble our own Solar System, without worrying about trying to make other solar systems. Theorists had blinders on. They were thinking about making planets around single stars, like the Sun, so they were thinking about making planets in regions of low-mass star formation. But now we realize that there are many other types of solar system. The shock of the first discovery,

CI   51 Peg, by Mayor and Queloz, was that this planet was roughly Jupiter-mass, on a nicely circular orbit that was about a hundred times smaller than Jupiter's orbit. Marcy and Butler weren't looking for something orbiting that quickly.

AB   There was a basic premise of looking for Jupiter-like planets with orbital periods of twelve years. Bill Cochran, at the University of Texas in Austin, and Bob McMillan at Arizona were doing this in the mid eighties. At a conference in the early nineties, McMillan and Cochran were asked, "You've been taking data for five or six years now, so you should start seeing something. Have you found any planets yet?" Their response was, "We're looking for twelve-year periods. We haven't started reducing our data. We're going to wait until we get ten years of data." [Laughs] Marcy and Butler had the same point of view. They had been taking serious data since 1987, and by 1995 they hadn't spent much time reducing it.

Michel Mayor came into the field of planet-hunting through the back door. He had spent most of his life looking for binary-star companions. Binary stars can have a wide variety of orbital periods; some of them are contact binaries, which orbit around each other in periods of a day or less, and others are so wide they take millions of years to orbit each other. Michel came into the planet search open to whatever he might find, and used his binary-star-search algorithms, which look for an orbital period of any possible length. He had his data analyzed quickly and looked for a short-period star. Instead, he discovered a short-period planet, which excited him immensely. The great thing was that it could be confirmed by other groups rapidly because its period was so short.

CI   What about you? As the discoveries came and the number of hot Jupiters grew, were you blown away by how little we knew?

AB   I was a reviewer on the 51 Peg paper in *Nature*. I was one of the people trying to find a reason why the data should *not* be believed, but they had done all the checks on the data to confirm that it wasn't due to some other oscillation of the star, and showed that the star was photometrically stable. I accepted that they had evidence for a Jupiter- or half-Jupiter-mass object on a short-period orbit around a solar-type star.

The significance of that came to me in the middle of the night. I woke up at 3 a.m. and stared at the ceiling for a while, and it came to me that the only explanation for the short-period planet must be that planets can migrate, because otherwise I had a hard time understanding how a planet could form so close to a star. I had the misfortune of publishing a paper earlier that same year, 1995, saying that we were going to find giant planets at 2 to 5 AU, maybe, but nothing much closer. That prediction was spectacularly disproved by 51 Peg's discovery. The part of the puzzle that I had not put into my paper was that I was talking purely about where planets *formed*, not where they might end up.

Planet migration was a serious worry for theorists since the late seventies and early eighties, when Scott Tremaine and Peter Goldreich analyzed Saturn's rings. They had shown that embedded moonlets in Saturn's rings can create some of the ring structure, and they would also interact with the rings through the gravitational force that the moons exert on the rings, and the rings exert back on the moons. Those gravitational forces were typically unbalanced, so there was a net torque, and the satellite would veer inwards or outwards on quite rapid timescales. It was still a problem for Saturn's rings, unless those rings are newly formed. Goldreich and Tremaine wrote an appendix asking what happens if the same analysis is applied to the Solar System instead of Saturn's rings. Take a disk of gas to form a planetary system, put in the mass of Jupiter instead of the mass of a satellite of Saturn, and see what happens. Essentially they said, "Oh my goodness, Jupiter should migrate like mad through these interactions on a timescale of maybe ten thousand years." [Laughs] They left that as an unsolved problem.

Jack Lissauer and I spent several weeks back in 1985 trying to figure out how in the world we could stop planets from migrating. In our Solar System, there was no good evidence for migration. At the time, we thought we knew how to make terrestrial planets form where they are right now, and how to make Jupiter and Saturn pretty much where they are right now, without migration. We had to figure out a way to stop migration. The best we could come up with was to say if you've got a long-lived disk and planets, somehow the torques must sum up to zero, that the nebula has just the right surface density and the right shape so that the total amount of torques pulling inwards are balanced by the torques pulling outwards, so you stay put. That was a pretty artificial way of solving the problem, but at the time it was the only thing we could think of to relieve our anxiety about losing the planets. We didn't publish a paper; we stuck notes in our file cabinet and forgot about them. Ten years later, along comes 51 Peg, and in the middle of the night I wake up and realize that the torques were *not* balanced in the 51 Peg system. The torques made a gas-giant planet migrate nearly onto the surface of its star. That's the only explanation.

CI   Did you or any of the other theorists working on these issues ever pinch yourself because you could have predicted this?

AB   Oh, my body is heavily bruised from all the pinching, and everyone else as well. Observers generally love to gloat about victories over theorists. They have a lot to gloat about. Paul Butler loves to say that not a single theoretical prediction has been borne out since the discoveries of extrasolar planets. That's true. If I had a score, it would be "Observers 450" and "Theorists 0" at this point. We've found an awful lot of these bodies, and theorists have not really been able to explain the formation of any of them.

CI   This type of science is different from experimental sciences. It's more similar to archaeology. How do you address all the possibilities of planetary evolution over such long timescales when you want to have an initial situation and propagate it forward, and do that in a unique way?

AB   You touched on the essence of it. The analogy with archaeology is a proper one. I am always amazed when someone looks at a few bone fragments or a portion of a jaw, and can state with a straight face that those are part of a *Homo habilis* that lived 3.6 million years ago, and this is how it liked to walk! While astronomy is hard, we have a much easier job than archaeologists in reconstructing the past; in large part it's because we're confident we know the physical laws that regulate how planets form and how the solar nebula evolved. We know the physics; it's a question of trying to find which of the possible paths our Solar System actually took.

CI   Can you talk about the two mechanisms for giant-planet formation?

AB   The conventional wisdom, and I include myself as having been a conventional thinker for many years, is that you make giant planets by the same process you use to make terrestrial planets – bang together increasingly larger solids through collisions. In the case of terrestrial planets, these collisions occur so late that the gas is long gone, so you end up with a rocky planet.

In the case of the giant planets, the idea was that you'd build some rather large planets, maybe ten times the mass of the Earth, quickly, within 10 million years, and the gaseous disk would still be there. Once you got up to ten Earth-masses, gas from the disk would fall onto the surface of a planet and form an atmosphere, and when the planet was massive enough the atmosphere would no longer be stable; it would collapse onto the planet, making an increasingly denser envelope on its surface. This was thought to be a dynamic runaway process, so a core of ten Earth masses could pull on 300 Earth-masses of gas, and end up as a planet like Jupiter.

The same sort of thing would apply to Saturn and to a lesser extent for Uranus and Neptune, although from the beginning, people realized that trying to make Uranus and Neptune by the collision–accumulation process was a difficult task.

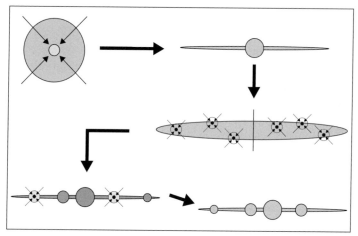

Figure 24.2. The core-accretion model for giant planet formation
A schematic view of the core-accretion model for planet formation. After the initial collapse of the protostellar nebula into a disk, random collisions of dust grains form small planetesimals. When they reach a size of about 10 km, long-range gravitational forces cause them to sweep up material until after a million years they have grown to 5–10 Earth-masses. Then runaway accretion of these planetary embryos or cores uses much of the remaining material in the gas disk and giant planets form in roughly 10 million years (courtesy Waqas Bhatti).

Safronov calculated that it would take 10 billion years to make Neptune, though we knew the Solar System was only 4.5 billion years old.

CI   The conventional model sounds sensible. What's wrong with it?

AB   [Laughs] Its strength and weakness is that people believe it, and they work on it. Core accretion has been worked on a great deal. One of the major problems is the assumption that the disk of the gas would last for 10 million years or more. Some gaseous disks probably do last that long, but on average, the gas is gone after 3 million years, and in some cases it's gone in less than a million years.

If you only had to make one solar system like ours, we might just have happened to form in one of the disks that lasted for 10 million years. But we know from the observations of other planetary systems that gas giants like Jupiter are common, and maybe 40 percent of nearby stars have gas giants we can detect, so maybe 50 percent of stars manage to make gas giant planets. We know that 50 percent of young stars do not have long-lived disks, so that's a mismatch. Either gas-giant planets have to be made faster in the core-accretion model, or something else is going on.

The people working on gas-giant-planet formation by core accretion have been busy. Some of them now can make a gas giant in as little as a few million years. But they make assumptions to get their planets to grow a lot faster, and that raises critical arguments. They sometimes assume that a planet migrates while it's trying to grow. That's an advantage because a migrating planet can eat those other bodies and become larger, faster. However, the migration process is also a great danger. Folks doing these calculations typically say, "Planets will migrate, but we'll only let them migrate at an optimal rate for eating other planets." When people calculate from first principles how rapidly planets should migrate if there's an interaction with a disk, the migration timescales can be a factor of 30 or 100 times faster than what people *select* as the right value when they want to grow a planet quickly. In reality, the planet will migrate so quickly that it will grow, but it will end up going into the proto-Sun, or perhaps be left in a parking orbit next to the proto-Sun, like 51 Peg. You'll end up with a whole bunch of hot Jupiters, but you won't have Jupiters at the distances that typify our own Solar System.

CI   It sounds like core accretion is too slow, and all attempts to fix it have their own problems. What's the alternative?

AB   There is a completely different way to get a gas giant-planet. The core-accretion model is a bottom-up model. You start off with small, rocky bodies, collide them, and pull on a lot of gas on top to make the gas-giant planet. The alternative is a top-down model. Imagine the gaseous disk around a star that is cool enough that if you have a clump forming randomly in the disk, that clump is self-gravitating. In other words it will have enough gravity to start pulling more gas onto itself. If that happens, the clump gets more massive, pulling in even more gas, which makes the clump even more massive, and it's a runaway process. When the disk gets cool enough, it is likely to be unstable to the formation of clumps.

As a second stage, the dust grains inside the disk would coagulate and sediment down towards the center of that clump to form a planetary core, much like the core that's at the center of a gas-giant planet. This alternative process is rapid and takes only a few hundred orbits to occur. You can make clumps within about a thousand years.

CI   That's fairly retro, almost Laplacian.

AB   Yes. [Laughs] The nebulae that Herschel observed were spiral galaxies, and at the time people thought they were stars in the process of forming. The theoretical models of this process have strong spiral arms, and look like spiral galaxies. It is extraordinarily retro. But it's not quite the same as what Laplace talked about: a rapidly rotating disk of gas, contracting down and occasionally spinning out rings of material which would then go on to form a planetary system. Spiral galaxies were the original misconception that launched Laplace's ideas, and the analogy's still pretty good.

CI   Was there physics missing 200 years ago that would have invalidated Laplace's idea?

AB   What was missing 200 years ago was an ability to calculate the evolution of the gaseous disk. There was no way of solving the three-dimensional equations of hydrodynamics without doing them numerically. There still isn't much you can do analytically. All people had were images of spiral nebula and a lot of imagination. Laplace used studies of the stability of rotating fluids, or of rotating solids, as his vertical model. These idealizations can be modeled mathematically fairly

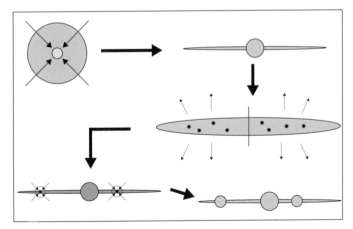

A schematic view of the disk gravitational instability model giant-planet formation. The protoplanet cools, dissipating heat by radiative and convective processes. Small parts of the disk become gravitationally unstable, generating fragments of 10–20 Earth-masses in as little as a thousand years. These fragments then generate fully-formed giant planets by runaway core accretion, a process that takes about a million years (courtesy Waqas Bhatti).

CI   well. When you don't know too much, that's where you start. But we now know that those analogies are not appropriate for a gaseous disk, because we can study these things in great, gory detail.

CI   You've made disk instability sound completely sensible. Why is it less popular as a theory?

AB   There are several reasons why disk instability tends not to win aficionados. The main one is that people are happy to continue pushing the core-accretion model. Core accretion almost certainly makes some of the planets; I think disk instability makes planets as well. What we're really arguing about is which fraction these mechanisms make. Is it 90:10, or 50:50, or 1:99? Since core accretion probably makes some planets, it's viable research. We know terrestrial planets form by colliding solids. The question is whether you can make the cores large enough and fast enough to make a gas giant.

    Many theorists have already spent a good portion of their lives working on core accretion and developing numerical methods for handling orbital dynamics and Newtonian mechanics. This idea of disk instability comes out of nowhere, which requires a different bag of theoretical skills. You're not doing orbital mechanics; you're doing fluid mechanics. When you're advanced in your career, you don't necessarily want to take two years to learn how to do a new numerical technique. Only a few crazy folks like me, coming from another field, are geared up to do disk-instability calculations. That partially explains the imbalance between core-accretion pundits and disk-instability fanatics.

CI   Does the huge amount of data give any sense of which mechanism dominates?

AB   Ten years ago, the feeling was that core accretion made everything and disk instability didn't work at all. From my own biased point of view, right now I would guess that disk instability makes more than core accretion does. The census of nearby planetary systems says there are a lot of gas-giant planets out there. Disk instability is almost impossible to prevent. Disk instability predicts that practically any star with a gas disk is likely to have some gas-giant planets, which seems to be the case. That's a strong argument.

    On the other hand, giant planets have been found that seem to be metal-rich on the inside. Some of them have much larger cores than you would expect disk instability to produce. Those objects look like they formed by core accretion, but there are only a handful of them. A core-accretion person would probably admit there are systems that only disk instability can form, like the planets with 10 or 13 Jupiter-masses, because it's hard for core accretion to make such a big object. But they say a Jupiter-mass object can be made by core accretion. We're still early in the game. We don't know what observations will rule out as the dominant mechanism.

CI   You mentioned conventional wisdoms and the way a field can operate based on assumptions, especially when the data are in scarce supply and there are a lot of people with clever ideas. You labeled yourself a heretic, albeit tongue-in-cheek. How easy or difficult is it to work in a minority viewpoint?

AB   I've always been a bit of a loner. I started working on star formation while in a physics department, where no one else in the entire department knew what I was doing. I spent twenty years understanding how dark clouds could collapse and fragment into binary-star systems. Eventually, a combination of theoretical work I had done, others had done, and observational data, showed that fragmentation really is how binary stars form. It has become the dominant paradigm for binary-star formation. If you really work on something hard and have a good idea, that may be how nature operates, and you can prove it to everyone else who might have doubted it.

   I didn't start off trying to be a heretic. By chance, I had run some models while studying how disks evolved. They made clumps, and I was intrigued, so I looked further to see if those clumps went on to make planets. Following that idea, the evidence became stronger and stronger, both theoretically and observationally. It's made me feel that I'm on the track of something positive, and with time it will be shown to be part of the truth.

   Meanwhile, you're still a heretic wandering in the wilderness. It can get hairy when you're trying to get a grant proposal accepted, because people reviewing it are going to be fairly negative about what you're doing. I have the benefit of working for the Carnegie Institution, and they give me a fair amount of support. Because of that, I feel brave enough to continue. Luckily, some folks are willing to support the heretic.

25

Geoff Marcy

Geoff Marcy was making do with moonlit nights on a medium-sized telescope, the kind of time that few astronomers want, but he knew that his dream of finding exoplanets was close to being realized. When the discovery came, it was the result of years of painstaking work to increase the precision of radial-velocity measurements of stars. He has a BA in astronomy from UCLA and a PhD from UC Santa Cruz, followed by fifteen years as a professor at San Francisco State University. He is currently a professor at UC Berkeley. Nearly half of the more than 450 exoplanets known were discovered by Marcy and his team. He has won the Henry Draper Medal of the National Academy of Sciences, the Carl Sagan Award of the Planetary Society, and the Beatrice Tinsley Prize of the American Astronomical Society. *Discover* magazine named him Space Scientist of the Year in 2003. He was awarded NASA's Medal for Exceptional Scientific Achievement and is a Fellow of the National Academy of Sciences.

CI   In the history of planet hunting, 1995 was an important year. As a pioneer in the field, you have an insider's view on the arduous process of discovering extrasolar planets. How long did it take to be in the position to make the first discovery? When did it really start?

GM   That question has two or three layers to it. The superficial layer is that I was a double major in physics and astronomy at UCLA in the early seventies. I loved physics and astronomy, and I remember thinking that the grand picture of the universe put humans in the proper perspective as small cogs on a great wheel. I remember feeling that there was some kind of importance associated with knowing our place – how we came to be here on Earth. It was kind of starry-eyed, but there's a grain of that in me, an excitement about the notion of the vast space and time of the universe.

CI   It sounds like you were already inclined to look at bigger questions, instead of focusing on a narrow research topic with the aim of becoming a world expert.

GM   No, I honestly wouldn't say that. This might be hard to believe, but I felt like I was struggling most of the time. I didn't feel comfortable enough to think that I could actually answer the big questions. I was feeling lucky just to be a part of the astronomy world as a graduate student. I was delighted and excited to be a part of astronomy research, but I didn't have grand notions about actually making a contribution. For most of my early career my mantra was, "If I can just carry out a career and make some tiny contribution, some little increment to the knowledge of humanity about the universe, I will be satisfied."

CI   That's a reasonable perspective for anyone who's a research scientist. We're lucky enough just to be able to earn a living. To have a grandiose and ambitious goal is icing on top, and possibly even unrealistic.

GM   Yeah, that was how I felt about it. It was just icing to be able to participate, and if I could actually make a little difference, that would be the best.

CI   Rewind a little – was George Abell at UCLA when you were there?

GM   He played a critical role in my development. He was my official advisor, so I went to him for academic advice. But he also taught me two remarkable things. One was an incredible love of science: what it meant to be inquisitive, to have the ability to determine the orbit of an asteroid from three observations. I can almost see him smiling and standing on his toes as he described how glorious it was that, even in the eighteenth and nineteenth centuries, people could work out orbits and so could we.

      The other thing he taught was care. You had to be extraordinarily careful. Using the Gauss technique to get the orbit required a hundred different arithmetic operations. You were taking differences of differences to get second derivatives, and if you made even the slightest error in the eighth digit in one of your hundred steps, the whole thing was wrong, and you would have no idea where the error was. He emphasized that attention to detail, and I imagine that the precision he promoted in that exercise is the same kind of precision that we need now to measure Doppler shifts to eight or nine significant digits. It's a labor of picayunish attention that pays off in the end.

CI   You're describing an apprenticeship – learning by direct experience and example.

GM   Absolutely, and I haven't forgotten it. I can still remember sitting in the astronomy library, poring over those interpolations and trigonometric tables, and thinking, "My God, what if I make a mistake in the eighth digit? I'm doomed!" It's still that way with planet hunting.

CI   What was the topic of your PhD at Santa Cruz?

GM   I did it on the Zeeman effect in Sun-like stars. The idea was to take spectra at very high resolution and look for the broadening of spectral lines that were sensitive to the Zeeman effect. It was the first attempt to survey the magnetic field on other Sun-like F, G, and K stars.

In 1982, I became a Carnegie Fellow, which was my first and only postdoc. I could tell that the Zeeman work wasn't going to go very far, because it was too difficult. I was feeling down about my future as a scientist, because I could see that my one area of expertise was not going to be very fruitful. I thought that I had already done what I could do, and there were many uncertainties. I remember lingering in the shower one morning in early 1983 thinking, "What in the world am I going to do for the rest of my career? Am I going to make it as an astronomer?" I didn't think that I was, but I thought that if I wasn't going to be very successful, the best thing I could do was go for broke. I would try to answer a question that was meaningful to me on a human, personal level, never mind what conventional science thought was a proper question for stellar spectroscopy. With the high-resolution spectroscopy experience I had in my pocket from the Zeeman work, I thought I might be able to measure Doppler shifts very precisely. I remember walking out of the shower thinking, "That's what I'm going to do. It may be my last gasp, but I'm going to try to measure Doppler shifts very precisely." I did that using the Mount Wilson 100-inch telescope, starting in 1983.

I got a lot of time because no one else wanted the 100-inch, except for Allan Sandage, who was doing metallicity work on halo stars. I got ten nights a month. I started measuring the Doppler shifts of stars, learning where the errors came from, and slowly began to recognize that no one had properly assessed the sources of errors in radial velocities. I started to learn about issues regarding the guiding of the star on the slit, the focus of the spectrometer, the point spread function of the spectrometer, and asymmetries in the spectrometer itself.

For the first time it made sense for me as a young person to ask, "What is limiting the Doppler shift? Why can't you measure Doppler shifts to arbitrary precision, and therefore detect Jupiters?" People were embarrassed for me when I started looking for low-mass planets at Mount Wilson. At that time, five Jupiter-masses was considered low mass. I remember one trip to Lick Observatory, where I told George Herbig and a few other well-known astronomers that I was going to hunt for planets using Doppler shifts. People looked at their feet, shuffled a little bit, and then changed the subject. It's hard to imagine, because it all seems so obvious since 1995, but in the eighties and early nineties, hunting for planets was not socially different than hunting for alien spacecraft or cheap energy sources like cold fusion or pyramid power. Hunting for planets smelled like looking for little green men.

CI   Was that due to the checkered history of trying to find brown dwarfs in the preceding decade?

GM   It was woven together. For one thing, people basically knew that you couldn't find planets. They were down by a factor of 10^9 in brightness, which meant you just couldn't see them. And as you point out, there was the stuffed graveyard of false claims. That added to the morose climate of planet hunting.

CI   You were exhibiting a wonderful mixture of bravery and foolishness in pursuing planets.

GM   I had nothing to lose. It maybe sounds odd to say, but I didn't think I was cut out to be a very good astronomer. As proof, I went into teaching in 1984 when my two years as a Carnegie Fellow were up. Instead of taking a job at a research institution like the Space Telescope Science Institute, I decided to go to San Francisco State, where I spent fifteen years as a professor. I taught three classes per semester: two full lecture courses and one full lab course where I graded the lab books and did everything. There were no graduate students or teaching assistants.

CI   You were teaching ten times as much as a Caltech professor!

GM   Exactly, and I still love San Francisco State. I'm very glad I went there, because that's where Paul Butler and I developed the Doppler-shift technique that's now so successful. I liked teaching, and I liked students.

CI: I'm sure pushing the limits of what you could do with radial velocities was a long tunnel of technical work. When did you begin to feel that the hunt was really on – that you were within spitting distance of the precision and errors that would get you what you wanted?

GM   What happened is important for understanding how you do a project like this. In 1986 or 1987, Paul Butler and I started trying out various ideas. We realized that we needed a wavelength calibration device, so we borrowed from the Canadian team that had used hydrogen fluoride. Then we decided to find our own molecule that would impose a wavelength standard right on the spectrum. To make a long story short, we found iodine. In the late eighties, using the iodine technique that we conjured up, we began to get good results. We would take ten measurements of the same star, over and over again over the course of a few months, and the variation was only $20\,\mathrm{m\,s^{-1}}$.

CI   Mount Wilson was closed at this point. Were you working at Palomar?

GM   I was using a very small 24-inch telescope at Lick Observatory. At San Francisco State, we had no telescopes, and we weren't formally allowed to use Lick Observatory, which was part of the UC system. I asked the director if I could use throwaway nights on the 24-inch telescope. We were in good shape, except that with a 24-inch telescope, we could only do fifth magnitude stars, which is so bright you could see them with your naked eye. The neat thing was that by the late eighties, using iodine as the wavelength reference, the accuracy of our velocities was a factor of ten better than anyone except that done by the Canadian team previously. Jupiter induces a wobble in our Sun of $12\,\mathrm{m\,s^{-1}}$, so we were within a factor of two of a Jupiter analog.

At that stage, theories were very uncertain; one or two people who worked on our own Solar System said that Jupiters were going to form at 5 AU and have periods of ten or twenty years. We felt that at $10\,\mathrm{m\,s^{-1}}$ we were in a position to make nondetections, and that was the key to our telescope proposals while we weren't finding a damn thing. I could say, "Look, we can rule out a universe that has planets bigger than Jupiter, and maybe right down to Jupiter, if we can get our errors down to maybe $10\,\mathrm{m\,s^{-1}}$."

CI   It may be counterintuitive, if you don't know about the Doppler technique, to find things as difficult and faint as planets with such a small telescope. Why didn't you use the biggest telescope you could get your hands on?

GM   There are luckily some Sun-like stars nearby. The very nearest of them – within light years – are bright enough that they're naked eye stars that you can name. You don't even need binoculars.

We started with a set of 120 stars that we tried to observe with this small telescope. Occasionally we were able to get a night on the 3-meter telescope, right at full Moon. No one else wanted full Moon, but we didn't care, because the stars were so bright that neither the Moon nor the San Jose lights killed us. Another reason that this project could work on a small telescope was that it fed a ten-million-dollar spectrometer. The spectrometer was far more precious than the telescope itself. So we were able to hunt for planets with the same integrity – just at a slower pace.

CI   Was it just a matter of time and patience? There's instant gratification in a pretty picture of a nebula, but you were gathering data points the hard way, knowing that it might be years before a signal. What sustained you through that phase?

GM   There's a plus and a minus. The plus is that everybody knew Jupiters were going to take ten to twenty years to go around their host star, so we had time to kill, and it was totally appropriate for us to take data as we were doing. I say that tongue-in-cheek, of course. Additionally, the work on the Doppler spectroscopic analysis, which was very challenging, was the heart of our effort. Paul and I spent almost all our time developing the spectroscopic analysis and we knew that patience was going to be part of the ballgame. There was just no other way. If Jupiter takes twelve years to go around the Sun, we were going to have to wait twelve years to see analogs of Jupiter go around their stars.

CI   You were involved in a long and painstaking process, but were there any "Ah-ha" moments or sudden epiphanies?

GM   Let me briefly give you three. One was technical, and as you know sometimes technical achievements are really the "Ah-ha" part. Once you've accomplished a technical goal, the science is going to happen. That occurred around 1992, when the Canadian team finished their effort. They had only studied 21 stars, but their precision was $10\,\mathrm{m\,s^{-1}}$, and they had found no planets at all. Paul and I were on a train together in the Netherlands going to a big meeting, when suddenly we knew that $10\,\mathrm{m\,s^{-1}}$ was not good enough. What was the point of doing 100 stars at $10\,\mathrm{m\,s^{-1}}$ if the Canadians had just finished 21 stars at $10\,\mathrm{m\,s^{-1}}$? We said on the train that we would have to go back to the drawing board and find out all the sources of error to get ourselves down to under $10\,\mathrm{m\,s^{-1}}$. We had a source of error in hand and we attacked it for two years. In brief, we concentrated on the point spread function of the spectrometer. When we did that, it was an "Ah-ha" moment, because our errors dropped to $5\,\mathrm{m\,s^{-1}}$ for the first time.

The next "Ah-ha" moment was the Swiss team's discovery of 51 Pegasus. It was a coincidence that six days after they made the announcement in Florence, we happened to have four straight nights on the Lick 3-meter telescope – a rarity for us. We took four consecutive measurements of 51 Pegasus and drove off the mountain with a sinusoidal curve, knowing that the Swiss were right. It became huge news; a bombastic splash that *Time Magazine* and others reported on. *Nightline* came to San Francisco State with cameras and reporters. And there we were, reporting how we had confirmed the Swiss discovery six days later.

CI   What an exciting time! Was there even a tinge of disappointment?

GM   No, absolutely not. A lot of people ask that, or say things like, "You lost the race." But what people forget is that there were ten teams out there hunting for planets, and most of them had dropped by the wayside. We're one of the two winners. At that time – early October – we still thought that the chances of finding a planet in our lifetime were pretty remote. It's hard to picture going through ten years and wondering if we would find a shred of evidence that there were planets out there. When the Swiss made the announcement it was really inspiring and exciting. We didn't feel like we had lost at all; we felt like the door had just swung open.

CI   And you had your own data set which hadn't yet been analyzed.

GM   Exactly. We had just perfected a technique at $5\,\mathrm{m\,s^{-1}}$ and had 120 stars sitting on a hard disk. We were only held back by the slowness of the computers of that time. It still took us six hours of CPU time on a good Sun Microsystems computer to just get one Doppler shift from one spectrum. We had to borrow Sun computers from all over Berkeley campus, and start running jobs here and there just so we could crunch through the stars that we had sitting on the hard disk.

Within two months, we had the third of these "Ah-ha" moments, which was the discovery of two more planets. One was the planet around 70 Virginis, which was really exciting, because it has a period of about 116 days. Paul and I stared at the computer screen for about an hour, absolutely speechless, when we saw the planet around 70 Virginis.

We had found the planet around 47 Ursa Majoris previously, but it needed more data for us to be really sure; we're very conservative. So basically, two months after 51 Pegasus, there were two more planets. These two planets played a very significant role. For about three years, 51 Peg became embroiled in controversy. We figured out very quickly that it was indeed a planet, and vigorously defended 51 Pegasus on web pages and in public talks.

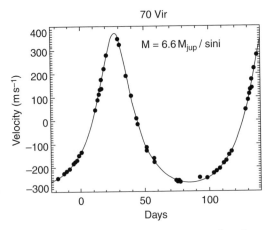

70 Vir

$M = 6.6\,M_{jup} / \sin i$

The Doppler curve of one of the first extrasolar planets discovered, a planet with about seven times Jupiter's mass orbiting the Sun-like star 70 Virginis. The planet is not observed directly, but reveals its presence by a gravitational tug on the host star, seen as a periodic Doppler shift in the spectrum of the star. The time taken for one cycle of variations is the orbital period of the planet. The precision of the data is no bigger than the size of the filled circles, so much smaller variations and much lower mass planets can be detected (courtesy Geoff Marcy).

CI   As you push a technique like this even further, do you run into astrophysical effects that limit the process?

GM   Yes, the dominant issue is the turbulence on the surfaces of stars that, in the parlance of solar physics, we call granulation. The turbulent motion of the gas over the hemisphere of the star has a velocity of hundreds of meters per second. Because each parcel of gas has a wildly crazy velocity which changes rapidly, it's just luck that there are enough of these cells over the surface that they average out. But it still constitutes the floor on the precision of the Doppler technique. We know that it's going to be very difficult to measure Doppler shifts more precisely than about ± 1 m s^{-1}. Which is, by the way, where we are right now.

CI   What are you looking for now?

GM   There are two areas. One is looking for short-period planets of about ten Earth-masses having orbital periods under a month that can be detected if your Doppler precision is 1 m s^{-1}. We're hoping to detect what we expect to be rocky planets, or miniature Neptunes, if they reside very close to the host star.

CI   Using the same targets that you've been using all along? You're just looking for harmonics?

GM   Ironically, we're going back to the original Lick sample of the brightest nearby stars. But we have also set aside a sample of 200 stars that we are observing with the Keck telescope. A good fraction of those stars are the same ones we did from Lick Observatory, but now at 1 m s^{-1} precision. The goal is to understand whether or not rocky planets are common, at least down to the level of super-Earths.

The other goal is the original one: the detection of Jupiter analogs. Amazingly, with all the successes, we've found very few Jupiter-mass planets with an orbit at 5 AU. For anthropocentric reasons, it's still compelling to find Jupiters in nearly circular orbits out at 5 AU.

Right now we're surveying 2000 stars using telescopes all over the world. At least a dozen of them show a clear Doppler signature of a Jupiter-sized planet out at 5 AU. In the next two or three years I suspect we'll find a handful of Jupiters; the massive planet we discovered orbiting 55 Cancri every fourteen years was the first. The interesting issue is the distribution of orbital eccentricities among Jupiters and Saturns at 5 AU. In other words, what fraction of Jupiters has nearly circular orbits?

CI   At what point did the discovery of closely orbiting, super-Jupiters become strange and puzzling? How did theorists react?

GM   Soon after 51 Peg was discovered, Douglas Lin, Peter Bodenheimer, and Derrick Richardson wrote a paper suggesting that migration would bring Jupiters inward, instead of dragging them outward as originally thought. Now, people more or less agree that migration is a fact of life.

The remaining theoretical question then was why they don't migrate all the way in. Doug Lin's answer, and my answer, is that they do – planets migrate in and fall right into their star. Another round of planets form, migrate in, and at some point the musical chairs stop. When the protoplanetary disk goes away, the planets are frozen in the chairs they last found themselves. So migration, planet dynamics, and the interaction of the planets are all very important.

CI   You've taken another step in the Copernican revolution by showing that planets are a natural consequence of star formation. But you've also found that even though other planets and other planetary systems exist, there is still something special about our own Solar System.

GM   You're right: there are two sides to this. On the one hand, we aren't special. On the other hand, even though we're just one of many planetary systems, the architecture of our planetary system is special in that it's a low entropy state. By that I mean a state where if you nudge one of the planets even the smallest amount – if you perturb Jupiter or Saturn or even Mars – the house of cards falls apart and it's vaulted into a realm of no return. This means that eccentric orbits and Darwinian planetary selection within a system is really the dominant activity. Planets, once disturbed by gravity, eject each other from the planetary system. At that point, you end up with eccentric orbits, maybe with only the most massive planets among them remaining as a final product. We don't know how often this happens. That's why it's so important for us to find Jupiters at 5 AU, and figure out how many are in circular orbits.

CI   You've bagged enough planets you could be forgiven for just putting your feet up on the desk and smoking a big cigar! What's the next step?

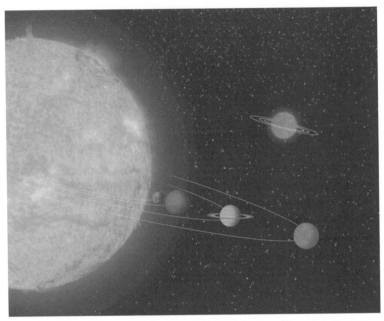

An artist's impression of the extrasolar planets around 55 Cancri, the first extrasolar planetary system found to have five members. The star is 41 light years away in the constellation of Cancer, and it has roughly the same mass and age as the Sun. It took over 300 observations to disentangle the signals of the five planets, which range in size from Neptune-like to four times the size of Jupiter, and orbital times of 3 days to 14 years. This is an exotic "cousin" to the Solar System (courtesy NASA/JPL).

GM   A few years ago, my team made a few discoveries that I think are extraordinary. One is a planet of 7.5 Earth-masses that we found around Gliese 876, which is far less than anything previously found. It opened the door technically, as well as inspirationally, to finding planets of 5 Earth-masses or 3 Earth-masses. Finding rocky planets is tremendously exciting.

We also found a planet orbiting the star HD 149026. It may not have made a big splash in *The New York Times*, but we're very confident that it has a large, rocky core. It's a Saturn-size planet orbiting close, but it transits. Debra Fischer was the lead on this. What Debra and her collaborators, including myself, found was that the planet is much too small to be pure hydrogen and helium. It has to have a significant core of rocky material. The reason that this is profound – at least from an astrophysical standpoint – is that this shows that planets form from the bottom up. Basically, heavy atomic material coagulates into planetesimals with the gas, and then gravitationally accretes onto that core. It's a very strong indication, though certainly not a proof, that the paradigm is right. The paradigm of how Saturn, Neptune, and Uranus formed is apparently operating for most of the giant planets that we're finding. Things like that still get me up in the morning; I'm not going to put my feet up on the table quite yet.

CI   It's possible that there's a clone of Earth with a 5- or 6-billion-year head start on us. That's staggering to think about.

GM   I think the probability is very high. We know of many, many stars whose ages we can measure; ages on the order of 8 or 9 million years. Some of those stars have heavy-element abundances like that of the Sun. So, bottom line, there almost certainly are planets orbiting those stars. I can't see any reason why there wouldn't have been dust coagulation and formation of cores around those stars. Presumably they have the elements on the periodic table that would enable them to form complex molecules, like amino acids. Maybe there are fewer such stars, but from the grand perspective, there are billions within our Milky Way galaxy. Billions of stars that are several billions of years older than our Sun have all the ingredients to make both planets and organic molecules.

CI   Is it just a matter of time before we find biomarkers on other terrestrial planets?

GM   That question opens up a huge can of worms. We need more discussion, not just among scientists, but with the public, because it is going to take funding from Congress. There are some very serious political issues going on, as well as international issues. We should be collaborating with the Europeans, as many of your colleagues have articulated.

The quick version is that to hunt for biomarkers around rocky planets will require an imaging, space-borne telescope that can take a spectrum. Biomarkers such as methane, water vapor, or the concurrence of oxygen plus methane, or ozone plus methane – molecules that are chemically not stable together – are all very exciting. But the Terrestrial Planet Finder as an optical challenge has not been proven – either the interferometric version of it or the coronographic version of it. Those are the two architectures on the table. Right now, the coronographs are favored, but we are probably fifteen years away from any kind of Terrestrial Planet Finder launch.

I think biomarkers are great science and great synergism because the study of them is so multidisciplinary. It's even spiritually exciting, but again may be farther down the road. It's analogous to the dream that we will someday put humans on Mars. It seems like everyone in the public knows that we want to put humans on Mars, but I don't think they have any idea just how far away from that we are. The same is true of the Terrestrial Planet Finder. But I am excited; it's a great goal. It might not happen within our lifetimes, but it's going to happen.

26

Debra Fischer

Debra Fischer earned her Master's in physics from San Francisco State University, where she was a member of the California and Carnegie Planet Search Project. Alongside Geoff Marcy and Paul Butler, she identified a large percentage of the growing number of exoplanets. She earned her PhD from the University of California, Santa Barbara, and is currently a Professor of Astronomy at San Francisco State University. Fischer has authored over eighty papers on substellar mass objects, particularly exoplanets. While she can take credit for modeling the first multiple-planet system, Upsilon Andromedae, as part of her graduate research, she gives a third-grade class from Idaho all the credit for naming the planets: Twopiter, Fourpiter, and Dinky. One of her current projects at Lick Observatory is an extremely sensitive search using the Doppler method for exoplanets around nearby star Alpha Centauri and its binary partner. It represents the best prospect for finding nearby terrestrial planets and she says she will be "stunned" if she doesn't find any.

CI   How did you get your start in astronomy?

DF   I was a pre-med major at San Diego State University, taking physics and math. Astronomy wasn't in my palette of possibilities, but I took an astronomy class and got hooked. Then I got a Master's degree in physics at San Francisco State University. I did my thesis with Geoff Marcy, looking at how common M-dwarf stars were in binary systems. From there I went to graduate school at UC Santa Cruz.

CI   This was before the breakthrough on exoplanets?

DF   Yes. I was finishing my PhD when the first planet was discovered by Didier Queloz and Michel Mayor. Paul Butler and Geoff Marcy confirmed the discovery. It was exciting. As soon as I finished, they invited me to come back as a postdoc. I was delighted; it was *very* good timing.

CI   The door was just opening on a huge new field. Were there times when you could have done something different?

DF   No, astronomy stuck. When I was in grad school, people said, "It's so hard to get a job in astronomy," and I was going back as a returning student, so they said, "What are *you* going to do? Nobody's going to hire you." I thought that was probably true, and told myself, "Well, that's okay, if I just get to do research while I'm a graduate student, that'll be enough."

CI   When did you start raising a family?

DF   While I was in the Master's physics program at San Francisco State. I was taking classical electrodynamics class from Susan Lee when I was *very* pregnant. No one sat next to me. I waddled into the room and all the guys in the class shifted to the other side of the room.

CI   As though you were an alien.

DF   Exactly, I was an alien! I went to class on Friday, had my daughter on Saturday, and was back in class on Monday. Mostly just to say I did it. [Laughs]

CI   That was your first?

DF   Yes. And my third was an unplanned pregnancy while I was starting my second year of graduate school at Santa Cruz. Part of the decision to go ahead with the pregnancy was that I thought I wouldn't get a job anyway. Things worked out.

CI   There must have been difficult times.

DF   Every week was almost impossible. Another student was pregnant right before me. I remember thinking when she got pregnant, "She's not a serious student." We helped each other through. Literally every week I would think, "I won't make it through this, but I'll just make it to the end of the quarter." I made it, one day at a time. I couldn't take a long view because it was too overwhelming.

CI   Are universities and other employers trying to make it easier for women to have families at an early stage of their careers?

DF   I honestly don't think they are making it easier; I don't think anyone is. You make a choice. I remember sitting at a "Women in Astronomy" luncheon in Sydney, Australia, at the International Astronomical Union meeting. Each of the women said, "I have small children. I don't want to leave them, and I'm making a choice not to put in the hours that my male colleagues put in." I chose to put in the hours. It was tough.

CI   There's a great story I've heard – kids naming a multiple-planet system.

DF   That's the Upsilon Andromedae system. I had been observing this star at Lick Observatory in 1997. It was exciting – Paul Butler and Geoff Marcy had found one planet in an orbit of 4.7 days and there was a clear trend of a second planet. I rushed up to the observatory every night; I'd work during the day, drive up, observe that star, work for a few hours, and then go back and do my regular day. Finally I had this great data set. It was my first job as a postdoc.

CI   Was it the first multiple-planet system?

DF   It was. Paul and Geoff were convinced that it was a double-planet system – and I was too, but what did I know? But I couldn't get it done at all. Finally, as I'm fitting a single planet and the double planet, I subtracted those two fits to see what was left, and there was a beautiful sine wave, which is the third inner planet. We published, and it appeared in the newspapers.

A fourth-grade class in Moscow, Idaho, sent me a letter saying: "Dear Doctor Fischer, we're studying astronomy and our teacher brought in your article about this new planetary system. We wondered if it had been named yet, because if not" – of course – "we have ideas for names." They thought that the one closest to the star, which is about three-quarters the mass of Jupiter, should be called Dinky; the second planet, which is twice the mass of Jupiter, should be Twopiter; and the third one, which is equal to four Jupiters, should be Fourpiter. Of course! They're such clever names, especially the latter two.

I gave a talk at NASA's Ames Research Center. My observer's talk was followed by a series of theorists, including Doug Lin, who was my professor when I was at Santa Cruz. I told this story at the end about Twopiter and Dinky and Fourpiter. We didn't have a naming convention at the time. The idea was the star would be A, the first planet would be little b, then c, and d. It was the first time that we'd had a multiple-planet system, so it was confusing. Doug Lin stood up and said, "Look, if c, no I mean if d – look, if Fourpiter gets too close to Twopiter, it's all over and Dinky will be ejected from the system." [Laughs] Everybody understood that really clearly. So it stuck.

CI   There are hundreds of planets beyond the Solar System, but this is still a fairly recent set of discoveries. People may wonder: why did it take so long? Why was it so hard?

DF   Before computer CPUs matched the demands of the data processing, we were in big trouble. Paul Butler gave a colloquium at Harvard in 1993 or 1994, before the first planet had been found. He said, "At Lick Observatory, we observe fifty stars a night." Someone asked, "How long does it take you then to analyze one of your stars?" And he said, "It takes about twenty-four hours for one star." You've got fifty stars, and then the next night you come back and you have another fifty, so you've got an incredible backlog of data. Sun Microsystems

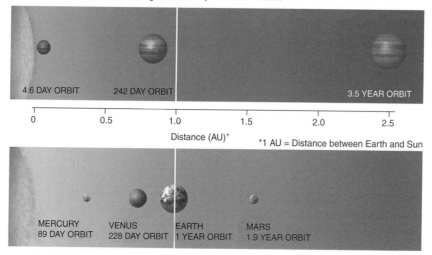

THE UPSILON ANDROMEDAE SYSTEM

Water has been detected in the region around Upsilon Andromedae

The Upsilon Andromedae system was the first time multiple exoplanets had been found around a Sun-like star. The architecture is totally unlike the Solar System, since there is a massive gas giant at a distance of 2.5 AU ("Fourpiter"), a somewhat less massive gas giant at a distance of 0.8 AU ("Twopiter"), and a sub-Jupiter-mass planet on a very rapid and close orbit ("Dinky"). None of the planets seems to be habitable (courtesy David Darling).

contributed three very fast computers to the project, and they had to run simultaneously.

Then there is the subtlety of the signal. If you look back at radial velocity projects, astronomers used to look at binary-star systems. Those shifts are maybe $10\,\mathrm{km\,s^{-1}}$, and they're several pixel shifts on the CCD; you can see the lines moving with your eyes. But with a planet, $50\,\mathrm{m\,s^{-1}}$ is a tiny fraction of a pixel on a CCD, so you're looking for lines with very subtle shifts. You have to have a huge number of lines, and that creates another problem.

CI There must be an extraordinary requirement on the stability of the instrument.

DF At Lick Observatory, we still have an instrument that is not that stable. It's sitting in the basement with a concrete wall around it. I put a little meteorology station in the spectrometer room; its temperature varies by 40 °C from summer to winter and there are diurnal cycles as well. What saved us there was the iodine cell. It was a clever idea. You can think of the Earth's atmosphere as the first iodine cell, where instead of iodine lines we have telluric lines, or lines imprinted in the spectrum by gases in the atmosphere, and those are not moving with respect to the spectrometer; we see the stellar lines then shifting back and forth with respect to the telluric lines. Gordon Walker and Bruce Campbell, who are Canadian astronomers, took this one step further. They used hydrogen fluoride, put it in a little cell, and used that as a reference spectrum for the stars. Paul Butler's Master's thesis project at San Francisco State was to come up with something else, and the answer was iodine, which is very stable. The iodine cell provides a rock-steady reference for measuring velocities.

CI   Why is having a reference spectrum in a cell in your telescope better than using the atmosphere?

DF   Because the telluric lines in the atmosphere are very broad, and they also tend to be in the red part of the spectrum. That doesn't matter so much if you have good quantum efficiency in your instrument to get a red spectrum. But at the time, we didn't. Iodine has thousands of sharp and shallow lines, and they're no deeper than a couple of percent and typically 1 pixel wide. We get a beautiful little grid etched into the spectrum.

CI   Are you limited by the brightness of the stars you're studying?

DF   We are, but we can observe the fainter stars longer. We lose precision when we have lower signal-to-noise in the spectrum, and that means that for fainter stars we have to go to a large telescope like Keck. We're getting $1\,\mathrm{m\,s^{-1}}$ precision there now.

CI   The Doppler method has been used to find the great majority of the exoplanets, and most of them are like Jupiter. Where does this technique run out of steam?

DF   We're going to go all the way to sub-Earth-mass planets.

CI   Really?

DF   That's what I believe. We'll have to see if the stars will cooperate, because they have turbulent motion. Stars have outflows at speeds of a kilometer per second, and then the material falls back as cooler gas, which can result in asymmetries in the line profiles. It's stunning; I look at the surface of the Sun boiling in the calcium bands, and it's amazing that it averages out such that we can measure dynamical motions of a meter per second.

CI   What does a meter per second correspond to, in terms of mass?

DF   It's a function of both mass and distance from the star. Maybe the best way to answer is to ask what's the amplitude the Earth would induce on the Sun: $9\,\mathrm{cm\,s^{-1}}$.

How are we going to get down to $9\,\mathrm{cm\,s^{-1}}$? Greg Laughlin and I are trying to go to the southern hemisphere and study the Alpha Centauri A and B stars. Our simulations show that if we sit on these stars all night, every night, for two years – as long as the noise is random, even if it's up at a level of a few meters per second – we can pull out a Mercury, a Venus, an Earth, and almost a Mars. All four come out of the data if there are terrestrial planets like ours in that system. But it requires thousands of observations of that star. That's completely different from anything we've done before.

CI   Would you be limited to a few nearby stars?

DF   I consider this a prototype. If it's successful, I suspect there'll be a lot of interest in building larger telescopes that can collect this data faster on many stars.

CI   Given that most of the exoplanets are gas giants, why are astronomers confident that there are terrestrial planets in abundance?

DF   If you look at the masses of planets that have been found as a function of time from 1995 to today, we're finding very few Jupiter-like planets today, because we harvested those massive planets in the first five or six years. What we're finding now are Saturn-mass planets going down towards Neptune-mass planets, with the lowest-mass planet at something like a couple of Earth-masses. The mass distribution of planets rises *dramatically*, or exponentially, towards lower-mass planets. That suggests that lower-mass planets are going to be common, unless there's some bizarre break in the planet-formation mechanism, which doesn't seem physically reasonable.

CI   There was a lot of data sitting on the shelf when the first planets were found. Now you're up at a Jupiter orbit's worth of data – are normal giant planets being found?

DF   That's a really good question. Our precision has improved dramatically – we refurbished the optics at Lick Observatory, and the Keck project started in 1997. We're finding more planets at wider separations than we do close-in. It's what we expect.

CI   How many hot Jupiters are being found?

DF   Truly hot Jupiters – with orbital periods less than ten days – are about 1 percent. They're interesting in their own right because they have a larger probability of transiting; in that case, they reveal the conditions of the atmosphere. We have a separate project looking for hot Jupiters because we can find them quickly and easily.

The most surprising thing about the Jupiters and Saturns at wider separations is that the eccentricities span a range, but they still tend to be noncircular. That has real implications for exobiology. Now we're only finding those Jupiters around a small percentage of the stars, about 20 percent. It's unlikely that there *aren't* planets around the remaining 80 percent of the stars. It's likely that a large fraction of them have planets. Imagine a Jupiter that induces a reflex velocity on the Sun of $12\,\mathrm{m\,s^{-1}}$ and give that system a bit of inclination; the velocity amplitude goes down to a few meters per second. Those systems are difficult to detect.

CI   Is all the data consistent with all Sun-like stars having both giant and terrestrial planets?

DF   We can't rule that out. If stars form, all the material in the star drains into a protoplanetary disk; the material that's left behind forms planets. That's a chaotic process that results in a huge number of possibilities in terms of planetary system architecture. In terms of each planet having a gravitational domain … it's a roll of the dice.

I did a scale model of the Solar System for fifth graders; the kids are running down the soccer field and they're only out to Mars, and I'm thinking, "Look at how much empty space there is in the Solar System. The Sun is a basketball and the Earth is a peppercorn. Why aren't there a lot more peppercorns in this system?" The amazing thing I learned when we discovered the Upsilon Andromedae system is that our Solar System is actually dynamically *full* of planets. When people who model the Solar System try to drop in an extra planet, the whole system goes into chaos – some planets are lost, some fall into the star, some are ejected, and then everything finally settles down. Each planet has its own gravitational domain and those domains are pushed up next to each other. Our Solar System resides on the verge of instability; it's stable, but only just.

CI   Is this related to the numerical coincidence of their nearly geometric spacing?

DF   Bode's law? Yes. They clear out disks; many lines of evidence suggest that core accretion is the correct model. Then they begin to migrate in until they come into a zone; again, if they get any closer, they're ejected. When I noted this back in 2000, Hal Levinson raised his hand and said, "No, no, that's not true – there's a place between Mars and Jupiter where a Venus-sized planet will survive." And I think, "How many simulations did you have to run to find that tiny little window? That doesn't count!" [Laughs] It may not be saturated, but it's pretty full. Upsilon Andromedae shares the same characteristic: in the inner 10 AU, we can't drop a Moon-sized object in and have it survive.

This characteristic is exciting and interesting: not only do stars form planets, but they have fairly full solar systems. That fullness may be a function of the surface density of the protoplanetary disk, which in turn is regulated by the metallicity of the disk and the star, and perhaps the mass of the star. Metal-poor stars may have fewer planetary systems, or weaker and thinner planetary systems. But for stars with solar metallicity and super-solar metallicity, the multi-planet systems we've found are full.

CI   How many systems have more than one planet?

DF   Around a third of them. As we keep going we see all sorts of trends. For about fifty percent of the stars with planets, we're going to find or have already found additional components.

CI   Detection of multiple planets sounds like "harmony of the spheres" – you fit one planet, subtract it out, and then you're looking for other harmonics?

DF   Right. We're in luck because of Kepler's law. It's true that we have all these different frequencies. For planets with larger orbits, and much longer periods for their orbits, we see different periodicities and they're easy to separate out. It can be difficult to disentangle the 2:1 resonance from a high-eccentricity single-planet fit, and there could be other resonances for which we're just not able to solve.

CI   What you've said about filling up solar systems is interesting. A wild hypothesis of a solar system with a dozen Earths is unlikely. If not a quota, there's probably a dynamical upper bound.

DF   Yes. There's a dynamical upper bound. People like John Chambers are working on this. He starts out with a thousand Moon-sized particles orbiting a star. They sometimes collide, and they're ejected. He always ends up with a few planets, six to eight – that's the typical outcome of these dynamical situations.

CI   The term "habitable zone" has become shadowy. What does it mean these days?

DF   It's completely bogus. We should be talking about habitable spots, maybe. Those are difficult to pinpoint; Europa is a habitable spot – liquid water, warm enough for life. The chance that there's something underneath the ice seems high. If we

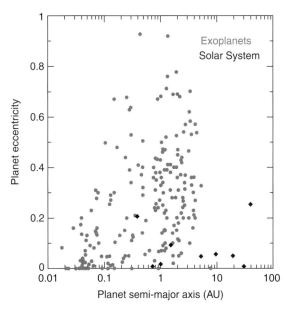

The eccentricities of most exoplanets (filled circles) are much larger than those of the planets in the Solar System (filled diamonds). In the formation process, interactions can easily act to create noncircular orbits, although it is not clear how and when they become circular (courtesy Lund Observatory).

go to Mars and there's subsurface water, it seems likely there'll be bacteria. I can't wait to see analysis of their DNA. Which amino acids are they using? Are they the same twenty we're using here? Does it have the left-handed chirality? That's going to be profoundly interesting.

CI   The hundreds of extrasolar planets found so far are unlikely to be habitable, but there are probably moons around many of those planets that are habitable, by inference from our Solar System. The next step is the hard part of astrobiology – inspecting atmospheres of modest-sized extrasolar planets.

DF   That's going to be doable. As I look at the requirements to form life, the biggest mystery is how you go from molecules to RNA and life.

CI   One of the ways our Solar System is still unusual is in the circularity of the orbits. That relates to planetary stability. Are all exoplanets wildly eccentric?

DF   No, they're not. A third or a quarter of them has eccentricities of 20 percent or below.

CI   Is it enigmatic *why* they don't circularize?

DF   I'm convinced that eccentricity is a natural outcome. Migration sounds gentle, but it's probably not a gentle process at all. There are torques as the planet clears a zone around the star, sweeping out material and building up a nest. There's probably a pileup of material at the outer edge of the disk. When that pileup and disk mass become comparable to the mass of a planet, they're going to exert a torque that causes them to migrate in. That may result in an impulse, rather than a gentle, gradual process. Any impulse is going to kick any circular planet into an eccentric orbit. There are also collisions – there must have been *thousands* of planetesimals forming in the protoplanetary disks to end up with what we have today. It's gravitational musical chairs, and those collisions are going to knock things into eccentric orbits. It's surprising that the planets in our Solar System ended up being in such tidy circular orbits.

CI   I want to return to your project to find terrestrial planets around Alpha Centauri. People are interested in the nearest Earth-like planet, because technology may enable tenth-light-speed travel, so we could get there in a few decades. When you can identify a terrestrial planet that close, someone will start planning.

DF   We go! I think we should go. We have to start planning it now. NASA's proposing the Terrestrial Planet Finder, a massive six-telescope array combining light by interferometry, phasing all the starlight out. It's great, it's ambitious, it's brilliant. But we still have big problems nulling out the stars. Let's go there.

CI   After the initial excitement to the Doppler technique, the public tends to say: "Where's the picture? I want to see something." We know how hard it is to do atmospheric chemistry of these exoplanets. But if we look hard for the closest example and it's not that far away, then yes, we cut to the chase.

DF   If Alpha Centauri doesn't have planets we will be *stunned*. It's a metal-rich star system, with slightly higher metallicity than the Sun. The two binary stars are separated by a large enough distance that any planets within 3 AU of either star would survive.

CI   Has anyone simulated the survivability?

DF   Absolutely. Plus, we have Doppler data going back for quite a while showing that there aren't any Jupiters there, which bodes well for little rocky planets. This is a binary system – the orbital inclination of the system is 79°, very close to edge-on, so we get almost the full Doppler signal; we don't have to worry about losing the mass of an Earth-like planet. Any protoplanetary disks would have been forced into the orbital plane of the binary system. These are the brightest stars – Alpha Centauri couldn't be better, it's *the* place to look.

CI   So you're doing that?

DF   We're trying. We found the telescope – there's a 1.5-meter telescope belonging to a university consortium. We think we have the spectrometer; if we can get the funding, we'll start this year.

CI   That sounds like something you could get private money for, it's a great hook: the nearest Earth. That brings me back to observing. Does the fun wear off?

DF   Every time I'm there, I'm happy. It's peaceful. On top of a high mountain you're completely removed from the hectic pace. As I'm going up I'm thinking, "I have five papers to write; I have to do letters of recommendation; I have to prepare for my classes." I get up there and all that disappears. I have to train myself; when I'm there, I focus. I have a hard time when students go up; they're observing, they start an exposure, and they want to go do something, and I have to say, "No! You stay, you watch. When the spectra come out you check the integrity, everything. Stay focused on what you're doing right now. When you're done, your job is to go to sleep." It's a discipline.

CI   You also get close to your data that way. Students are so facile with computers and data they wonder what being at the telescope adds. The act of observing is not really threatened, but there are larger numbers of astronomers who don't know how to do it. It couples to the data – what you find, what might go wrong, and how would you know if it did go wrong?

DF   Exactly, the validity of it, and the interpretation. In our group of senior scientists – myself, Steve Vogt, Geoff Marcy, Paul Butler – we *all* still go up to the telescope to collect our data. We take graduate students to train them, but we go on almost every big run.

CI   Why let students have all the fun?

DF   Exactly! The only thing I don't like is the long plane ride.

CI   Your Alpha Cen project is exciting, but where do you see your research going more generally in the next five or ten years?

DF   We have our sights set on Earth-like planets. At Keck we're doing high-cadence, high-frequency sampling of a set of stars we've deemed "rocky targets" by virtue of the chromospheric stability and the brightness of the stars. We know we can get high signal-to-noise and $1\,\mathrm{m\,s^{-1}}$ velocity precision. We're pounding away on them. We're finding a *lot* of amplitudes that are $3\,\mathrm{m\,s^{-1}}$ – that corresponds to 10 Earth-masses roughly – in about 60-day orbits. We're working our way out to the habitable zone.

We know that Jupiters, giant planets, form. The next question is: what about planets like Neptune? Those aren't gas-giant planets; they have relatively thin atmospheres. Neptune is 17 Earth-masses and Uranus is 14 Earth-masses. As we get down to the 7 to 10 Earth-mass regime – which we'll do easily – what's the frequency, how often do those planets form, and where are they, at what sorts of separations? We're taking it one step at a time, trying to flesh out the possibilities for planets around stars.

CI   I wanted to finish by asking you about talking to the public. You've been involved in a revolution – not just in astronomy, but in our understanding of the universe. Has that revolution percolated into more general awareness? Extrasolar planets still get occasional covers of *Time* and *Newsweek*, but is the scientific implication of what we've found out about other planets penetrating?

DF   It's the usual thing: the people who are interested and have a natural inclination towards science are tuned in to what's happening, and there are the usual people who are *really* interested. But no – if you confess you're an astronomer on your way to Hawaii (which means you won't get to sleep for five hours), a lot of people are still surprised that planets have been found around other stars. The big flash that happened in the nineties has been attenuated, forgotten, in the general knowledge of people.

CI   There's probably a *Guinness Book of Records* mentality; unless you break the record, it's not newsworthy.

DF   Yes, and I think that's fair.

CI   If Alpha Cen turns out right, you'll have your covers.

DF   Top of the front page, full spread in *The New York Times*, that's right! [Laughs]

27

Sara Seager

Sara Seager had started a PhD thesis in cosmology when the first exoplanets were discovered and she felt the pull of an exciting new field of research. After a BSc in math and physics at the University of Toronto and a PhD at Harvard, she spent time on the research staffs at the Carnegie Institution of Washington and the Institute for Advanced Studies in Princeton before taking her current job as a faculty member in Planetary Sciences and Physics at MIT. She has won the Bok Prize while at Harvard and the Helen B. Warner Prize from the American Astronomical Society. Seager does theoretical modeling of the atmospheres and interiors of exoplanets, leading to the first detection of an exoplanet atmosphere. She was part of the teams that made the first discoveries of reflected light from an exoplanet and the first spectrum taken of an exoplanet. It has become a maxim in her research that "For exoplanets, anything is possible under the laws of physics and chemistry."

CI   What path led you to your current research?

SS   I was a Harvard graduate student in cosmology, working on recombination in the early universe. That problem was becoming a dead end, so I wrapped up my calculations. The basic physics had been worked out by other people, and I was just dotting I's and crossing T's.

Extrasolar planets had just been discovered. 51 Peg was discovered in the fall of 1995, and the next summer three more close-in or "hot" Jupiters were found. It became an option for my thesis. The problem was that a lot of people didn't even believe they were planets at that time. It was exciting. The whole paradigm of planets established by our Solar System – the Jupiters far from the star and the terrestrial planets closer – was shot. There was a new class of giant planets close to the star, and a new opportunity to understand how these planets were heated by the star and how the radiation affected them. So I switched topics.

CI   Did you encounter any skepticism early in your career when you moved into a maverick field with not much subject matter?

SS   Definitely. My thesis committee kept asking when we'd get data on exoplanet atmospheres or if getting data was ever going to be possible. Other people said, "So-and-so thinks that they're not planets but some kind of stellar variation. You should talk to him." But I didn't worry about it.

CI   You were following your gut.

SS   I was taking a risk. But I could always go back to cosmology. What I was learning would still be applicable to other things.

CI   At what point after the first discoveries did you get a strong sense that you could bet your career on this, that it was a rich place to do research?

SS   I didn't have a lot of time to develop my computer code and to do calculations for my thesis, since I had switched topics. I had thought it was going to take me two weeks to do the introduction, but my advisor said, "You don't have that time; it should only take you two hours." So I stayed up all one night writing it. What I wrote in that introduction is coming to pass now. At the time it was still far-fetched to think of being able to detect atmospheres of these planets. Planets close to the star have a high probability to transit, so when I was graduating, I knew that one would transit soon, and I knew that once transits were detected it would open up a lot of possibilities for detecting the atmospheres. Right around the end of my thesis, I knew it was going to be good.

CI   Why did it take so long to find these planets? Everyone was hunting so hard.

SS   The planets are very small and a thousand times less massive than the star. You cannot see the planet directly. The star is so much brighter than the planet that the planet is impossible to see. Instead, you have to look for the tiny effects that the planet induces in the star itself. People had been working on this for a long time. But neither the hardware nor the software had been developed. It was also a problem of expectations – everybody was thinking about Jupiters far from the star which have a tiny signal, and not big planets close to the star, which actually have a much greater signal. If they had known what to look for, they could have found hot Jupiters at least ten years earlier.

CI   Astronomers were also working down in stellar mass, and there had been some famous brown-dwarf detections that went away. Was there a stain from that?

SS   Yes, partly. There was a lot of skepticism about the whole field. People who had worked quietly for so long, including Paul Bulter and Geoff Marcy, were ridiculed. It was a matter of technology, and patience.

CI   Flash forward less than twenty years, and the body count is over 450. How would you summarize the properties of those systems?

SS   The planets include Jupiter-mass planets at a range of distances from the star, from ones that are close-in to ones at about the same distance as our Jupiter. It takes twelve years for Jupiter to go around the Sun, so the surveys have to be in operation for twelve years to find planets like our Jupiter. The planet hunters are starting to find those right now.

Closer to the star, they're able to detect even lower masses. In the last few years they've found over a dozen smaller planets, of Neptune's mass, and even down to about 3–5 Earth-masses; they've been dubbed "super-Earths." We see a range of eccentricities as well, and it's showing us that planets are born in a random way. The transit search method has uncovered around eighty transiting planets, including one that is Neptune-size and Neptune-mass.

CI   The hot, tight-orbiting super-Jupiters that were found initially must have caused a theoretical debate. How could you get massive planets so close in?

SS   Everybody likes to come out of the woodwork and say they predicted this, but there was never a clear-cut prediction. People knew that planets are born in disks of material, and that they can interact with that disk, which exchanges angular momentum that can move them around. While there were papers about migration, no one anticipated the planet being so close to the star, and today the mechanism for how the planets stop there remains a mystery.

CI   Is it a transient stage of evolution?

SS   Definitely not. The planets in tight orbits are just there, and they're stable. This is the resting place of some giant planets. People wonder whether some planets continued to migrate right into their stars, or whether some planets were more massive when they formed and their atmospheres partly evaporated. These hot Jupiters have raised many unanswered questions. We think they didn't form close to the star because we observe young stars with disks around them, and there's not enough material near the star to form a big planet.

CI   So there's a reason to believe they didn't form there, and a plausible mechanism to get them closer in?

SS   There are also ideas about how to stop them migrating in, but it's still puzzling.

CI   The Copernican revolution told us not to expect ourselves to be unusual. Is our Solar System? And is it destined to stay in its current arrangement?

SS   It will stay like this for billions of years. Originally people thought all solar systems would be like ours. We can now safely say that about 15 percent of solar systems are not like ours, because that fraction of Sun-like stars has big planets close to the star. But the jury is still out on the other 85 percent of stars. There could be lots of Jupiters at 5 AU, or Jupiter-like planets at Jupiter-like distances, that we haven't been able to detect yet. There's a caveat. The microlensing surveys for planets work on a statistical basis; if 50 percent or more solar systems had a Jupiter, they would have had more detections. They've hardly seen anything.

CI   How different is the incidence rate of giant planets for a more-or-less-massive main-sequence star than the Sun?

SS   It's a good question. I'm not sure how much it's been worked on. People expect planet formation to be different for low-mass dwarfs or M stars. That has to do with the mass of the disk, how much the planet can interact with the disk and move. They've found only a couple of giant planets close-in after monitoring several hundred M stars. It's already clear that the M stars don't have as many close, big planets as the Sun-like stars. Theories aren't much help because the physics is nonlinear; it's a complicated, difficult field. Theories are more useful for interpreting than for predicting.

CI   Is it fun to be a theorist when there are a lot of ideas on the table?

SS   Yes. The field of extrasolar planet atmospheres is taking off. The Spitzer Space Telescope measured the infrared flux from two different hot Jupiters. From those two planets there were three data points, and already there are four papers, two of which contradict each other. It's fun, but we should be careful. We waited so long to get data, we get overenthusiastic.

CI   Do close-in super-Jupiters wreak havoc on terrestrial planets?

SS   It depends when exactly they moved in. If the Jupiters form and move in quickly, then they disturb the disks, but there might still be time for an Earth-like planet to form out of the remaining material. But if the giant planet came in more slowly as the Earth was forming, it could potentially break things up. It depends how close the planet is. If the planet is around 1 AU, then an Earth-sized planet probably cannot exist in a place where the temperatures are suitable for liquid water. These giant planets are useful because we can look at the stars with them, and we can tell which ones could have an Earth there and which ones couldn't. Generally they're bad news, but there's hope if they're close to the star.

CI   Even if they don't prohibit a terrestrial planet from forming, do they have a big effect on the debris environment of a terrestrial planet?

SS   The debris environment is affected, and if they move in later after the planet has formed, they could force the planet to move inwards, too.

CI   How far away are we from finding planets like Earth and Mars?

SS   We're a ways from finding Earth-mass planets at Earth-like distances from Sun-like stars. But we're not far from finding Earth-mass planets close to an M star. If some of these M stars have Earths close to the star, the planet would be in the habitable zone of the stars. A lot of people are interested in finding those.

CI   And this is still by the Doppler technique?

SS   Yes. It's quite hard to do now, because only a few hundred stars would be bright enough. They've searched and nothing has been found. If you had the Doppler technique at near-infrared wavelengths, where stars are brighter, it might be possible to find something. You might not get quite down to an Earth-mass, but you could get down to several Earth-masses in the habitable zone.

CI   Let's talk about transit detection. How does the transit technique work, and what are the odds of finding it?

SS   A transit occurs when a planet goes in front of the star as seen by us, just like a partial eclipse. We expect a random set of orbital inclinations for exoplanets around their parent stars. Some of them are going to pass directly in front of the stars, and the probability for a planet to pass in front of its star as seen from Earth is related simply to the ratio of the radius of the star to the semimajor axis of the orbit. The closer the planet is to the star, the more likely the planet is to go in front of the star. We were lucky that the close-in planets exist, because there's a 10 percent probability that any particular one will transit its star. That's why we were so enthusiastic. That 10 percent probability for these closer planets was great for our research. Out at Jupiter's distance, eclipses are rare, about 0.1 percent probability.

CI   And you'd have to wait a decade or more.

SS   To see repeat events you'd have to wait a couple of decades. People thought of doing decade-long surveys, but it wasn't realistic. These hot planets, plus the fact that they have a 10 percent probability of a transit, and the fact that the transit

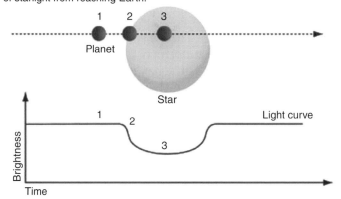

In Transit
A planet (1–3) crosses in front of its parent star, creating a mini-eclipse that blocks a small amount of starlight from reaching Earth.

When extrasolar planets have their orbits aligned so that they pass between us and the host star, the light is dimmed in a partial eclipse. The amount of dimming is given by the ratio of the cross-sectional area of the planet to the cross-sectional area of the star. Only a small fraction of distant planetary systems have their orbits suitably aligned (courtesy ESA/Charbonneau *et al.*).

dropped the star's brightness fractions of a percent in a partial eclipse, made everybody – myself included for a while – crazy to try to find more of them with a different technique. If you have a field with thousands of stars, you can monitor them and look for a momentary drop in brightness, like one star blinking. The star drops in brightness because a planet crossed in front of it.

CI The size of the effect would be about a percent?

SS Yes, because the effect is related to the area of the planet compared to the area of the star, and Jupiter has about 1 percent of the area of the Sun.

CI Which to the layperson sounds small, but in astronomy, especially from space, it's fairly easy to do, right?

SS It's not too hard to do from the ground, but it's hard to do it on thousands of stars in the same field because the stars have a range of brightness, and there are other practical issues. There's a new class of planet with even shorter periods. Radial-velocity surveys don't detect periods of less than three days, whereas transit surveys found planets at a little over a day.

CI These outrageously short periods remind me that the first two extrasolar planets were pulsar planets.

SS We often forget that, and the pulsar people get mad at us.

CI It was like a jigsaw-puzzle piece that didn't fit in anywhere, so people put it aside. Is it hard to understand orbits this close and rapid? Is there some bound at which you expect them to be sucked swiftly into the star?

SS There is. Such systems may have a different migration history. The planets could have migrated close to the star, lost mass, changed orbit and moved back out. No one really knows, but it seems like they must have a different history.

CI What can transits tell you?

SS Transits can tell us so many fundamental things. With the radial-velocity method, we only see the effects of the star. You see the star moving around on the sky, but you don't see the planets at all. We can learn about the minimum mass of the planet, the semimajor axis or eccentricity of its orbit, but we can't learn anything about the planet itself. Transits let us learn more about the planet. If it goes in front of the star and the starlight drops in brightness, we can use the ratio of the planet-to-star area to learn about the size of the planet. This is important in many ways. First, it confirms for skeptical people that these are really planets, because you saw a transit when you expected to see one. The size of the planet tells us it has to be light, it has to be low density, and it has to be made of hydrogen and helium, just like Jupiter. It's not a huge rock.

CI So calling them super-Jupiters was an educated guess for a while?

SS Always expect the unexpected. People had trouble imagining a massive rock. Transits confirmed that they're big balls of gas, and that was good. They also tell us something about the evolution of the planet. Planets are born big and hot, and they contract and cool as they age. The only difference between these planets and Jupiter is that they're close to a hot star, so we're trying to understand these planets in a new environment.

CI   Do they have trouble retaining such a large atmosphere that close to a star?

SS   No, because of their gravitational well. The planet's atmosphere is about 1000 K, but the planet is also heavy, so it retains everything. The complication is the radiation from the star, and UV radiation can potentially heat the upper atmosphere of the planet. Some people think the upper atmosphere temperature can be as hot as 10 000 K. This could lead to gas escaping. Over the lifetime of the planet, anywhere from 0.1 to a few percent of the planet's mass might leak out because of this hot upper atmosphere. But we don't think the planet is evaporating.

CI   How many stars do you have to monitor to expect to find even one transit?

SS   It depends. From space, maybe one in three thousand stars has a transiting hot Jupiter. But from the ground, with the day–night cycle and other problems, maybe one in ten thousand. You have to be able to monitor all of those stars for about a month. Astronomers like transiting planets around bright stars for the ease of follow-up observations. The statistical method requires many stars in your field of view, so they tend to be fainter. But the bright ones are great, because we can take spectra and learn about their atmospheres. There are two different ways we can do that.

When the planet goes in front of the star, it's like shining a flashlight through a fog – the light of the star goes through the planet's atmosphere, and imprinted on that light should be some information on the planet's atmosphere. Astronomers take a spectrum of the star by itself, when the planet isn't in front of it, and a second spectrum when the planet is in front, and they compare those two and divide out the starlight. They're left with the planet's light. It's hard to do, because it depends on the size of the planet's atmosphere. If you think of the annulus of the atmosphere projected on the star, it's about 0.1 or 0.01 percent of the surface area. You need the stability of space to do this because you're subtracting two sets of data, and even that data is not perfect; there are small changes all the time.

CI   What was seen the first time this experiment was tried?

SS   Hydrogen was the primary ingredient of a huge extended exosphere. Sodium was also detected. The planet's atmosphere was found to be well over 1000 K. It's nice to confirm the basic picture – we expect sodium to be in a solar gas, and it is. But a lot less sodium was detected than expected, so the simple picture is not totally right.

CI   Can you learn about the profile of a planet atmosphere by the shape of the transit, or is that too difficult?

SS   That's too hard to do, because the atmosphere's not that thick. The second way to learn about a planet's atmosphere is a secondary eclipse, when you see the planet just before it goes behind the star. When the planet disappears behind the star, you don't see the planet at all, you just have the star, and you can subtract those two observations. That gives you the whole planet either reflecting or thermally radiating light, not just the atmosphere annulus.

CI   It sounds like hard work. What are the best prospects for getting more eclipses?

SS They're extremely high, due to the successful operation of the Spitzer Space Telescope. The hot Jupiters and their host stars have only a few thousand times difference in brightness in the infrared, so the infrared is ideal for observations. The method has already been successful on two bright host stars with transiting planets. The first time astronomers proposed to use Spitzer, the time allocation committee called it an extremely risky proposal but they accepted it, because the potential payoff was high. Next time around, they accepted every single proposal on transiting hot Jupiters.

CI What atmospheric tracers or biosignatures does the infrared bring into play?

SS Infrared observations can help determine the temperature of the planet, help say whether the planet has water vapor, or even methane and carbon monoxide, and help determine whether the planet has clouds. Most giant planets have huge gas atmospheres without a surface as we know it. That said, we don't expect to see any life, so we aren't looking for oxygen or ozone. On hot Jupiters, the dominant form of oxygen is water, but we don't expect to see oxygen or ozone as gases.

Spitzer has transformed the study of extrasolar planets by identifying planets with strong day–night temperature variations and those without, and by

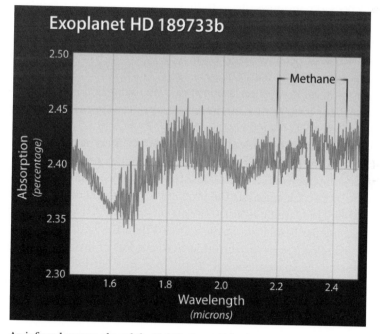

An infrared camera aboard the Hubble Space Telescope was used to detect methane in the atmosphere of the hot Jupiter that orbits HD 189733; water vapor or steam had previously been detected in the atmosphere of this same planet. Although methane is known as a biomarker, it also occurs naturally and it's unlikely to indicate life on a planet whose atmosphere has a temperature of 1000 K (courtesy NASA/STScI).

measuring the vertical temperature structure of transiting planets. Both Spitzer and Hubble have identified molecules in two transiting exoplanets, including water and methane.

CI   Earth-finding is one of the biggest hooks that scientists can lay into the public or Congress. The Kepler mission will hopefully suceed in doing that. Does that make a difference when funding is tight and all the priorities are being juggled?

SS   It does. Even if they don't like science in general, most people on the street are interested in whether there are other Earths out there, and whether we're alone. It speaks to people from all walks of life.

CI   How will the James Webb Space Telescope (JWST), the successor to Hubble, and Terrestrial Planet Finder (TPF) play a role in searching for extrasolar planets and in understanding their properties?

SS   JWST will be able to do all the things we talked about for transits. It will be able to do this for hot Jupiters and hopefully for smaller objects, maybe Neptunes. If it retains its coronagraphs, JWST will also search for planets 10 or 20 AU from the stars, and get spectra of them. But JWST itself can't do the kinds of things TPF is going to do, because it's so hard to get rid of the starlight, and JWST is not being designed or optimized in any way to do that.

CI   How does TPF suppress the starlight?

SS   It uses a visible wavelength coronagraph. With a traditional coronagraph, you put a disk somewhere in your optical path and you block out the light. But TPF uses a special peephole where the light interacts with itself so that part of the image is bright and part of it ends up very dark. There are clever ways to get the light to cancel itself in certain locations on the image.

CI   Have these been tested in the lab at that rejection level of one part in ten billion?

SS   They're getting close. There are a number of different methods, so now it looks doable. We have made dramatic improvements in our technology.

CI   What's the trade-off between doing this experiment optically or in the infrared?

SS   In the infrared, it's not just the problem of getting rid of the starlight, it's having the ability to spatially separate the planet and star. A telescope of a given size has worse angular resolution in the infrared than the optical. You can't put a huge infrared telescope in space; you have to put lots of little ones acting as an inter-ferometer and space them out, acting like one huge telescope. In the visible, we can do what we already know how to do: set up a big telescope, something like Hubble, but bigger. But in the infrared you'd have to do formation flying, and that's something we haven't done yet in space.

CI   It sounds like you're spreading your bets. Some of your projects will be paying off right now, but you're excited enough about these ambitious missions to put real effort and time investment into them.

SS   They're worth the investment. TPF has to know what it's looking for in order to be built properly. It's important to work on what Earths might look like as seen from a great distance, what properties they could have and what properties the early Earth had, in order to make sure we design TPF so that we can detect Earths and also the unexpected. The future of TPF is uncertain, but the super-Earths have given us a short-term Holy Grail, attainable with existing technology: the discovery and characterization of super-Earths in the habitable zones of small stars, and the eventual search for atmospheric biosignatures with JWST.

CI   After the first phase of extrasolar-planet hunting, the public might be forgiven for losing the thrill each time there's a new discovery. How close are we to identifying the full range of architectures of planetary systems?

SS   We're not too far for giant planets, but we're far away from doing that for smaller-mass planets. We're going to reach the limit of what we can do from the ground, and we'll need one of these new missions to fly in order to reach the next level. The radial-velocity surveys tell us about architectures down to low mass *close* to the star, but they can't tell us anything further from the star for lower masses.

CI   So we have to wait for those missions to know how unusual our Solar System is, or what the incidence of terrestrial planets in habitable zones might be?

SS   Right. Maybe ten years from now we'll know how many planetary systems are similar to our own.

CI   What might happen in the next decade that would be most exciting to you?

SS   Finding that Earth-like planets are very common, and having TPF characterize a number of them and show a diversity of planets, including those with spectra that are completely unlike anything we expect. For the public, the most exciting thing will probably be something we can be more sure about, and that would be finding Earth twins, planets that look a lot like Earth and have oxygen and water. We'll never be able to say unequivocally that a planet has life. But we can be pretty sure if a planet has an atmosphere like ours. That will be the next Copernican revolution.

28

David Charbonneau

David Charbonneau is having the time of his life as one of the astronomers who is leading the pioneering phase of exoplanet research. He got his undergraduate degree in math and physics from the University of Toronto and his Harvard PhD earned him the Robert J. Trumpler Award from the Astronomical Society of the Pacific. He joined the faculty of Harvard University in 2004 and the next year he made the first direct detection of light from a planet outside the Solar System. He has developed several innovative methods for characterizing exoplanets and he is a founding member of the Trans Atlantic Exoplanet Survey, which uses an array of small, automated telescopes to search for periodic eclipses that indicate the passage of orbiting planets. Charbonneau has already won numerous honors in his brief career, including an Alfred P. Sloan Fellowship, a David and Lucille Packard Fellowship, the NASA Exceptional Scientific Achievement Medal, and the Alan T. Waterman Award from the NSF. *Discover* magazine named him "Scientist of the Year" in 2007.

CI   How did you get into your field?

DC   Both of my parents were government scientists.

CI   So you were doomed, basically? [Laughs]

DC   Not true. My sister works in a different job. We grew up in Ottawa, Canada. I was interested in astronomy from a young age. I'd go canoeing on scout trips and bring my star chart, so I could figure out the constellations. I was able to glimpse the Andromeda galaxy by eye. I made out a fuzzy patch and realized it was much farther than anything else I could see.

      Then for a long time I had an interest in marine biology. When I was twelve, we went out to the West Coast and I saw tide pools for the first time, rich with life. I was set on marine biology until the end of high school, when I was exposed to some deep physics, in particular to big ideas about quantum mechanics and special relativity. That seemed so exciting. It's what got me back into astronomy for college. At the University of Toronto, I studied mathematics and physics and put in as much astronomy as I could on the side. I applied to graduate school and got into Harvard, and moved down to Boston.

CI   Research there is so diverse that you could have gone into anything.

DC   I was absolutely certain I wanted to study theoretical cosmology. My background emphasized analytic mathematical techniques, but I had never designed an experiment where I could get real data. I assumed that the challenging, exciting stuff would be working with complex analytic equations. Then I was inspired by one of the regular afternoon talks over beer and snacks that the faculty gave. A professor named Bob Noyes told us about planets that had just been discovered on close orbits of nearby stars.

CI   This was 1996?

DC   Yes, the year I arrived in graduate school. Bob thought that since the planet was so close to its star, we might be able to detect the planet directly by measuring the light it intercepts and reflects. Reflective starlight would prove that these were planets. At the time, there was a debate between the astronomers who felt that the Doppler method was finding planets, and astronomers who felt the star was pulsating and we weren't diagnosing that pulsation correctly.

CI   That's a bit of the history that people forget. It was a real issue for the first year.

DC   It was a real issue for *several* years. Prominent papers were published on both sides. If we could measure the reflective light, we could prove beyond doubt that there was a body there, a planet that was reflecting light. Of course we never succeeded – nobody has managed to detect reflected light because they don't reflect much light. That proof ultimately came with the first transit. When the planet goes in front of the star, it makes a major eclipse. It happens at exactly the time you predict from the wobble observations, so it's excellent proof that it really is a planet. I did that later in my thesis.

CI   Exoplanets were hot! Once you got your teeth into them, did you have regrets about not becoming a cosmologist?

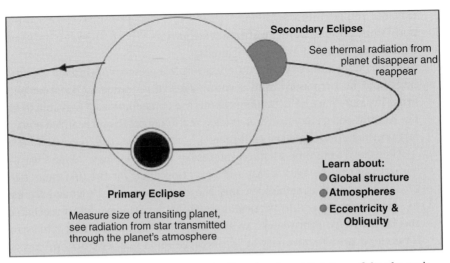

When a distant planetary system is aligned so the orbital place of the planets is along the line of sight, a planet in a tight orbit of the distant star has a significant chance of crossing in front of the star, dimming it slightly for a short time. Eclipses give data on the sizes of exoplanets and, combined with the mass and orbital information from the Doppler detection, the structure of the exoplanet can be estimated. When the planet passes behind the star in a secondary eclipse there is another signature as the thermal radiation of the planet disappears and then appears (courtesy NASA Ames Research Center).

DC   Not at all. That was the great joy – I didn't have to read textbooks, or thirty years of papers. The basic questions were unanswered, so there was an opportunity for students to make an important contribution. When I wrapped up this first project to look for the reflective light, I approached Bob and told him I wanted to work on exoplanets for my PhD thesis. The folks doing the "wobble" technique, and measuring the Doppler shifts, had moved very rapidly. It wasn't likely that a student could make an important contribution. But if one of these planets should eventually transit and make an eclipse, that would be extremely interesting. It would tell us these really were planets, and more importantly, it would provide an estimate of their physical size. The wobble technique gets the mass, so we can combine those two and calculate a density to figure out what these planets are made of.

CI   So eclipses were promising because there were so many orbits with the planet close to the star?

DC   Yes. The 51 Peg discovery meant there were bodies with mass similar to Jupiter much closer to their stars than Mercury is to our Sun. That meant that there was a much greater probability that they would transit. Orbits are randomly aligned to our line of sight, so planets as far from their stars as Jupiter is from the Sun only have one chance in a thousand of transiting. But these hot Jupiters had chances of one in ten.

CI   You also didn't have to wait long.

DC   These close planets complete an orbit every three days, instead of every twelve years for a planet like Jupiter. It's a great project for graduate students, because nobody wants to be in grad school for twelve years. Bob Noyes acquainted me with Tim Brown, who was working at the High Altitude Observatory in Boulder, Colorado. I bought an old car and drove out there to work for Tim, who had built a telescope that would do a survey to search tens of thousands of bright stars for these little transits. Up until this point, no one had found any transits, regardless of the detection technique.

      I proposed that we use that telescope to follow up on a few systems that had already been discovered by the wobble, but had not yet been examined to see if that planet had transited. Before leaving, I talked to another scientist at Harvard-Smithsonian Center for Astrophysics named Dave Latham. He had been doing a wobble survey at Keck with the Geneva folks, and they had inferred that there was one such planet, which had a boring catalog name, HD 209458. In August we started using Tim's telescope to follow this one star. Opportunities happened once every seven days. The weather in August in Colorado is affected by the monsoon, so when we set the telescope up it was often cloudy and rainy.

CI   This is big science you can do with a small telescope?

DC   It was a 4-inch telescope in a woodshed behind Tim's office. It was right near the end of a landing strip for small aircraft. At the beginning of the night, we would see all these airplanes passing through our images. Finally, in September, we got some clear nights on which the planet was expected to pass in front of the star, and ultimately, when we analyzed the data, we saw that it had. We made the first measurement of a transit.

CI   Was it competitive to find transits?

DC   Yes. Most of the Doppler teams were doing their own follow-up before they announced planets. This star had been independently detected by Dave Latham and his group. Another group headed by Geoff Marcy in California had a guy named Greg Henry doing the photometric follow-up. He also observed some of the transits. This came to a head in November 1999, when we both announced our results.

CI   You announced at the same time?

DC   Yes. There was a push to publish. With that discovery we realized all the other things we could do with transiting planets. We immediately estimated the density, and concluded that this planet, HD 209458b, was similar to Jupiter: it was made of hydrogen and helium. It was not a ball of water, and it was not a rock. Then we proposed to get data with the Hubble Space Telescope, and were granted that time on fairly short notice. We went from a 4-inch telescope to arguably the most powerful telescope ever constructed.

CI   Once you find transit, you can mine it for as long as you want?

DC   That's what we've been doing. On that one object alone, we gathered data with many different observatories and were able to look for the reflected light. We also studied the atmosphere. When the planet passes in front of the star, some of the

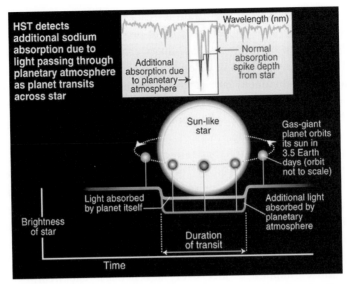

When an exoplanet is in eclipse, a spectrograph can be pointed at it, and the spectrum of the star will show additional absorption imprinted by the atmosphere of the transiting star. This reveals the composition of the planetary atmosphere. This star and its hot Jupiter-like planet was the first where such an observation was successful (courtesy NASA/STScI).

light from the star passes through the outer layers of the planetary atmosphere. Imprinted on that light is the fingerprint of atoms and molecules present in the atmosphere. That's been done successfully now for many different atomic and molecular species.

When the planet passes behind the star, we can see the emission of the planet in the infrared disappear as the planet goes behind the star, then reappear. When the planet's behind the star, we can study the star in great detail, and then subtract those data from data gathered at any other time when both star and planet are in view. That's one way to study the spectrum of the emitted radiation from the planet.

CI   The Hubble observation was exciting. Can you talk about that?

DC   The way you get that data is to write what's called a Director's Discretionary Proposal. You are writing a brief description, and asking to be granted time on very short notice. Otherwise you'd be waiting a year, and there'd be a committee, which is how most of the time is awarded. I put the proposal in before I left for Christmas break, and we heard shortly after that we had been granted time. We got the data in the spring of 2000.

The data were exquisite. As the planet passed in front of the star, we could see in clear detail exactly how long it took for the planet to cross the limb of the star, and therefore how big the planet was, independent of other constraints like the size of the star. That data had the sensitivity to tell us whether the planet had rings similar to Saturn, or if the planet had a satellite similar to the Earth in its physical size. It doesn't have rings and satellites. But here was an observation made from a 4-inch telescope in the fall; by the spring, we were talking about looking for systems of satellites, perhaps akin to what we see around Jupiter and Saturn.

CI What were the special attributes of Hubble that made that possible?

DC The key attribute was that it's in space, so it's incredibly stable for gathering and measuring light intensity. We didn't need resolution. We were using Hubble as a light-bucket.

CI What did you find in the atmosphere of that planet?

DC The element we chose to target was sodium. It wasn't that we thought the planet atmosphere was made of sodium, it was the fact that sodium has a prominent spectroscopic feature, which is broad and has a yellow-orangish tint. Several theoretical calculations predicted which features would be prominent in the atmospheres of these planets, and they all predicted that sodium would be easy to see. We had to pick which wavelength, which colors of light, to look at. We chose to target the sodium region and were able to make that detection. We could see that when the planet passed in front of the star, all of the light was suppressed by about 1.5 percent, but at the wavelengths where sodium would absorb light, there was additional suppression. The planet appeared bigger in that color, meaning that there was an atmosphere which was opaque in this wavelength of light. That technique has since been used to observe many other planets.

CI Discoveries are always amazing and unique. How have the statistics gone over the intervening years? Was the expectation of the rate of eclipses from the hot Jupiters supported?

DC That has been borne out. The radial-velocity-wobble teams have continued to find many hot Jupiters. The correct fraction of those planets, roughly one in ten, do present transits, and some exciting systems have been discovered. One planet was found to have a mass much less than Jupiter. Jupiter's about 300 times the mass of the Earth. They've found a planet with a mass roughly 20 times that of the Earth. Based on its mass, we expected it to be much more similar to Neptune than to Jupiter. It transited, and indeed it's similar in size to Neptune, so it's not made mostly of hydrogen and helium. It has a big core of ice and rock, and an envelope of hydrogen and helium. We can see that directly.

We got involved in the transit surveys, trying to find planets first by eclipses and then measure the wobble as a follow-up. For many years, we deployed humble telescopes at various mountaintops around the world, and the recovery rate was extremely low. We'd look at tens of thousands of stars and wouldn't find anything. It's not that there was a problem reconciling the wobble resolves and the transit surveys. It was simply that we had many challenges in terms of analyzing the data, and being certain that if there had been a planetary transit, we would have seen it. Recently, that detection method has matured in the same way that the Doppler method matured in the fall of 1995. By late 2008, there were over forty transiting planets, and the large majority of those were found first by transits and later confirmed by wobble.

CI   Since your detection efficiency in the transit surveys is lower, is the motivation that you can get a lot of information from each system?

DC   Yes. The wobble technique alone tells us a great deal about the architecture of those planetary systems. We often find several planets. We can learn whether their orbits are circular or eccentric, and whether the planets are close to the star or far from it, and we can compare that to our Solar System.

The transit method gives us the ability to study these planets as physical bodies. We can learn what they're made of, what their internal structure is like, and study the chemistry and dynamics of their atmospheres.

CI   Are these exoplanets throwing up any surprises to tell us that our giant planets are abnormal or atypical?

DC   Absolutely. The central lesson of exoplanet studies, both from wobble and transits, is that the diversity of planets is far greater than what we had predicted, both in terms of their orbits and physical properties. There were sophisticated predictions, and we thought they were grounded in a basic understanding of planet formation. But we were subject to the bias of having grown up in our Solar System. Nature produces diverse planetary system architectures, and diverse properties of planets themselves.

Hot Jupiters vary immensely in their density. Some of them have a mass and size similar to Jupiter, and we think we understand those. Some, for the same mass, are much puffier. That's a basic physics problem: we don't understand how they can maintain such low density, because under that much gravity they should contract rapidly down to the size of Jupiter. There are basic puzzles in understanding these planets and why there's such large diversity in their masses and radii.

CI   Do the structure and composition of these hot Jupiters tie in to the debate about how and when they got into those parked orbits?

DC   Certainly. We think Jupiter formed far from the Sun, where the temperatures were cold enough to allow ices to condense – particularly water. It would have enough solid material to build up a core of ice and rock many times the mass of the Earth. Once it was big enough that its gravity attracted the surrounding nebula of gas, it could balloon out to 300 times the mass of the Earth. We cling to that picture so dearly that we think *all* these hot Jupiters weren't born where we find them, but formed like our Jupiter, and then migrated in subsequently through some mechanism that brought them close and miraculously parked them right next to the star.

Encoded in the properties we can observe about the planet is its formation process. We'd like to know if there is a core at the heart of these planets. If there's a core, the density is affected. I described some of these planets as puffy; in that case, we don't have a good sense of whether there's a core or not. But in many cases, we can see that the planets do have a core in their center. If they were made entirely from hydrogen and helium, they'd be bigger than we observe. That fits well with this picture, because it's consistent with our view that they formed far from the star and migrated in subsequently.

CI   You're inverting the nature of our knowledge and ignorance. I was surprised to realize how poorly determined the core masses of our own Jupiter and Saturn are, but you're making those measurements for exoplanets?

DC   Right. Saturn has a core, but Jupiter falls in a range where there's an ambiguity. Because of its massive radius and our incomplete understanding of physics of hydrogen, we aren't certain whether Jupiter has a core or not. For less-massive planets, the core has a stronger effect. That's what's going on with Saturn, and with what we're calling hot Jupiters, but which in some cases are more like hot Saturns. We can see that core because it doesn't fall in the regime where there's ambiguity.

CI   Let's shift back to chemical fingerprints. We want to know if there are water worlds out there, even if the only ones we can detect right now are more massive than terrestrial planets. Can we get a handle on that signature?

DC   Yes. In 2007, a team has used the Spitzer Space Telescope for transmission spectroscopy, which is a basic way to see starlight filter through the atmosphere of a planet. They made detailed measurements of one of these planets as it went in front of the star, and saw what looks like a clear signature of water absorption. We expect there to be quite a bit of water. On the Earth, what's exciting is that the water's in liquid form. That's not the case with these hot Jupiters, where the water is in vapor form.

CI   That's a great discovery. Steam. Is there more of that on the horizon?

DC   Yes, much more. There's a productive combination of many groups using humble telescopes to survey big swaths of the sky. They're finding planets at a growing rate. In 1999, we had one. It was many years until we found a second. Two years ago, we only had a dozen. In the past year alone, we've discovered the bulk of them. We can use powerful, general-purpose observatories, such as the Hubble and Spitzer Space Telescopes, to compare the properties of different exoplanets, and compare those properties with the planets of the Solar System.

CI   What about pushing to lower mass? Is there any prospect of studying terrestrial planets with eclipses?

DC   I think so. The first and best way we learn about terrestrial planets and their properties may be with transits. The NASA Kepler mission was launched in 2009 and it will study 100 000 stars. Because it's in space, it can search continuously. By waiting for four years, it can detect planets with orbital periods of longer than a year. The precision is enough that if an Earth-sized thing goes in front of a Sun-sized thing, they can detect that. They will find these systems, and most importantly, they will tell us this precious number: the rate of occurrence of Earth-like planets in the habitable zones of those stars.

CI   It's a full factor in the Drake equation, determined by one mission.

DC   It is. Do we live in a *Star Trek* universe, where every star has a habitable planet, or do less than 2 percent of Sun-like stars have planets? If the Kepler mission doesn't find *any*, that will affect our plans down the road, in terms of what we might build to study the atmospheres of those planets.

There's an opportunity to do this from the ground, because most stars are much smaller than the Sun. The small stars are called M dwarfs. Just south of Tucson, I'm building an array of small telescopes called the MEarth project. We're going to pick a whole bunch of M dwarf stars and study them carefully. We're cheating: essentially, we're shrinking the stars, so the signal is bigger, and the ratio of the planet to the star is big enough that we can measure it, despite the atmospheric effects of the Earth.

CI   Despite the nice work you can do on Hubble and Spitzer, it sounds like you're still drawn to the small telescopes and getting data in a hands-on way?

DC   Absolutely. It's great fun. These are telescopes you can build yourself, and we persuade students to help us.

CI   You're training a new generation of planet hunters.

DC   Yes. Hubble and Spitzer aren't planet-detection instruments. They're not survey instruments. They can't look at many stars at once. If you want to do a survey from space, you have to build your own mission, and that's hundreds of millions of dollars. On the ground, we're talking about things that can be done for a tiny fraction of that budget and be taken on by one or two interested individuals.

29

Vikki Meadows

Vikki Meadows has a job that sounds straight out of a science-fiction movie: she runs NASA's Virtual Planet Laboratory. Her team does not actually make planets, either real or imagined, but they simulate the full range of plausible atmospheric and geological properties in a self-consistent way so that various biomarkers can be correctly interpreted when astronomers finally get spectral information for the terrestrial exoplanets. Meadows got a BSc in physics from the University of New South Wales and a PhD in astrophysics from the University of Sydney. She is a research scientist at NASA's Jet Propulsion Lab and a Principal Investigator in the NASA Astrobiology Institute. Until its recent demise as its cryogens ran out, she also served as the lead Solar System Observations Scientist for the Spitzer Space Telescope. Her early experience in remote sensing serves her well in the Virtual Planet Lab work, which extends to making models of the early Earth when its biological signatures were quite different than they are today.

CI   You changed continents in your career. What prompted that move?

VM   I was born in Sydney, Australia, and I went to the University of Sydney to study astrophysics. I was working with the Anglo-Australian Observatory, so I had an advisor there as well. My PhD thesis was jokingly called "Infrared Observations of Just About Everything." It was split between high-redshift galaxy surveys and studying water vapor in the lower atmosphere of Venus. I helped build an infrared camera with David Allen, my advisor. He died in the last year of my PhD.

CI   That was tragic. I have common history with you – as a grad student in Edinburgh I cut my teeth on observing at the Anglo-Australian Telescope. I knew Dave Allen from that time.

VM   His death was premature and incredibly sad. We were observing the impact of Comet Shoemaker–Levy 9 with Jupiter, which he had been looking forward to, but he knew at the time that he was dying. So we called him. I spoke to him last on the impact of fragment A. Then he went into his final coma, and he died three days after the impact of fragment W. I turned in my thesis three days after that. It was a dramatic week all round.

CI   Between distant galaxies and Venus, when did you decide you wanted to work closer to home?

VM   David Crisp and David Allen were collaborating on the Venus project, and David offered me a job at JPL. I went over there nominally for two years, to see what it would be like, and stayed for the rest of my career. [Laughs] In 2005, I transferred over to Caltech. I work at the Spitzer Science Center.

       I've always been interested in planetary science, but in Australia it was difficult to get a mentor in that area. The Venus study, which introduced me to David Crisp, was a side project, but it ended up being my thesis. The work I do now is closely related to that. We use remote sensing to look for signs of life or habitability on other planets.

CI   How did the concept of your Virtual Planetary Lab come about? It's a beautiful name. It conjures up video games and the Sorcerer's Apprentice.

VM   JPL wanted to get into astrobiology. I was working at the time on the limits of the habitable zone. I was approached about leading a proposal and said okay, not knowing what I was getting into. We were told it had to be interdisciplinary. We had a lot of models available, and we thought, "Nobody's ever put these together. Wouldn't it be neat if we could make a model of a planet?" I went into this project understanding that it was extremely challenging.

       Once we had the team, we started ironing out all the ideas, and it fell together. It was a way to synthesize a lot of different fields by focusing on one long-term goal and one very challenging project. It forced people from different disciplines to work together, because the product was common. Having this core concept that everybody could contribute to is a nice way to do astrobiology, rather than having people do separate research on their own and then sew it together after the fact. Along the way, I thought, "This has to have a catchy name." I came up with the Virtual Planetary Laboratory, which is jokingly called Vikki's Planetary Laboratory.

CI   Is it virtual in location too? Is it a loose collaboration of people at different places?

VM   Yes, eighteen institutions are covered in the VPL. Politically it was astute, because NASA's Astrobiology Institute was stressing this idea of the virtual institute. We put together a laboratory to make virtual planets, and we're the virtual planetary laboratory because we are distributed. The double entendre was intentional.

CI   It's only a problem if NASA says, "We have virtual funding for you."

VM   [Laughs] That's right. Virtual funding, virtual results. There've been lots of jokes on the virtual aspect. We meet through video conference, telecon, and email. We try to get together once a year, but most of our interactions are not face to face.

CI   Spectroscopy is a powerful technique. Why is spectroscopy of planets elsewhere so hard?

VM   We hope to find a small, terrestrial, rocky planet like the Earth around a star other than our Sun. It's an extremely challenging goal to begin with, because reflected light from a planet is swamped by light from the star it orbits. But when we find that planet, if we're able to directly detect the photons coming from it, we'll have a chance at being able to characterize it. The first thing to look at is its distance from its parent star to understand its orbit, whether it's circular or elliptical. Does it get changing starlight throughout its year? Then maybe we'll measure colors by doing photometry at different wavelengths. That only gives us some basic, crude information.

The most powerful way to get information about a planet is to take a spectrum. We break reflected light from the planet into as many constituent wavelengths as we can. There's a trade-off between the number of photons we're able to collect in a given channel versus how much detail we would like to see in the spectrum of the planet. When we get the spectrum, it's going to be from the entire face of the planet. All the continents, oceans, clouds – everything's going to be crushed together into a disk average. Even so, it's extremely powerful, because within that spectrum will be signatures of atmospheric composition, and perhaps what the surface is like, whether or not there's vegetation, and whether or not there's life. All of this can be collected from telescopes orbiting the Earth. We don't go there.

CI   We have over 450 extrasolar planets, and a lot of them are Jupiters or similar, but we're working our way down towards terrestrial-planet masses. How far are we from applying these spectroscopic techniques?

VM   We're already pushing into the realm of what we would call terrestrial planets. Anything up to 10 Earth-masses we consider to be a terrestrial planet. Beyond 10 Earth-masses, you have a core for growing a gas-giant planet. We've already got down to 2 to 3 Earth-masses. To go much beyond that, we need technology that NASA's developing for the Terrestrial Planet Finder missions and also SIM, the Space Interferometry Mission, which will use an indirect detection technique to look for the wobble of planets around their stars. If you combine the data of the SIM mission with the TPF mission, you might be able to push down from 3 to 1.5 Earth-masses. We're keenly anticipating TPF-C, the first Terrestrial Planet Finder. That will be our first direct detection of Earth-sized planets around other stars.

CI   Why do we have to go into space to do this work?

VM   For two reasons. One, you need the stability of space. The C in TPF-C stands for coronagraph, a method for blocking out the light from the parent star to see the much fainter planet nearby. You need a stable instrument and a stable mirror system, and space is good for that. But the real scientific reason is that once we get the direct detection of these planets, we're going to take spectra of them, and look for constituents in their atmosphere that are present in our atmosphere. We don't want to look through our atmosphere when we're trying to detect these incredibly weak signals elsewhere. It's important to get above the atmosphere so that we're not trying to separate extraterrestrial oxygen from terrestrial oxygen.

CI   That's imaging, and the spectroscopy comes later?

VM   Yes, there are two missions. The coronagraph flies first, in part because we already know how to build large optics and a big mirror system in space. TPF-I is a much more technologically challenging project, and it involves using free-flying spacecraft to form an interferometry system with a changeable baseline. That's at the edge of our technological capabilities. TPF-C does the initial detection and then does spectroscopy in the visible; TPF-I detects, and then does spectroscopy in the mid-infrared. TPF-C will have fewer targets to look at but it will get visible light

Terrestrial Planet Finder is an ambitious two-phase NASA mission to characterize terrestrial planets down to near Earth-masses. The second phase will be this interferometer, which combines the information from multiple spacecraft to achieve very sharp images. Infrared wavelength observations can reveal biomarkers, or spectroscopic signatures of gas composition that might indicate a metabolism at work. TPF is under study but not currently funded (courtesy NASA/JPL).

spectra of them. TPF-I will be able to do almost all the targets they want C to do, plus it will have the capability to go further. It's not only a complementary wavelength, it's also a next-generation move.

CI   Are people brave enough to venture when TPF-I will launch?

VM   It all depends on NASA budgets. TPF-I is supposed to come five years after TPF-C, so 2019, or 2020.

CI   Your research has value in its own right, but the payoff from these ambitious missions is a long way away. Is that difficult?

VM   I've worked with NASA for fifteen years now, so I'm used to it. The Hubble Space Telescope took twenty-seven years from concept to launch, so we're used to long gestation periods. We're trying to lay the theoretical foundation for these missions, so we're happy to have some time to do that, too.

CI   In an uncertain funding climate, your research will help motivate support for the missions.

VM   Absolutely, and not just politically, but supported scientifically as well, because we're exploring a parameter space that either of the two TPF missions could do observationally, but we're doing it theoretically. We're determining the range of observable properties of terrestrial planets. Can we imagine worlds that are physically and chemically self-consistent? They may be out there, so what would they look like? Are there any other, different biomarkers we might look for? A lot of things are produced on the surface, but not all of them are visible in the global spectrum. What if we had a habitable environment where sulfur bacteria or methanogens dominated rather than denitrification and photosynthesis? What will we look for, and can we design TPF so that it could detect those different metabolisms? We can explore a whole array of planetary and life characteristics. The ultimate goal of VPL is to generate synthetic spectra that say "these are the things we should be looking for; these are the things that might happen on distant planets."

CI   What are the best biomarkers? Are any of them so striking that they amount to a smoking gun?

VM   Our overall philosophy is that a biosignature has to be interpreted in the context of its environment. A smoking gun is a nice idea, but as a scientist I have trouble with it because so many conditions must be met. Biomarkers have a probability of indicating life. You see a series of spectral features and you ask, "what is the probability that this is produced by the planet itself and not by life?"

On Earth, methane in the presence of oxygen is considered to be a biomarker, and combined with oxygen, an extremely strong biomarker. The gases destroy each other if left to their own devices. But because there are strong sources of both gases on the surface of the planet, they coexist in chemical equilibrium. You would have to know something about the rest of the atmosphere to see that they are in chemical equilibrium. If you looked on Titan, you'd see lots of methane. But methane on Titan isn't a biomarker; it's a main constituent of the atmosphere. You have to know something about the bulk composition of the atmosphere, the chemistry of the planet, and its physical conditions, before you can say whether something is a biomarker or not.

Having said that, oxygen is our best biomarker. If it's above a certain abundance in the atmosphere, it's probably produced by life. People have explored other mechanisms for producing large amounts of oxygen in the atmosphere – by large, I mean over 1 percent – and it's difficult to do by planetary or abiotic processes alone.

CI   Since each potential biomarker can also have a purely geological origin, you have to look at the ensemble of tracers and make a model that is consistent with biological mechanisms.

VM   That's right. We can do it rigorously. The claim to have detected life elsewhere in the universe sets a very high bar. You can always bring out the Carl Sagan quote about extraordinary claims requiring extraordinary evidence. That's one of the crucial reasons why we want both TPF-I and TPF-C: with both of the wavelength regions, the visible and the mid-infrared, we have a much better handle on the environment of the planet, and we'll be more able to distinguish between false positives and actual, true life.

It would not be good *just* to detect oxygen or *just* to detect ozone, because you would want to know what was going on with the rest of the planet. But even if you just detect oxygen or ozone, there's still a high probability that the planet is inhabited, because other things that might produce oxygen are not that common. But that is our Earth-centric view. We don't know what we're going to find. TPF is a mission of discovery: finding planets exactly like the Earth would be incredibly exciting, but it would also be very disappointing. I want to find planets that are *not* like the Earth, that are maybe habitable or inhabited by life.

CI   Does that philosophy extend to targeting? Is TPF only going to look at stars like the Sun?

VM   At the moment, the sample is limited to Sun-like stars and stars a little bit hotter and a little bit cooler than our Sun. They're considered higher probability stars for having a habitable planet around them. The hot limit is based on the lifetime of the star. If you go hotter, the lifetime of the star is 2 billion years or less. Planets might form, but is that enough time to develop a strong enough biomarker to be able to detect it remotely? On the cool side of the sweet spot, we worry about the lack of UV radiation and the lack of overall flux, but they're very long-lived, and that's good. That also means that the planet itself has to be close to the star to be within the habitable zone, which makes them much harder to separate. A few M stars might be included in the sample, but with the angular separation problem and the slender habitable zone, we're limited.

CI   You're placing bets. Does that amount to a strong assumption?

VM   That's still a lot of stars. We're eliminating the stars that live fast or die young. But we are missing the large M dwarf population. Most of the stars in the galaxy are cool dwarfs. They're much more common, but they're also faint, which makes it difficult to detect a planet around them because the habitable zone appears very close to the star.

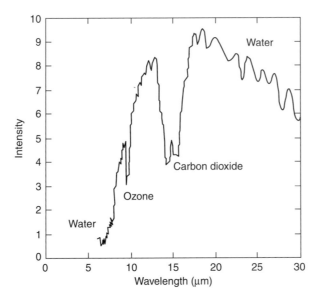

This infrared spectrum of Earth was taken by Mars Global Surveyor as it headed towards Mars. The level of spectral detail is similar to what will be possible with TPF, but the data quality is high because Earth is bright and nearby compared to exoplanets. Water vapor creates many absorption features and the ozone feature is so strong that it's ten times easier to detect ozone than oxygen (courtesy NASA/JPL).

CI   When you interpret a spectrum, there must be a range of habitable planets that might produce a spectrum that, given the data quality, is identical to the one you measured. How do you deal with the concern about uniqueness?

VM   That's exactly what we're trying to explore with these models. I can generate a lot of variations. For example, we've taken an Earth-like planet around a G star, and we've cut down the amount of oxygen by successive factors of ten, down to one part in 100 000, and looked at what that does to the strength of the oxygen and ozone absorption bands. TPF won't be able to discriminate minor details. With our models, we can generate a unique result, but also determine how sensitive we are to changes in the gross planetary properties. Are we ever going to detect them from TPF-C or TPF-I? That's part of our research.

CI   I like the idea that you're creating a landscape of hypothetical terrestrial planets. For each one, you predict something that TPF might detect, and you'll let the data decide what you see.

VM   Ultimately we're trying to create a catalog of potential or plausible spectra. We don't want things hanging in the air, and we don't want to create a planet that's physically unrealistic. We're taking it one step further and saying, if I pull out the oxygen, what happens? This: the chemistry and temperature structure of the atmosphere change, which changes the mid-infrared detectability. A whole bunch of attributes reorganize themselves, shuffle around, and come into equilibrium when you make a gross change to a planet. With VPL, we can create a library of plausible spectra. None of them may exist in the real universe, but some of them might. We then have a resource not only for planning what we potentially might look for, but also a framework for analyzing the data when it comes back.

CI   Astronomy is full of surprises. A spectrum might not match anything in your grid.

VM   Exactly. But we'll have the tools to start working on it. By pushing the models to be as thorough as possible, we're trying to make sure that they can encompass the majority of what we might see with TPF. I would be surprised if we modeled alien planets that turned out to be *exactly* like planets we find with TPF.

CI   Let's use Mars as an example. We might one day find microbial life under the surface. Could we deduce the presence of life on Mars as seen from afar?

VM   I don't think we can, because Mars has about 60 parts per billion methane. That's incredibly hard to detect in a spectrum without both high resolution and high sensitivity. But you could imagine another planet that was larger, closer in to the habitable zone, and for whatever reason never developed an ozone shield, and so ended up with life under the surface to avoid UV radiation on the surface. Gas emitted by the microbial community would escape into its atmosphere and build up, and we would be able to detect biomarkers globally. But in the specific case of an extrasolar planet like Mars, the signal is too weak.

CI   So there are going to be situations where there *is* microbial life, but its signatures are too subtle to detect?

VM   Yes. We won't be able to get all of them. In this first phase of discovery, we'll find planets where life is absolutely obvious. When you take a spectrum of the Earth, it is obviously inhabited. We see water vapor, oxygen, ozone, methane, nitrous oxide, and chloride. We see ingredients that are generated by biology and have built up to significant levels – out of chemical equilibrium – in our atmosphere.

Having said that, there's a mission beyond the Terrestrial Planet Finders. NASA's talking about sometime after 2025, and that's probably optimistic. This mission is called Life Finder, and we know very little about it, except that it's supposed to have a much higher spectral resolution and sensitivity than the missions before it. TPF-C and TPF-I are the first reconnaissance missions. They're bringing back the first, best-possible, but admittedly crude characterization of the planets. Life Finder can not only look for subtle features in the spectrum, perhaps indicating continents and oceans, but will be able to do better time sampling as well, and have a larger range of planets to study.

CI   TPF-I is already big, expensive, and unfunded. What would motivate you to do even more difficult work?

VM   The infrared is a rich region for finding spectral signatures due to molecules. In particular, the by-products of metabolism are more likely to absorb in the infrared than in the visible. We're also much more sensitive to some of the more gross features of the atmosphere in the mid-infrared. The classic example is CO_2 – the CO_2 signature in the mid-infrared is the most obvious thing about terrestrial planets in any part of the spectrum. CO_2 shows a temperature inversion in the middle of the band. This big, broad CO_2 band helps you understand whether the planet has heating in its stratosphere, which could be indicative of a UV shield. In our atmosphere, it's ozone; on other planets, it could be something else. The UV shield has been important in the evolution of complex land-based life on Earth.

Just shortward of that band in the spectrum is the 9.6-μm ozone feature. According to our models, that ozone is a more sensitive indicator of oxygen in an atmosphere than oxygen is itself. As you go to successively lower amounts of oxygen in an atmosphere, there's a point where it becomes hard to detect in the optical spectrum; yet the ozone produced from that oxygen via photochemistry is still visible in the mid-infrared, and it's a factor of ten difference – we can detect oxygen down to 1 percent concentration in the optical, and 0.1 percent in the mid-infrared. If we had a tentative detection of oxygen on the planet with TPF-C, we would want to follow it up with TPF-I, because we would see it in the ozone, even though it's only at 0.2 percent or 0.3 percent concentration in optical data.

From 9 μm all the way down to 6 μm, there are other features that are metabolic by-products: nitrous oxide, methane, and methyl chloride. There's also water vapor, which is not a bioindicator, but is perhaps indicative that the planet could have surface water. Sulfur compounds like SO_2, dimethyl sulfide, and even more complicated molecules, also absorb throughout that wavelength range.

CI   What's the most complex molecule you can detect? There's been controversy over people taking infrared spectra of comets and claiming that they saw viruses. At what point is the signature ambiguous?

VM   The short answer is chlorophyll. In the visible part of the spectrum, it is part of a signature called the "red edge". Plants are highly reflective longward of about 0.7 μm, and they absorb shortward of that due to chlorophyll. I've heard two explanations for why they're highly reflective beyond 0.7. The debate is between whether it's the leaf cell structure changing the refractive index, or it's the optical properties of the chlorophyll molecule that produce this particular type of behavior.

CI   Is the search for the biomarkers predicated on photosynthesis? Are other types of metabolism amenable to this technique?

VM   Yes. We don't have to only detect photosynthesis. Methane, which was probably put out in larger quantities by life in the early history of the Earth, is something we could look for, and it's not photosynthetically generated. Some sulfur bacteria-output products are generated outside of photosynthesis. TPF is a discovery instrument. We shouldn't assume that biomarkers are going to be the same as on Earth. Perhaps there is subsurface life, or you have life only in the oceans, so these gases have to escape through the rock. But even if other modes of metabolism are possible, radiation coming from the star, at least within the traditional habitable zone, is such an enormous energy source that it's almost inconceivable that whatever life evolved on the planet would *not* exploit it.

CI   What's common to all brands of metabolism? Is it some aspect of nonequilibrium chemistry or of energy usage? What's the most general way to describe what you're looking for?

VM   We're looking for chemical equilibrium in the atmosphere, or surface signatures like the reflectivity and absorption of chlorophyll – that discontinuity in the spectrum. We don't know where that break might occur in the spectrum, but it's a characteristic pattern to look for. A lot of mineral signatures also have sharp rises in reflectivity. We have to be careful not to conclude this planet's been covered in a particular type of basalt or rhyolite that happens to have a reflective edge to it. We'll be looking for spectral edges, but trying not to be fooled by minerals. We'll also be looking for many different molecules in the atmosphere, and they'll be trace molecules, or near the limit of detection. In the context of the environment itself, we'll be looking for something out of place or something we know is derived from life on our planet. Methane, nitrous oxide, oxygen – we'll look very carefully for them on other planets. The key is energy usage; we're seeing waste products most of the time.

CI   I've heard it said that on the Earth, 90 percent of the biomass could be subterranean or deep in rock where it might not leave a tracer visible in an atmospheric spectrum.

VM   It depends how efficient the microbial community is. When the communities get together, one microbe's waste product tends to be another microbe's food, so they tend to be quite efficient. Very little escapes from the top because these are layered communities like stromatolites, where one uses the products of the one above or below it. Seeing these products in the atmosphere could be difficult. But in the case of photosynthesis, a lot gets out. Animals use it, and we use it, but we don't make as big an impact on the biosphere as the microbial plant community.

CI   You're participating in the next step in the Copernican revolution: the question of whether we're biologically alone. That must be a great motivator as you face long time frames for the missions. Is it exciting to be part of such a big question?

VM   Yes. It's great because it's so relevant to people. You can sit next to people on a plane and start a discussion, and everybody's got an opinion, or is interested in what you're doing, or thinks it's amazing. You talk about the long timescales; I turn it around. With these NASA missions, within our lifetimes, we may have the answer to whether or not we're alone. That's a question we've had for several thousand years, but it's only in the next few decades that we'll have the ability to address it. This is a very special time in history.

Part V FRONTIERS

30

Jill Tarter

As a young researcher, Jill Tarter faced a fair amount of opposition to her chosen path. Now Director of the Center for SETI Research, she both encourages young girls to pursue science and encourages scientists to persevere in the search for intelligent life. Tarter earned her PhD in astronomy from the University of California, Berkeley. She holds the Bernard M. Oliver Chair for SETI Research and was Project Scientist for NASA's High Resolution Microwave Survey. When SETI lost NASA funding, Tarter stepped in as Director of SETI's Project Phoenix. She is a fellow of the American Association for the Advancement of Science and the California Academy of Sciences, and has served as President of the International Astronomical Union Commission 51. Tarter has earned two NASA Public Service Medals and the Lifetime Achievement Award from Women in Aerospace. In 2004, *Time* magazine listed her as among the 100 most influential people in the world. In 2008, she won the prestigious TED prize for her dream that is "big enough to change the world." She was the major inspiration for Jodie Foster's character, Ellie Arroway, in Carl Sagan's book (and movie) *Contact*, and hopes that she has served indirectly as a positive role model in the popular consciousness.

CI   How did you become an astrobiologist?

JT   It was a total accident. In my first year as a graduate student at Berkeley, my job was to program the world's first mini-computer to run a spectrometer on an optical telescope that Berkeley was using as a teaching tool. The computer didn't have any language, and I had to program it. As I was finishing up, that old, defunct equipment was donated to Stu Boyer. He wanted to use it to search for extraterrestrial intelligence. The radio telescope measures both the amplitude and the phase. He realized he could take some of the signal and amplify it back; basically, split up the photons and reproduce them. He could steal some of the astronomer's signal and analyze it in the background for technological signals, rather than just astrophysics. This was in the early seventies.

CI   Did you know about Project OZMA and Frank Drake when you started?

JT   I learned about it by reading the Cyclops report, which was the engineering-design study done in 1971 by Barney Oliver. Stu suggested I read it and join his group. I'd never thought about SETI at all. It was a dense report, but I got hooked. I read it from cover to cover. I was so excited that I happened to live in the first generation of human beings who could try and answer this question. For millennia, all we could do was ask the priests and philosophers; the answers we received were in tune with someone's belief system. This was an opportunity for scientists and engineers to study the question experimentally, and I thought that was fantastic.

CI   There's fear associated with being a pioneer. People must have told you not to go into this. What was the reaction from peers and colleagues?

JT   It was the science equivalent of "What's a nice girl like you doing in a place like this?" Many of my colleagues told me it was a fine *avocation*, but a terrible choice as a *vocation*.

CI   But the hook was set so deep that it didn't worry you?

JT   Exactly. It was a wonderful opportunity to combine my engineering undergraduate work with the physics I did later to find out what the universe would look like at high resolution. When you get a chance to work on a project that could make a difference for everyone, it's a pretty special opportunity.

     After I left Berkeley, I did an NRC postdoc at NASA Ames, continuing to work on the brown dwarfs from my thesis. That was great, and I had a lot of fun, but I always wondered: why should the taxpayers support my doing this? Then I worked on SETI as a volunteer; I walked across the Ames campus and knocked on John Billingham's door and said, "I've got extra time; the NRC doesn't own my nights and weekends. What are you doing, and how can I help?" We laid out a program and I learned radio astronomy in order to do SETI.

CI   Exobiology and SETI had been part of the research that NASA embraced. Then it got twitchy. How many times did the appropriators pull the plug?

JT   There was the Golden Fleece Award – awarded by Senator William Proxmive for research that "wastes taxpayer dollars" – and that you can live with. Following that, the funds were terminated in 1981; we had to write a termination plan. We stayed alive on carryover funds from 1981 that hadn't been spent in other projects. In 1982, NASA put SETI back in its budget very boldly, and said, "This is not something you can make fun of; it's something we want to do." So we survived. But in 1993, Senator Bryan terminated it, and it was clear there would be no surviving that.

CI   It's pretty ironic. He's from the district in Nevada that includes Area 51 and he shamelessly lobbied for congressional funds to name the "extraterrestrial highway," right?

JT   Yes. It's such hypocrisy; it's hard to even talk about.

CI   Before those roadblocks, wasn't SETI part of NASA's road map?

JT   Yes. Since the beginning of the astronomy decade reviews, starting with the Greenstein report in the seventies, SETI was listed as something worth doing – not something big money should be spent on, but it was supported scientifically. It was important to have that stamp of approval from the National Academy.

CI   Through those fluctuating fortunes of SETI in terms of funding, how did that affect your career, or your commitment to stay the course?

JT   When I started volunteering with John Billingham, I said, "I'll give this five years, let's see what I can get going." Five years later, I was given an award for women in aerospace for my work in exobiology – and the same day, I found out the budget had been cut. It was the first hump of a rollercoaster, and I realized it was going to take more than five years.

    I was told not to give up my day job when I started SETI; I should keep doing some traditional astrophysics – if nothing else, as a way to measure my own progress, my own ranking. As long as I could compete with the big boys and get my papers published and do credible science, then I could justify myself and figure I was doing a good job. I did that for a while, because I was still interested in the brown dwarfs. But I gave up the "traditional" science, because SETI was a lot of fun, and a lot of work. I found another way of determining whether or not I was doing a good job, and that was looking at how rapidly I was able to search and improve the instrumentation and the parameter space we explored. There was a small community – Paul Horowitz being one a prime example – of very smart people starting to do SETI. If you could keep up with Paul Horowitz, it was a good indication that you were doing the right thing.

CI   This is a nonscientific sampling, but in my experience, physical scientists and astronomers tend to be much more optimistic about the prospect of widespread life and intelligent life than paleontologists and biologists. Do you notice any disciplinary divisions between the ways people think about SETI?

JT   Yes. It stems from the dichotomy between the views of Stephen Jay Gould and Simon Conway Morris – contingency versus convergence.

CI   They look at the same data on Earth history and draw different conclusions.

JT   Right. The convergence school doesn't have any problem with the idea that we'll get intelligent creatures elsewhere. The contingency school has a big problem. It depends on who you ask.

CI   But do we know enough about the role of either of those factors in Earth's history to set the expectation for a search for life beyond our Solar System?

JT   No. The wisest thing that's ever been said on the subject was by Cocconi and Morrison in that seminal paper in *Nature* in 1959. The last sentence was: "The probability of detection is difficult to estimate – but if we never search, the chance of success is zero." That's it. We're stuck with twenty-first century technology and our current understanding of the universe. We can't do anything we can't yet conceive of. Beyond those limitations, we should try to search in as many ways as we possibly can.

CI   There aren't many fields of science in which, after decades of searching, you can have *no* detections and still be convinced that the search is worth doing.

JT   People are still looking for magnetic monopoles.

CI   Does part of that perseverance come from the growing capability of electronics and detectors that make each new search better than the sum of everything that came before?

JT   Yes. And "exponential" is pretty common in technology. If we got out of bed every morning and expected instant success, we'd all go to bed at night disappointed. That's not the way to live your life. You get up and say, "I'm going to see what I can do today better than I could last week." We were doing

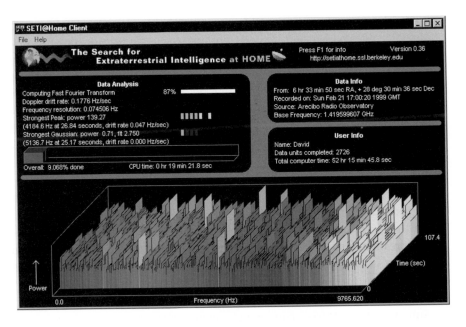

Part of the nine-dimensional parameter space of a radio SETI search, in this case a small amount of data running as a screen saver in the popular SETI@Home program. Time, frequency, and power are shown in three dimensions. Packets of data analyzed as part of SETI@Home formed the largest distributed computing project in history (courtesy NASA).

fourteen orders of magnitude better than we had done with OZMA – that's a lot of improvement. On the other hand, it's still a nine-dimensional search space, and we may not yet be looking the right way.

CI   What are the nine dimensions?

JT   The standard three of space and one of time, that's four. Then frequency. Two polarizations for an electromagnetic signal. Modulation. Lastly, sensitivity. We don't know how strong the signal might be or how far away it is, so we don't know how sensitive we need the search to be. We could have been searching in the right way for the right thing and have to do it all over again with greater sensitivity because the signal was coming from too far away and we missed it.

      We started with the radio searches and we're continuing with that. We're building our own telescope rather than getting time on other people's telescopes. We're also doing optical searches; that was something we were aware of early on. We hadn't done it before because the photon counters were too bloody expensive. We needed nanosecond time resolution in order to cut down the background. The military had some, but we couldn't afford them. Ten years ago they became available, and we started doing optical SETI.

CI   There was a strong rationale for radio searches – the photons are low energy to produce and the universe is fairly transparent if you pick the right frequency. But the universe is noisy and busy at optical wavelengths. What's the rationale for optical SETI?

JT   The big-picture rationale is that you can encode a lot of information in a way that makes it obviously *not* an astrophysical signal. The universe is only noisy if you're talking about averages. If you have a meter-class telescope and you're looking at a solar-type star a few hundred light years away, you only expect to get about a million photons from that star in a second. If you count photon arrivals at much shorter timescales, such as nanoseconds, then you don't expect a stellar background photon in any nanosecond. The arrival of a clump of photons in one nanosecond would clearly be an artifact. A lot of information could be encoded on those photons, because it's a broadband signal.

CI   The rationale is similar to the filtering done at radio wavelengths for artificially produced pulses, but you can also encode a lot of information.

JT   Right. Basically, all of SETI is looking for signals that have a characteristic where the time–bandwidth product of the signal is close to unity. In the radio: a lot of compression, a narrow frequency range for carrier-wave signals or narrow pulses. In the optical, the compression is in the time domain. It's a broadband signal, but short, so the time–bandwidth product is close to one. The uncertainty principle says that you can't get any smaller than that, and astrophysics as a natural emitter has time–bandwidth products that are enormous, primarily because there are so many emitting atoms or molecules. Even masers have large time–bandwidth products.

CI   Is optical SETI best done as a targeted search, or can it be done blind?

JT   Good question. Paul Horowitz isn't doing the targeted work anymore; he's been working for years with a graduate student to make a sky survey. He's built his own dedicated observatory with a large mirror, which is spun, so he doesn't need good optics – he can make a big, inexpensive mirror. Now manufacturers are making essentially linear arrays with photodiodes. Paul can do a drift scan with this telescope. It's very clever. The students get excellent training in all kinds of instrumentation building this.

CI   We survey the universe at different wavelengths and always try to go deeper, but the synoptic universe is still poorly understood.

JT   Exactly.

CI   None of the strange, fascinating phenomenon – gamma ray bursts, quasars – that do go bump in the night was predicted. Is there any confusion with any astrophysical source based on the physics we understand?

JT   No. There's a paper looking at what could be natural sources of nanosecond optical phenomena, and there's no rationale for a natural background. Phil Morrison predicted this. Andrew Howard did a study looking at what might be there and was able to put upper limits on it.

CI   Philip Morrison is the perfect example of someone who could think deeply about multiple subjects.

JT   He's wonderful and I miss him. He had a lovely phrase for SETI: the archeology of the future. It's archeology because of the finite speed of light. Whenever we get a signal, it will be telling us about their past. But if we do get a signal, it tells us that it's possible for us to have a future. In order for two technological civilizations to coexist in space and time in this Galaxy, technologies have to last for a long time. If the technology isn't long-lived, then the probability of two technological civilizations detecting one another is zilch. If SETI succeeds, it tells us that it's possible to survive our technological infancy.

CI   That's a positive message. The simple extrapolation of our current capability seems to give a civilization the ability to explore, colonize, or at least view most of the Galaxy in a short fraction of the Galaxy's age. Why haven't they done that? Is the Fermi paradox just a side-show argument?

JT   No, paradoxes can be extraordinarily powerful arguments. Logically, this isn't a paradox. For it to be a logical paradox, you'd have to believe they're not here. I don't mean that they're showing up and abducting Aunt Alice for fallacious medical exams. But "they're not here" would say that we have so thoroughly explored our Solar System that we *know* there's nothing here. We know a bit – we can tell you there's nothing as large and shiny as the Starship Enterprise sitting in the Lagrange points of the Earth, Sun or Moon systems. There's also a bias that they're colonizing, which means that some big, wet biology is traipsing around the universe. Colonization might be done with nanotechnology. We can't defend the "they're not here" statement, so it's not a paradox, yet.

CI   What about another possible answer to the Fermi question? Arthur C. Clarke said that a sufficiently advanced civilization is indistinguishable from magic. It's hard to imagine our own capabilities in a thousand years, let alone a civilization

Part of the Allen Array, a set of 42 radio dishes in northern California that can do general radio astronomy, but also the most powerful SETI experiment in history. The eventual goal is 350 dishes, all acting in concert to look for artificial radio signals from many thousands of stars like the Sun at distances of up to a few thousand light years (courtesy SETI Institute).

that could have a five-billion-year head start on us. Does that constrain how you approach the detection?

JT That argues for the most aggressive program of astronomy and astrophysics possible. When we detect something that doesn't look right, we can't throw it out to make the grass look prettier – we need to think about the possibility that it's an indication of astroengineering. Early in the history of gamma-ray bursters, when there wasn't a good theory, someone took the bursters that had been detected to date and tried to line them up into four-dimensional space–time. Maybe those were half-MeV photons resulting from the accelerations of the matter/antimatter spaceship! They didn't know. For things we don't understand, and can't build, we need to keep our eyes open.

CI That's a sophisticated echo of Jocelyn Bell's *Little Green Men One, Two, Three*.

JT Exactly.

CI Back to radio. When the Allen Array is completed, what will its capabilities be?

JT When am I going to get the money? [Laughs] It's totally dependent on that. We have our production line up and running and we can build one of these telescopes and get it up in the field in just over a day.

CI How many do you have?

JT Forty-two. It's frustrating, because the longer it takes, the more expensive it gets. We envisioned this telescope being really cheap. But when we buy ten at a time, instead of three hundred, we don't get anywhere near the price break we wanted.

But now we know how to do it, and we can do it really well. Three hundred and fifty telescopes, each 6.1 meters – it's actually 20 feet, because they don't do metric in Idaho Falls. The array has the capability to change a lot of what we know about the sky. It's fantastic for transient studies. At the lower frequencies, we can survey the entire sky in a day or two. At the higher frequencies, it takes longer because the beam is smaller.

CI   You're inverting history and have traditional astronomy piggybacking SETI.

JT   It's more than that. We're building it as a combined and joint instrument; it's been conceived as doing radio astronomy and SETI simultaneously from the outset.

CI   Without compromising either project.

JT   We have to share the sky. But since the beam at the 21-cm hydrogen line is 3° across, it's a big piece of the sky. I need a catalog of stars with a few million entries. Right now I've got one with a quarter of a million stars. Any beam of sky the radio astronomers want to look at has targets. There are four simultaneously tunable IF paths out of the telescope, so nominally, the radio astronomers can select the frequency on two of those and they have correlators for those IF chains, and I can select the frequencies on the other two. I can use the astronomy frequencies as well, if those are interesting to me. We can both make an image of the sky over that whole field of view with correlators, or we can phase up the output of all of the antennas sixteen different ways.

CI   You said the parameter space of the search has increased fourteen orders of magnitude since OZMA. What gain will the full Allen Array give you?

JT   That depends on whether I can build the back-end processors. We're finally at the point where we can get away from special-purpose hardware. With SETI, we built everything ourselves, including the chips that did the transforms, because nothing was fast enough. We're now into a mode where we're using rack-mounted PCs, each of which has a single accelerator card – special hardware we built. Now we're at a point where we can go into a cluster, and it can all be done in software. We made the conscious decision that the output of the telescope, which originally was going to be a custom output, synchronous format, is going to be IP packets instead; those are a standard where you can bring anything back into the telescope you want. The processing will get better in time because we're putting out IP packets – it can be an open telescope for which people develop and implement their own algorithms. We've spent the last decade looking at a thousand nearby stars in the 1–3 GHz range. We're hoping that in the next decade we can look at a half a million to a million stars over 1–10 GHz.

CI   That's an enormous gain. Our grad students wrote a spoof paper about detector technology advancing exponentially. There's a logical argument to procrastinate. It's the slacker mentality that says you should keep waiting, because better technology is right around the corner.

JT   Seth Shostak has a good comeback to that, as he does for most things. He said, "Columbus didn't wait for a 747 to cross the ocean."

CI Some people may find it surprising that you're still struggling to finish the Allen Array. They'd think, "The guy spends a few hundred million dollars a pop on sports teams; those are some deep pockets."

JT You hit the nail on the head. It's hard to raise the funds. I ask somebody who could give me a million-dollar gift and I get ten thousand, because they say, "Paul Allen has more money than I do. He can do this if he wants." Paul made us a deal where he would put in $25 million and we'd have to raise the rest. We thought we could do that easily, but that hasn't been the case. He wants partners; he wants other people validating this as a good idea.

CI How much time do you spend on fundraising?

JT A whole lot.

CI In addition to struggling with astronomers who raise their eyebrows at SETI, the public tends to believe that we've already made contact and UFOs are real. How do you convey the scientific aspect of SETI to the public, given this backdrop that what you're looking for is already known, and the government is hiding it?

JT I address it up front and say, "If I ever claim that I have detected a signal, you need to demand from me incredible data and proof, and I have to demand the same level of verification and validation from anyone else who claims to have seen something or been abducted." Unfortunately, the other part is that I have to make myself available to validate or discredit other people's claims, which takes time and has not been fruitful. For amazing claims, we should use the Sagan criteria – it should come with amazing amounts of evidence, and if it doesn't, you shouldn't believe it.

CI Is it possible to do SETI without anthropocentric views?

JT As far as possible, but we can't get away from it. Our methodology for thinking about things is obviously anthropocentric. We back off things and say we're looking for signals that have a time–bandwidth of unity. Technology can create that; astrophysics can't. Maybe someday we'll end up detecting a signal that is in fact an astrophysical manifestation. I'm keeping an eye on the giant pulses from the Crab pulsar. That signal might have some nanosecond structure and a time–band that's approaching unity.

CI These issues are similar to those of the gravity-wave field. Kip Thorne has said that in a noisy situation, digging out signals, they worry that they could throw the baby out with the bathwater. They can't anticipate all the signals the universe might produce. The act of detection and filtering becomes profound because you impose your prejudices and expectations on the analysis.

JT Absolutely. We use two widely spaced telescopes simultaneously. Since we're looking for narrow-band signals, we can look at the differential Doppler of that signal as seen at both telescopes and use that as a discriminant against satellites or airplanes. But still – how do you know that wasn't an ET satellite? We make pragmatic choices; we do as much as possible, but we can't do everything.

CI There's been a lot of exciting progress in astrobiology – extremophiles, the prospect of detecting Earths, and potentially the alteration of their atmospheres by life. Do the advances in astrobiology give you a heightened expectation as those early terms in the Drake equation get defined for the first time?

JT   There's more cosmic real estate for life out there than we would have claimed when I was a graduate student. It's a question of whether those Drake equation terms are in fact all independent. If we find ozone or some other nonequilibrium chemical signature in the atmosphere of a nearby terrestrial planet, we're not going to know if it's microbes or mathematicians. SETI is our only way of finding *intelligent* inhabitants.

CI   Let me ask about your personal experiences. You are a pioneer, not just in your chosen field, but as a woman. Have you seen any change from when you were coming up?

JT: Yes. I make it a point to get out in front of every group that's likely to have young girls, because if you can't see somebody doing it, then you can't conceive of doing it yourself. It's important to have role models.

CI   You've penetrated the popular consciousness indirectly as a role model through the movie *Contact*. Can you talk about your connection with the Ellie Arroway character?

JT   Carl Sagan was on our Board of Directors at the SETI Institute. He had been a colleague for a long time; I was back at Cornell for some occasion and Carl invited me up to the house for a cocktail party. Ann Druyan and Carl got me in a corner and said, "Carl is writing a science-fiction book." I laughed, because the weekend before, *The New York Times* revealed the price of his advance, and everyone was jealous as hell. Ann told me I might recognize someone in the book, "But I think you'll like her." I just laughed and said, "If she doesn't eat ice cream cones for lunch, who's going to think it's me?" I was so wrong. When Carl sent me a pre-publication copy to read, I was flabbergasted – "Carl doesn't know this about me – how could this be?" It turns out he'd read a report about me and other women in science by the American Association of University Women. For most women in that study, their fathers had been the centers of their universes, the people who encouraged them. Many, including me, had lost their father at a young age; that left a lasting impression and stubbornness, and also gave us the *carpe diem* philosophy: "Don't take anything for granted, because it may not be here tomorrow." Carl used that. But the character is mainly Carl. It's Carl in drag.

CI   Didn't Jodie Foster use you to pattern her role?

JT   Yes. That was a wonderful experience. We had some phone conversations before the movie started shooting and then I went down and worked with her at Arecibo. She's a brilliant, wonderful, amazing woman. It was a great deal of fun, and it formed a friendship.

CI   Even though it's fictionalized and manifested through a great actress, you must be proud of having such a visible role model patterned on yourself.

JT   It's great. Maybe it will help encourage some other young women to go into science. They'll understand that science isn't just about memorizing – it's about trying to solve puzzles, and it's fun. The science in *Contact* is good because Carl wrote it.

CI   The capstone on your career would be a bona fide signal with the fully functional Allen Array. Are you taking personal bets on whether you'll see it in your lifetime?

JT   Chris, I don't know the answer to that question. I go back to the last sentence of the Morrison-Cocconi paper. I *want* to succeed, but I don't know when or if we ever will.

CI   You're in it for the long haul, and the journey itself.

JT   The journey is very interesting. It's not always easy. But it is interesting.

31

Seth Shostak

Seth Shostak is the "public face" of the Search for Extraterrestrial Intelligence, giving over 50 public talks a year and hosting his own weekly radio show called *Are We Alone?*. After a BA in physics from Princeton and a PhD in astronomy from Caltech, he did research work on radio observations of galaxies in the USA and Europe before coming to the SETI Institute, where he is Senior Astronomer and in charge of all outreach activities. He has written hundreds of articles for newspapers, magazines and the Space.com website, and three books, including a popular textbook on astrobiology. He was awarded the Klumpke-Roberts Prize by the Astronomical Society of the Pacific in honor of his contributions to the public understanding and appreciation of astronomy. Shostak is aware that even though SETI has been met with nearly fifty years of the "Great Silence," gains in detection sensitivity and bandwidth mean that the search is just starting to get interesting.

CI Have you always been a SETI man?

SS The tragedy of my life – and there are many – is that I have been interested in too wide a range of things. I've worked for the railroads. I had a company that did computer animation and made movies when I was younger. I got interested in astronomy at age eight, which is pretty typical. That's the age at which you develop the interests you carry with you forever. In the back of my parents' atlas, there was a diagram of the Solar System. I asked my mom, "What the heck is this?" She said, "Those are planets." I'd never heard the word. After that I went to planetariums all the time, particularly the Hayden Planetarium in New York.

CI Is that where you grew up?

SS No, I grew up in northern Virginia, in Arlington. I had relatives in Brooklyn, so during vacations, my parents put me on the train to New York. My aunt always sent me to the Hayden Planetarium, I think to get me out of the apartment. By eleven or twelve I had built a telescope and was trying to take pictures of the Moon, and make 8-mm time-lapse movies of Jupiter.

I did my undergraduate degree in physics. Then I went to grad school at Caltech. I entered in physics, but I wandered around the astronomy department and they had all these wonderful drawings of the construction of the 200-inch telescope. Astronomy seemed so much more interesting than particle physics. So I switched majors for my degree program. I did radio astronomy because I had been interested in electronics when I was a kid.

CI Were you a tinkerer?

SS That's my strong suit, tinkering. I did radio astronomy. I made maps of galaxies in the neutral hydrogen line. I spent weeks trying to get the rotation curves to go down in the outside areas of these various galaxies. They would *not* go down; they stayed flat. I was the first to find that with radio techniques in a convincing way. I had experience with radio astronomy and a postdoc in radio, so I went to university in Europe, in Holland, Groningen.

CI You discovered dark matter and then got rid of it! How long were you in Holland?

SS Thirteen years. I was still interested in the idea of life in space, because as a kid I'd read all these books about UFOs. There was a book with photos of flying saucers. At thirteen years old, looking at those photos, I thought, this looks like a hubcap – I could make a better photo than this! In 1981, Jill Tarter came to the university where I was working. I got to know her a bit as a friend and suggested we do a SETI experiment with Westerbork in Holland. We could point it at the center of the Galaxy. That's a logical place for a beacon because everybody's going to be looking at the center of the Galaxy every now and again.

CI Had there been some radio SETI before that?

SS Frank Drake did the first experiment, Project OZMA, in 1960. He was unaware of a paper that had appeared the year before in *Nature*, by Morrison and Cocconi, which laid out the basis of the whole idea.

CI Was this the first time a radio-interferometer had been used?

SS   As far as I know, it was the first time this had been done with interferometry. By using an interferometer, we thought we could get rid of all the natural radiation in the center of the Galaxy, so we could look for small sources of intense radiation, and also look for signals that were narrowband, which is the hallmark of a signal coming from a transmitter as opposed to coming from some natural process.

We had trouble getting this proposal past the observing committee at Groningen University. They gave us just four hours worth of observing. There's something interesting about the fact that the Dutch were so reluctant to give us the time. I gave a colloquium on SETI in the same university, and it was standing-room only. The first thing I did was ask the audience, "How many of you think there might be intelligent life out there?" They all raised their hands. These people were faculty, students, members of other departments, and the general public. Then I asked, "How many of you are willing to spend one guilder a year to look for it?" A guilder a year buys a cup of cheap university coffee. The hands all went down. They *all* went down. I asked them later, "Why is that? You guys have big telescopes, you have the expertise, you have the technology, and you think it's out there. But you're not willing to spend a guilder a year?" The answer was, "We're too sober for that." It's cultural.

CI   Interesting. Alien hunting is cultural.

SS   Yes. But it isn't just the Dutch. When you look at which countries do or have *ever* done SETI research, there are only three or four. The UK doesn't do it. They've got big antennas sitting around near Manchester; those are not doing much. Australia did; they're out of business right now. The United States does. Italy has a small experiment in Bologna.

CI   Was that because of the Marconi and Tesla connections?

SS   Nice thought, but that's not true. It's because some Italian astronomers happen to be interested. They want to develop new technologies, like new receivers. The Italians are smart, and they've developed mathematical transforms that would allow you to find different classes of signals. Americans are way behind in that. In the old days it was the Soviet Union – they were the real competitors. But as is so often the case with the Soviets, they couldn't afford to build any equipment that worked. They had some very good theoreticians. Many of the seminal ideas in this field came from the Soviet Union.

CI   That was the early sixties?

SS   Mid sixties. Shklovsky and Sagan's *Intelligent Life in the Universe* was the first popular book in this field. That's what got me interested in SETI. I was working on my thesis and reading that book at three in the morning. I was using antennas to study galaxies, but I realized that the same hardware would be useful for communicating between the stars.

CI   You had standard astronomer fare, and the wild stuff you think about at three in the morning.

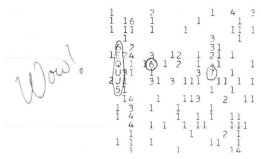

The "wow" signal, recorded at a radio telescope operated at Ohio State University in 1977. In this early SETI experiment, this computer chart shows numbers and characters representing the strength of a radio signal in a narrow frequency band. This strong peak was 30 times the level of the radio noise, but it never repeated and it could not be fixed to a particular star, so it was presumed to be an artifact (courtesy Jerry Ehman, Ohio State University).

SS   Not only at three o'clock. For years afterward, I thought about it every time I was on the telescope studying galaxies. I'd come to a hole where there weren't any galaxies for a while, but I still had some telescope time. I'd look up some nearby star-system positions, and point it at those.

CI   The famous "wow" signal was detected by a radio telescope at Ohio State in 1977. The signal processing power and techniques have improved so much since then. It must have been hard to know what your data were telling you.

SS   That's a problem. The "wow" signal is famous because it has good nomenclature. [Laughs] It was fortuitous that this guy came in in the morning. In those days, the telescopes printed out big computer-paper printouts.

Thousands of signals were found in the early days. You would observe, you'd record all the data on computer tape, and you would take that back to wherever you worked. Even in those days you had receivers that would monitor at least hundreds of channels, sometimes thousands, so of course you're going to pick up signals. But the question is, what was it? Was it ET or AT&T? It may have been the radar down at the local airport. How do you deal with that? In the old days, you just hope that you can go back six months later when you have more telescope time, and look at the same spot on the sky at the same frequency, and see if it's still there. If it is, then you get interested.

CI   So a pure, nonrecurring transient has to be consigned to limbo.

SS   Exactly. ET might not be on the air for six months just for your convenience.

CI   Some astrophysics works that way. Microlensing events are nonrepeatable, but people have managed to do statistics with the phenomena.

SS   That's because the universe for astrophysics tends to be quiet. The local airport doesn't interfere with you. If it did, you might not believe a lot of that stuff. You might not believe any of it.

SETI should have some way of being able to find one-off events, but it doesn't yet. In the case of the "wow" signal, they had two receivers on the telescope, and two horns. They were set up so that 70 seconds after the sky was focused into the first receiver, it got focused into a second one. So they got a follow-up observation just over a minute after the first one. And the "wow" signal was dead.

CI   The methodology for filtering or rejecting terrestrial phenomena is fairly clear. But the universe is such a magnificent place. Nobody anticipated gamma ray bursts and quasars. Jocelyn Bell wrote "LGM 1, 2, 3" for "Little Green Men 1, 2, 3" beside the first few pulsars while everyone was still figuring out what they were. How do you reject all astrophysical phenomena when you don't necessarily know what all the astrophysical phenomena are?

SS   The answer is: we don't. We accept all narrowband signals. We're not looking for the value of pi.

CI   It's not like *Contact*. It's not a prime-number sequence.

SS   There's no "Ah-ha!" moment. We look for narrowband components. We look for a signal that might be a hertz or less wide.

CI   Why is that so special?

SS   Nature is not good at making signals that only show up at one narrow spot on the radio dial. Consider a quasar. It doesn't matter where you tune in for a quasar, you're going to get radio static. It's stronger in some bands than others, and it changes slowly, but if you go from 1422 Hz to 1423 Hz, it doesn't change much.

CI   What about spectral lines, or molecule features?

SS   Even molecule features are pretty wideband, because Doppler motions spread them out; intrinsically, they wouldn't be. If you point a telescope at the center of the Milky Way and look at the hydrogen line, it's hundreds of kilohertz wide. That's hundreds of thousands times wider than the signal we're looking for. As far as we know, nature doesn't make signals as narrow as a hertz. Pulsars are regular in time: *zap! zap! zap!* But you can tune the radio knob just about anywhere and hear them. They're wasteful in terms of energy. That's nature. Nature's not a good engineer. Even masers are hundreds of hertz wide, which is hundreds of times wider than what we look for. Could there be astrophysical phenomena of which we're unaware that would mimic this signal? Maybe.

CI   It would be a discovery. Maybe not be as spectacular as ET, but it'd definitely be something new.

SS   It would be so interesting. If somebody asked me thirty years ago whether SETI will find something interesting in astrophysics, I would have said sure. Every time you open up a different phase space with a telescope, you find something you didn't expect.

CI   Radio has been the dominant technique over the history of SETI. Is that simply for energy reasons?

Despite the continued work on traditional radio approaches, optical SETI is becoming more popular. This Japanese 2-meter telescope is being used to look at nearby stars with very high time resolution, so that any artificial signal can be separated from the more uniform signals of the light from the star (courtesy Nishi-Harima/NAYUTA Telescope, and SETI Society).

SS   The first SETI was done at radio frequencies, and more specifically, microwave frequencies. SETI today is mostly done exactly that way. Part of that is history; we developed radio before we developed the laser. But there was also a long-standing argument against the idea of extraterrestrials flashing lights at us. An optical photon, a photon of visible light, has a lot more energy than a radio photon. It's many orders of magnitude more expensive to send bits of information around at optical frequencies than at radio. Energy costs are something you're always going to consider in a communication system.

CI   The Galaxy, since we're in it, is also much more opaque.

SS   Exactly. It's hard to do long-distance communication, because it gets absorbed and scattered. Those arguments didn't make any sense then, and they're not considered to make a whole lot of sense now. First off, once you go a little ways into the infrared, the Galaxy becomes transparent. So much for that argument.

The argument about energy is true, but for rather little money you can make a mirror a meter or two across and you can direct those photons. This more or less compensates for the cost of the photons. It turns out there's no clear preference in terms of the cost of setting up a communications system at optical or at radio. If you want to broadcast to the whole Galaxy, radio makes sense, because photons are cheap. But if you know there's somebody over there you want to get in touch with, it's just like direct broadcast satellite or microwaves here on Earth – you can use the highest frequency and beam it.

CI As opposed to looking for narrow bandwidths, you're looking for a quick pulse; something that will outshine the parent star for a tiny fraction of a second.

SS Exactly. People thought it was going to be difficult, because the Sun's putting out 10^{26} watts – how are we ever going to outshine that? Well, we *can* outshine that.

CI We have that technology?

SS We can do it. We take the biggest lasers, which are for fusion, not for signaling the Klingons. They make very short pulses, a nanosecond long. You work out how many photons that produces, and aim that into a mirror the size of a desk, and you aim *that* at a star like the Sun a hundred light years away. In a second, it's collecting on the order of 10^8, a hundred million photons, from the Sun. But in a nanosecond, that's less than one photon. This laser will put a few thousand photons down that telescope tube in that nanosecond. For a tiny fraction of a second, you've got something orders of magnitude brighter than the star.

CI So someone doing rapid time-sampling would see an enormous signal, if they were actually out there?

SS Right. That's easy to do. Point it at nearby stars and look for a whole bunch of photons in a nanosecond.

CI Is optical SETI becoming ascendant?

SS I would like it to be. All SETI suffers the same malady, and that is lack of funding. But optical SETI is cheap. Anybody could do it; you could build the equipment you need for ten thousand bucks or less. If you've got a telescope sitting around, it doesn't matter what the aperture is; you don't know how big the aliens' lasers are, anyhow.

CI You're riding the rising exponential growth of information technology, processing power, bandwidth, and computer power. Each new capability dwarfs what came before; that's probably an argument for procrastination. How do you decide when you're getting traction, when you're in interesting parameter space?

SS Building something, and spending time looking. I don't have a specific answer. Christopher Columbus's wooden ships were slow, uncomfortable, and dangerous. If he had waited five hundred years, he'd have crossed the Atlantic in a couple of hours, eating bad food. Wooden ships *were* good enough in that case. You don't know where the threshold is, but you can guess. In fact, I'll guess. We've looked at fewer than a thousand star systems carefully. I don't think one in a thousand star systems has ET. I'm pretty optimistic, but that's too optimistic.

I think the right number is probably a few million searches if the total number of societies in the Galaxy is about ten thousand. That doesn't sound unreasonable. When can we build something that allows us to look at that number of targets in a reasonable length of time? We're there with radio; that's why we're building the Allen Telescope Array. Within a couple of decades that will check out everything to a distance of a thousand light years. That's worth the effort.

CI The landscape of research is starting to point to widespread microbial life, but it's so hard to make sensible projections of intelligence and technology. Orcas are pretty smart, but I don't think they'll ever make a telescope.

SS   Well, that's just you. [Laughs]

CI   It's consistent with all our data that there's an extreme bottleneck between microbes and civilizations. How often do you think evolution generates complex intelligence?

SS   This is probably the most controversial aspect of the whole SETI enterprise. The Drake equation was just an agenda for a meeting, but it's been a durable idea. The last term is how long you last once you build a radio transmitter. Do you blow yourselves up? That's sociology. People think, now that we have the H-bomb, we don't have long to last, so we'd better enjoy life now. The term before that is what fraction of biological worlds are ever going to cook up intelligence that can build radio transmitters. We don't know anything about that.

The record that led to us can be discouraging, because there are plenty of forks in the evolutionary story and, had they gone the other way, the smartest things in Tucson would be the saguaro cactus. We don't know how intelligence got started on Earth. Was it a mechanism you would expect in lots of ecosystems, or was this a very lucky shot? There are two approaches to this question. One is to try to figure out why humans arrived, why we got smart, and what pressures caused this leap.

Other people feel we may never know why our ancestors didn't keep hanging from trees, but we can look at other species. If other species were also moving down this road towards greater intelligence, it suggests we're not so miraculous. We don't expect insects to get very smart, because they're too small, and they don't have neurons. It's expensive to be smart. Lots of other creatures have gotten smarter in the last 50 million years, not just the simians. Octopuses are reputed to be fairly clever – they'd give you a good game of Scrabble. Lori Marino has plotted the encephalization of various animals, and a certain fraction of them evolved to greater encephalization; it paid off for some of them. That suggests that, if it hadn't been us, it would have been some other species – but sooner or later, you would get a species that was clever enough to build telescopes.

CI   Even if it's highly contingent, it's still a possible outcome. If there are hundreds of millions of Petri dishes out there, that's a lot of time to work with. What about the timing argument that our conception of possibilities is too modest, because anything that got a four-billion-year head start on us will be to us as we are to pond scum?

SS   I'm sure they wouldn't care about us. I get emails all the time that the reason we haven't found the aliens is because they're dismayed by our despoiling of the environment.

CI   That's awfully anthropocentric.

SS   My God, is it ever! Completely nonsensical. We may be prosaic in assuming that they produce clues to their existence that we can recognize. This is Arthur C. Clarke's argument that advanced aliens might look like magic to us.

CI   If I was super-intelligent, I'd be making baby universes and all sorts of exciting stuff. I wouldn't waste my time on radio transmitters.

SS   Exactly. And it's hard to detect those baby universes.

CI   That's not a counterargument for SETI, but it doesn't make it any easier to come up with a rational strategy.

SS   We were talking earlier about optical SETI. We've had lasers for fifty years; what if they've had lasers for fifty thousand years? Their high school kids have better lasers than we've got – it's a science-fair project for them to signal a bunch of worlds. But all they'd have to do is crank up the brightness a few orders of magnitude, not an enormous number, and now we see it flashing in the sky with our naked eyes.

That's not so far beyond what we can do. Why limit ourselves to something they're going to need a telescope and photomultipliers to find? Why not just flash light, sequentially target all these stars around, and make one flash a year. From their point of view, it will just be a flash, but bright enough to be seen with their eyes, even though it was only a nanosecond long. They know there was a flash in that part of the sky. Next year, or maybe ten years later, whatever period may be short compared to their lifetime, they notice it again. Now they know there's something special in that part of the sky.

You can be sure that part of the sky will be on all of their star charts. Whatever instrumentation they've got, whatever telescopes they've got, are going to be looking at that place extraordinarily carefully. Then you have an omnidirectional transmitter beaming in all directions, because you don't know where, of all these places you've pinged, the intelligent creatures are, but now you have their attention. An omnidirectional unit doesn't have to be strong, because you can count on them building something big to look. Why don't they do that? Why don't we see astroengineering of these super-duper civilizations? Why don't we see stars being blown around?

CI   Maybe *we're* the astroengineering. I like the simulation hypothesis, that we're the pure recreation of such civilizations.

SS   I think we're really here.

CI   That's annoyingly hard to refute. The optimist–pessimist dichotomy amuses me. SETI scientists work hard to do their science and slough off the anthropocentric mantle that others would throw around them. But people still talk about being optimistic or pessimistic. Where are you?

SS   I'm optimistic. I'll bet everyone a cup of coffee we find ET within two dozen years. But it is a guess, and it's based on what we were talking about earlier: how much of the Galaxy you have to look at before you've given it a reasonable shot. Radio, that's just where we are now. I don't think we'll be using radio if they're a million years or a hundred million years ahead.

CI   We sent LPs into space. That's so seventies.

SS   Yes, it's true. On the other hand, technology can have a long lifetime, depending on physics. As far as we know, you can't send information faster than the speed of light. That could change, and then all bets are off.

Suppose that doesn't change. Electromagnetic radiation is a good way to get bits around. It's fast, it's cheap, and maybe you use that forever. We continue to use the wheel – it's an ancient invention, but it still turns out to be *the* solution for that particular problem. As far as we know, light and radio waves will always be in use.

CI If there's a phase of evolution and technological evolution that passes through that capability, you'll make a snapshot of it.

SS Yes, but you need big numbers to be sure you've got a few open windows.

CI Then the timing argument works against you, if it's a transient technology?

SS If it's transient, then you'd better have an awful lot of ETs out there at different stages, so you have some windows open to you. It also depends on what you think the window is. Several people have argued that we're looking for ETs like us, maybe a little more advanced, but within a hundred years of us.

CI What's your gut reaction to Fermi's question, or the puzzlement over where all the alien visitors are?

SS I think it's a great conversation topic. [Laughs] But it's a big extrapolation from a local observation.

CI We shouldn't necessarily be surprised by it?

SS I don't think so. We shouldn't read too much into it. Michael Hart wrote a paper in the eighties in which he dug up the Fermi paradox and said we should see evidence everywhere, including here. Since we don't, they're not out there. That's stimulated a cottage industry of people trying to reconcile the possibility that there's a lot of intelligence out there with the fact that we don't see it so obviously. Clever ideas came out of this. I'm sure most of them, maybe all of them, are wrong. But only one of them has to be right to explain why we don't see anything obvious.

In the history of Earth, colonizers always run out of steam quickly. They don't get far. The Mongols colonized a big hunk of real estate, but they lasted less than a generation. Even the Romans ran out of steam, and they were more successful than anybody else. That happens on Earth for reasons that may not apply in the Galaxy, mostly competition. Your lines of communication become long, and that is a real issue. If you're talking tens of millions of years, which we are, then you've become different species, and the original motivation is long gone.

CI There's a lot of unbounded speculation when you work on SETI.

SS Keeping focus on this requires single-minded effort. It takes years. By that time, you've invented machine intelligence. The real question to me isn't why ET isn't everywhere, but why ET's machines aren't everywhere.

32

Ray Kurzweil

In his work as an inventor and entrepreneur, Ray Kurzweil has created numerous technologies, which he then turned into nine businesses in various areas of AI. He helped develop the first print-to-speech reading machine for the blind, the first music synthesizer capable of receiving orchestral instruments, and the first CCD flat-bed scanner, among other inventions. *The Wall Street Journal* described him as "the restless genius," and *Inc. Magazine* called him "the rightful heir to Thomas Edison." Kurzweil has been inducted into the National Inventors Hall of Fame, he received the National Medal of Technology and the Lemelson-MIT Prize, and he has been the recipient of many other awards, including twelve honorary Doctorates, honors from three US presidents, and seven national and international film awards. His books include *The Age of Intelligent Machines, Fantastic Voyage: Live Long Enough to Live Forever*, and *The Singularity is Near: When Humans Transcend Biology*. His website, KurzweilAI.net, has over a million readers.

CI You're a polymath. How do you describe yourself?

RK Inventor, entrepreneur, and author.

CI Writing seems important to you – conveying ideas that are quite complex for a popular science book.

RK It's another way of inventing the technologies of the future. We don't have the computers of 2020 yet, but I can write about them – their implications and what will be feasible. My interest in anticipating the future stems from a practical project of anticipating technologies to time my inventions. Most inventions fail because the timing is wrong, not because people can't get the thing to work. About thirty years ago, I started tracking technology. I have a group that helps me gather data about information technology in many different fields and then builds mathematical models to anticipate what they will do. I use this to time my own technology projects, but an offshoot is to anticipate what will be feasible ten, twenty, or thirty years from now.

   Technology has benefits and downsides. It's done a lot of great things for human civilization. Consider where we were as a species only a few hundred years ago. There were no social safety nets; there were no antibiotics. If somebody got an infection – they were rampant then, because there was no sanitation – it was a disaster for a family, most of whom lived precariously on the edge of disaster. That was the fate of virtually all humans not long ago. But technology is a double-edged sword. I've written a lot about the downsides.

   My book, *The Age of Spiritual Machines*, led Sun Microsystems Chief Scientist Bill Joy to write his famous cover story in *Wired* magazine, "Why the Future Doesn't Need Us," about the dangers of genetics, nanotechnology, and robotics. He and I also coauthored an op-ed piece in *The New York Times*, criticizing the web posting of the 1918 flu genome. I've testified before Congress proposing a program to create a rapid-response system for new biological viruses. There is a reflexive anti-technology sentiment in some quarters; we see it in the movement against genetically modified organisms. Not all genetically modified organisms are necessarily safe. They should be studied for safety and efficacy; but there's a movement against them regardless of any testing. African nations are pressured to refuse vitally needed food aid or genetically modified seeds that could resist local blights because of irrational concerns. That's a modern version of an anti-technology bias that goes back to the Luddite movement in 1800.

CI People may be concerned that we're in a race against time – that there's always the possibility of microbes winning the evolutionary arms race.

RK We *are* in a race against time. The primary danger is not the microbe evolution, but terrorist use of bioengineering. Bioengineering has a positive side. It will enable us to reprogram the information processes underlying biology. We're designing drugs that can undercut cancer and heart disease and other major killers, as well as progressing in a dramatically more effective way than before. Biology used to be hit or miss. With new drugs, we discovered something that happened to work but had no real model for *why* it worked, and it had lots of side effects. We're now

able to precisely turn off a key gene or enzyme, or add an enzyme, and reprogram these biological processes away from disease. That's the positive side of the
equation, but the negative side is that a destructively minded individual could in
a routine college bioengineering laboratory create a new biological virus that is
deadly, communicable, and incurable.

CI   Could people without state-of-the-art resources do this?

RK   It's much easier to create a destructive biological virus than it would be, say, to
make an atomic bomb. An atomic bomb is fantastically complex, industrial, and
technical. But it's easy to create a new pathogen – several companies can send it
to you mail-order. You need some expertise, but creating a new biological virus is
much easier than making an atomic bomb, and it could be much more destructive because its effects are not local.

CI   So we ride the exponential curve of change, but risk making bad choices as a
society or as individuals?

RK   I don't think our choices will disrupt these exponential progressions. But they can
cause a lot of suffering. In the twentieth century, for example, 180 million people
were killed in about a hundred wars, 50 million in World War II alone. That suffering represents a lot of mistakes. We had a lot of misguided movements, but
there was a smooth double-exponential improvement in the price performance
of computing through the entire twentieth century. These exponentials continue
regardless of human folly, but how much suffering we will encounter is impacted
by the choices we make.

CI   You're talking about exponential improvements across engineering as well,
so it's not just computing.

RK   It's much broader. It's anything to do with information. Information technologies double their price-performance capacity, their bandwidth, every year. We see
it in computation, in communication, and in biological technologies. Biology is
now becoming an information technology. Fields like biology have pre- and post-
information eras. Due to the Human Genome project, the amount of genetic data
has doubled every year. The cost came down from ten dollars per genome in 1990
to a penny today. It took fifteen years to sequence HIV for the first time; we can
do it now in about seven days.

Our knowledge of the brain is undergoing similar exponential growth. That
used to be hit or miss – we didn't have brain scanners with enough resolution to
see what was going on. An MRI can only see clusters, not individual neurons or
neural connections. But new scanning technologies can see individual interneural
connections in living brains, and watch them signal in real time. The spatial resolution of brain scanning and 3D volume is doubling every year, as is the amount
of information we're gathering about the brain. We're turning that information
into workable models and simulations.

Ultimately, information will underlie everything of value. With full molecular
nanotechnology, we'll be able to bring the programmability of software to the
world of physical objects. We'll be able to create any physical object through

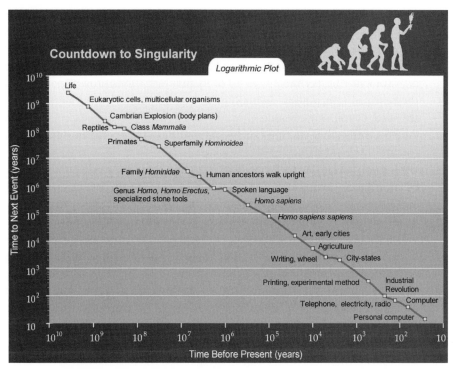

One way to represent the "singularity," the point at which accelerating technology moved humanity into a post-biological era, is to show the progress on a logarithmic plot. Advances on ever-shorter timescales are the hallmark of a crescendo of change, the convergence point of which is called the singularity (courtesy Ray Kurzweil and Kurzweil Technologies, Inc).

programmable software processes acting on organic raw materials – we'll be reassembling those materials intricately at the molecular level. Even within computation, exponential growth is broader than Moore's law, which refers to the shrinking of transistors on an integrated circuit. If we put twice as many on an integrated circuit every two years, and they're also faster because they're smaller, that technology doubles in price performance every year. Computing has ultimate limits, but the limits aren't very limiting – one cubic inch of nanotube circuitry would be a hundred million times more powerful than the human brain.

CI   There's plenty of room for improvement. Is an exponential a great model for this?

RK   Yes. Information content, data rate, bandwidth, capacity, and price performance of a technology all grow exponentially, not linearly. Technologies that are in a pre-information era, like energy, aren't subject to this law of accelerating trends. But once we can fabricate inexpensive, efficient solar panels and fuel cells using nanotechnology, energy will also become an information technology. Today, we only need to capture one percent of one percent, one ten-thousandth, of the sunlight that falls on the Earth. It's manageable within twenty years, even if we need three percent of a percent by then. Then energy will become an information technology. One of the messages of my book, *The Singularity is Near*, is that not only does information technology in all of its manifestations grow exponentially, but the reach of information technology is expanding and will underlie everything we need, everything of value.

CI   Perhaps things that we think of as commodities, like energy or information, will essentially become free.

RK   Well, information isn't free, though there are open-source forums of every type of information. Proprietary forms of information are the major underpinning of our economy. A substantial fraction of the world's GDP is in proprietary forms of information – whole industries, like publishing, movies, music, software, in which there's no physical substrate, just information. That isn't *free*, but the ratio of price to performance of information comes down every year. There's fifty percent deflation, which I think is good.

CI   What about the interface problem – the fact that humans and their sensory capabilities change on a biological timescale? How are we going to manage the future as far as not being left behind by our clever machines and computers and technologies?

RK   The short answer is we're going to merge with our technology. There are innate advantages of nonbiological intelligence. We can handle about two hundred calculations per second with our interneural connections, and electronics is a million times faster. Our interneural connections communicate inside our brains at a few hundred feet per second using chemical signaling; electronics is a million times faster. Machines share their knowledge at electronic speeds, which are a million times faster than language. They can share, communicate, and think about a million times more effectively than biological intelligence.

     We can already send blood-cell-sized devices into the bloodstream. That's today. If we take what can be done today and apply these exponential progressions, they mean a factor of a billion increase in the capabilities of electronics in twenty-five years, a factor of tens of thousands smaller in size because we're shrinking technology at a rate of over a hundred per 3D volume per decade. The nanobots of 2025 will have computers and communication networks. We'll be able to send millions or billions of nanobots into our brain at low cost, they'll be able to communicate wirelessly with our biological neurons, and we'll be able to directly augment our biological thinking with a nonbiological prosthesis, which augments our memories, our pattern recognition, and our analytical thinking.

CI   And subvert the built-in obsolescence of the organism itself?

RK   This is how we're going to evolve. Biology evolves at such a slow pace that it's completely irrelevant compared to technological evolution. We will pass our own intelligence by merging with our machines. People think that's undesirable

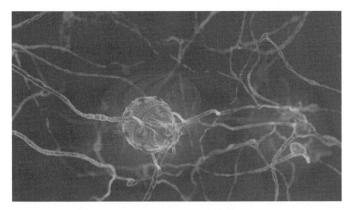

Nanobots are tiny mechanical devices or robots that have a partial biological function. Implanted medical devices are already in use, but in the future they will be miniaturized and have more flexible capabilities. Nanobots may be able to maintain and repair our bodies from within (courtesy NEWSin3D).

because they're thinking of the machines they've met, like their cell phones, which are not machines you'd want to merge with.

CI You're talking about merging in the grain of our being.

RK Yes, I'm talking about sublime machines that are truly as subtle and intelligent and complex as human intelligence. We're going to integrate this technology into ourselves. It's part of a long progression. Computers used to be remote. They're not integrated into our bodies and brains yet for the most part, but they're in our clothing, they're in our pockets. I hear parents say, "It may as well be implanted in his brain because he carries it everywhere. It's an extension of him." We're already augmenting what we can do with our computers. Few professionals could do their jobs without technology. We will get more intimate with it, we'll move it inside our bodies to keep us healthy, and inside our brains to provide full-immersion virtual reality from within the nervous system and augment all of our cognitive and emotional capabilities.

CI The gradual merger of the machine and the organism, even if it's accelerating, gives us a level of control. Do you discount the dystopian idea that our machines could do without us?

RK I am concerned about that. I call it, "The Intertwined Promise and Peril of GNR." G is genetics, somebody designing a pathological, biological virus. The downside of N, nanotechnology, would be a self-replicating entity that replicated out of control, a so-called "grey goo" scenario. What you're articulating is the downside of R, standing for robotics, but it really has to do with a strong AI, a nonbiological, intelligent entity that was more intelligent than "us," and would have no use for us. That's called the unfriendly AI scenario.

There's no technological defense against something more intelligent than you. History shows that civilizations with more advanced, more intelligent technology generally prevail. A scenario in which there were AIs with superior intelligence that saw no need for us would be a bad situation. But it's not likely to be that separate; it's not "Humans on the left side, machines on the right." They're going to be integrated together: biological humans will have lots of nonbiological processes running inside their bodies and brains; even nonbiological systems will be based on emulating biological systems. Machines are going to be deeply integrated into our civilization. It's more likely that there will still be some conflict between different philosophies. The real danger would be a fundamentalist force that had a regressive philosophy; if they happened to have more intelligent AI, they could dominate. The encouraging thing is that such civilizations tend *not* to have the most sophisticated technology.

CI   What if a strain of separate robotic intelligence improved itself exponentially, and we became irrelevant, or insignificant?

RK   I don't believe a runaway phenomenon would be that fast. It's going to take time for nonbiological intelligence to improve its own designs. That's why this trend can continue past the point of unenhanced human comprehension, the issue you brought up earlier. This is how our civilization, which is already an integrated human–machine civilization, continues to progress at an ever-accelerating pace.

CI   You have an avatar, Ramona, who sometimes does your bidding remotely. How do you envision yourself interacting with the world in twenty years, professionally or personally?

RK   We will have virtual reality within the decade – not *inside* the human nervous system, but visual–auditory virtual reality with images written directly into our retinas from eyeglasses. Not just an image on a PC screen, but a full-immersion virtual reality where we feel like we're together in some virtual environment. We'll be online all the time with the electronics we carry around or wear in our clothes. We'll have augmented reality. People and buildings will have pop-up displays that give information about them.

CI   And presumably alongside this, some vast processing power of intelligent agents assimilating information and adjusting to our preferences and goals.

RK   Exactly. By the late twenty twenties, this technology will be inside our bodies, and our brains will have radical extensions. Nanobots will be keeping us healthy from inside, and they'll be interacting with our biological neurons. We'll be spending increasing time in full-immersion virtual reality environments that are created from inside the nervous system, and will incorporate all five of our senses.

One of the features of virtual reality is that you don't have the same body in a virtual environment that you have in real life. You can be a different person, so we will pick different bodies the way we now pick different fashions; we can be different people in different situations. We'll be expanding our mental horizons through this merger with our technology.

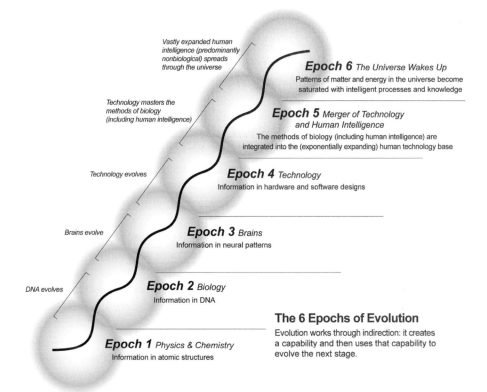

Vastly expanded human intelligence (predominantly nonbiological) spreads through the universe

Epoch 6 *The Universe Wakes Up*
Patterns of matter and energy in the universe become saturated with intelligent processes and knowledge

Technology masters the methods of biology (including human intelligence)

Epoch 5 *Merger of Technology and Human Intelligence*
The methods of biology (including human intelligence) are integrated into the (exponentially expanding) human technology base

Technology evolves

Epoch 4 *Technology*
Information in hardware and software designs

Brains evolve

Epoch 3 *Brains*
Information in neural patterns

DNA evolves

Epoch 2 *Biology*
Information in DNA

The 6 Epochs of Evolution
Evolution works through indirection: it creates a capability and then uses that capability to evolve the next stage.

Epoch 1 *Physics & Chemistry*
Information in atomic structures

Possible stages of evolution based on information. Up to this point, the universe has evolved from information stored in atoms and molecules to information stored in genetic material and neurons on at least one planet, and possible many more. One species on Earth has progressed to the capability of storing vast amounts of information via technology. Of the last two stages of evolution, one is imminent and the other is hypothetical, and if they have occurred elsewhere it poses the question asked by Enrico Fermi, "Where are they?" (courtesy Ray Kurzweil and Kurzweil Technologies, Inc.)

CI   This radical advance of technology and replacement of biological function brings to mind the arguments by which we might already live in a simulation. Perhaps this has already happened, and we are the playthings of some more advanced civilization.

RK   That becomes a philosophical issue as to what is a "simulation." If we're living in a simulation, it's pretty good; we see a lot of consistency, not too many glitches. There are some strange phenomena at the very limits, the details of quantum mechanics, for example, but overall it's fine-tuned. It's so good, we can't tell whether it's really any different from what we assume is reality.

CI   Let me move on to astrobiology. Our civilization faces post-biological evolution. What do you think the prospects are in a universe where chemistry is universal, where microbes could well be widespread, and where a fraction of those living planets will have produced intelligence and technology? Could post-biological evolution be a universal phenomenon?

RK   Once biological civilization has the primitive stirrings of technology – even horses and carriages, buildings, simple automatons – it's only a few centuries until that civilization can penetrate matter and energy in its vicinity with its intelligence at a molecular level, and spread out at the speed of light. This argument is based on exponential progression. A few centuries are very little time on a cosmological scale. The SETI assumption is that perhaps intelligent civilizations aren't likely, but there are so many billions or trillions of stars that even if it's one in a billion, there should be still billions of them out there. Some would be ahead of us, some would be behind us. Lots of them should be ahead of us, and not by forty years – they're going to be ahead by thousands or millions of years.

CI   Colonizing the galaxies would be small steps in this progression.

RK   Yes. Exponentially, it doesn't take that much time. So there should be millions of civilizations, but we don't notice them.

CI   This is the Fermi paradox.

RK   Yes. If we were talking about *one* civilization we don't see, the explanations for the Fermi paradox would be plausible. But it's not plausible in my view that *every single one* of the billions that should be out there have all either destroyed themselves or come to the *Star Trek* ethical decision to remain invisible and not disturb us. For *none* of them to use electromagnetic communication, even as a by-product of some industrial processes, seems unlikely.

      Someone might say, "Isn't it unlikely that of all the billions of stars, we're in the lead?" But that's the anthropic principle: *somebody* has to be in the lead. If we weren't here, we wouldn't be talking about it. For the universe to have all of its constants set to such precise values that complexity can evolve is extremely unlikely. If you took any of the thirty constants in the "standard model" of physics and changed them very slightly, there'd be no atoms and no galaxies and no telephones. How likely was it to have turned out so well for us? Again, if it weren't the case, we wouldn't be here.

CI   The other way to explain it is that there's a huge bottleneck between microbial life and complex, multicellular, intelligent life. Is that plausible?

RK   Microbial life started billions of years ago. It is possible that there are many primitive life forms elsewhere. But as soon as a technology-creating species develops, it's not far to the singularity. Our technology only started a few tens of thousands of years ago, and even that's a short time on a cosmological scale. From technologies like primitive radio, it's only a few centuries. So technology like ours might not be out there in abundance.

CI   Let me understand: based on the timing arguments, the multiplicity of sites for life, and some degree of inevitability to evolve to intelligence and technology, either the Fermi paradox is explained through various reasons, or we're the first?

RK   I'm arguing that we're the first. The various explanations that have been proposed are only satisfactory if you're trying to explain why one, two, or three civilizations didn't make it; but according to most interpretations of the SETI assumption, there should be billions of those civilizations. It's not credible that every single one is invisible to us.

CI   What about a bottleneck between intelligence and technology? Cephalopods have intelligence and impressive capabilities, but they are not technologically capable.

RK   There are two enabling factors for technology. One is intelligence; the other is the ability to manipulate the environment with a certain amount of dexterity. A chimpanzee's hand isn't quite good enough – the thumbs are slightly different than ours and lack the power grip, and they don't have fine motor coordination. Cephalopods have limited ability to move objects around, but they don't have our fine motor coordination. We were able to take our mental models of how things could be and do experiments in our mind, and then change the environment to create those things.

CI   Are you persuaded by anthropic arguments? Apart from the fact that we may be the first to reach a singularity, various aspects of the universe – from fine details of physics, to the cosmological expansion itself – are concordant and coincident with the evolution of complex biological organisms.

RK   That is something we observe. The philosophical issue is, why? Perhaps our universe was designed by an intelligent designer that was actually a considerably advanced civilization in some other universe that designed or created *our* universe so as to expand its own computing capability. We can always speculate.

CI   Let me get back to singularity as an issue for our species and our culture. The concept presumes us to be at an extraordinary time in cosmic history. How do we manage this transition well and not have one of the Doomsday scenarios come true?

RK   That *is* one of the biggest issues. Each step is more daunting than the last. We face the existential threat of a bioengineered pathogen right now. That's a grave threat. Nanotechnology will solve that. Once we have effective nanobots, they will be able to combat biological viruses. But they introduce their own peril. Nanotechnology devices can be very dangerous, even without self-replication. Ultimately, strong AI will close the window on that peril – strong intelligence can solve the problem of pathological nanotechnology. But we then confront pathological AI. The answer is even smarter AI. And so on, ad infinitum.

         My solution, which may seem vague and unsatisfying, is instilling the values – respect for knowledge, diversity, tolerance, and democracy – from our own civilization, because the integrated biological–nonbiological civilization that's going to emerge will be the same civilization. If we maintain those values, they will hopefully exist in the future. That is not a foolproof plan, but I think it's the best we can do. The future civilization is not disjoint from our current civilization – it *is* our current civilization, as it will evolve into the future. But these are the primary challenges. We can control the peril while we focus on the promise.

33

Nick Bostrom

Nick Bostrom doesn't claim to know the future, but he does have a few hunches. He cofounded the World Transhumanist Association and the Institute for Ethics and Emerging Technologies, and is currently the Director of Oxford's Future of Humanity Institute. Bostrom has a PhD in philosophy of science from the London School of Economics, as well as an MSc in computational neuroscience, and a second Master's in physics and philosophy. He was awarded a Postdoctoral Fellowship from the British Academy. With his skill of making the most abstruse topics comprehensible, he has given hundreds of interviews on cloning, cryonics, AI, the anthropic principle, nanotechnology, and the simulation argument. He has over 130 publications, one scholarly monograph, *Anthropic Bias*, and several edited books. His writings have been translated into sixteen languages. Bostrom has worked for the European Commission, the European Group on Ethics, and the Central Intelligence Agency.

CI   You have eclectic interests. What has been your academic path to this point?

NB   I have an educational background in physics and computational neuroscience, as well as philosophy. I did my PhD in philosophy of science and foundations of probability theory, and I've been employed by philosophy departments since.

CI   You were at Yale for a while, and now you have a new institute that sounds exciting.

NB   It's called the Future of Humanity Institute, and it's a multidisciplinary research institute at Oxford University. Our aim is to look at some of the big questions for humanity. There are four main research areas. One is human enhancement, which looks at how anticipated technologies might be used to extend human capacities or change various aspects of our biology, and the practical and ethical implications of that. The second is global catastrophic risk, which includes threats to human survival, but also other, less extreme risks. The third is rationality and wisdom, where we address certain methodological questions. And finally, future technology, which is looking at potentially transformative technologies, such as molecular machines and artificial intelligence.

CI   What's your general impression from a philosophical perspective as to the state of astrobiology? What expectations are justified about life elsewhere?

NB   The first question is whether we will find anything elsewhere at all. My view is that it's relatively unlikely that we will find life anywhere else. I certainly hope that we won't, and if we do, then I hope that it will just be some primitive form of life rather than anything even moderately advanced.

CI   Interesting. You don't think the rapid emergence of life on Earth and it radiating into a wide range of environments implies any inevitability to biology?

NB   No. There is an observation-selection effect in play, which a considerable portion of my work has addressed. I developed the theory of observational selection in my PhD thesis, and have continued to publish in that field. Take two different hypotheses about how common life is: hypothesis A says that life evolves whenever you have a suitable planet; hypothesis B says it's extremely unlikely for any planet to give rise to intelligent life, and maybe it happens once in a trillion trillion. What do these two different hypotheses predict about what we should expect to observe? Both of them predict that we should find ourselves on a planet where life evolved. The fact that we have evolved here doesn't give us any discriminating evidence for or against either of those hypotheses.

CI   Presumably, we also must discount the fact that we've developed a technological capability no other species has: the ability to understand the universe and look for other life forms.

NB   That doesn't give us much information about the probability of it happening on a random planet. We have to look at other planets. Where we've looked, we've found absolutely nothing. We might also look at details of our own evolutionary development, and see whether it looks like a steady flow of inevitable progress, or more like big leaps that might have been very improbable.

As far as our current data is concerned, everything's consistent with the second hypothesis. The step from prokaryotic to eukaryotic cells took about 1.8 billion years, during which time, not much happened. It might have been a matter of randomly trying lots of different combinations and getting lucky. There are a number of candidates for difficult steps where it might just have been enormous luck that made it happen, rather than inevitability. The case for inevitability gets stronger the farther down the road we look. For example, the step from apes to humans looks much easier – it didn't take too long to happen. That little bit might have a high probability, whereas some of these earlier stages – whether it's the emergence of the first self-replicators, or the step from prokaryotes to eukaryotes, or from there to sexual reproduction – might have been ridiculously improbable.

CI   If we were to find microbial life on Mars or subsurface Europa and it wasn't a shared origin with Earth, then our way of thinking changes, presumably. The difference between one and two examples of biology in a single solar system is a major conceptual change.

NB   It changes something, and exactly how much depends on how advanced this life had independently developed. Think of the line that starts from the lifeless planet and goes on to more advanced life forms, and eventually intelligent life forms such as our own. Shortly thereafter, it will give rise to a civilization that can travel through space and send out replicating Von Neumann probes. We haven't seen any; no aliens have visited our planet as far as we know. Somewhere between having an Earth-like planet and the endpoint of this line – a space-colonizing civilization – there must be a great filter, one or more steps that are so improbable that very few trajectories make it through.

There are lots of planets, but as yet no observed space-colonizing civilizations. This great filter that reduces the huge number of planets to the number of space-colonizing civilizations – what is this filter? There are two possibilities: either it occurs before or after our stage of development. From our point of view, it would be better if it was before us, because if it tends to happen after us, we're unlikely to reach that colonizing stage ourselves.

CI   A priori, isn't it more likely that it's before us, since we already make ourselves known across space and time with our transmissions?

NB   Yes. There is certainly a small window in which, if there had been civilizations at our own stage of development, and if they are near enough, we could have detected their signals. But beyond that there's a huge amount of time when they could have colonized the universe, and they would be here by now.

Say we find simple prokaryotic life on Europa. That would cut off a part of this line. It would show us that the great filter is not located between a lifeless planet and prokaryotic life, because we'd have an independent instance of something that made it past that initial state where there is no observation-selection effect to account for it. That reduces the remaining points for the great filter. If it's only prokaryotic life it's not extremely bad news – but suppose it was something more advanced, like some extinct form of squirrel. That immediately removes a lot of evolutionary distance where the great filter might lie, and only leaves a bit just behind our own point of development. It therefore increases the probability that the filter is after our developmental stage – which would be bad news.

The absence of visits from intelligent aliens, who should exist given the large number of habitable planets and long timescales available for evolution since the big bang, suggests that there may be a "great filter" that stands between us and truly advanced technological capability. It also features in SETI formalism as the factor for a civilization lifetime in the Drake equation (courtesy Mike Ivy and Wikipedia Foundation).

CI   That's ironic – the more advanced the first life forms we find beyond Earth, the worse it is for us.

NB   People were excited about this possible Mars fossil – everybody thought it would be great news if it turned out to be real. Whereas I think it would be very bad news.

CI   Is Rare Earth another type of argument affected by observational selection? The evidence is always the same, but some people look at particular aspects of the Earth – its environmental and geological conditions, the stabilizing affect of the Moon – and argue for the specialness of Earth-like habitable conditions based on that list.

NB   It's one way to figure out how improbable step zero was, step zero being a planet with the right properties. Step one is the evolution of simple self-replicators, and there are other steps after that. We can try to determine the likelihood of each step. With the later steps, like the step from prokaryotic to eukaryotic life, it's difficult to figure out the probability from first principles, given what we know in evolutionary biology and genetics. But, with step zero, it's slightly different. It's easier to calculate how many different planets there are of different sorts, and to make some guesses about what conditions would have been necessary for life to arise. It certainly seems worth trying.

CI Although it's still bound by the chronology of our biology. The flavor of the Rare Earth argument is the Goldilocks principle, but there is a counterargument – that extreme radiation environments or fluctuating climates could clear out ecological niches and spur faster evolution. There isn't any way to decide until we find out how many terrestrial planets have Earth-like conditions.

NB That's right. There are some constraints we can be confident about, and others seem very shaky. Perhaps because it's easier to do the research, there's this huge emphasis on step zero, having a suitable planet. The steps after that might be much more improbable, but it's harder to show rigorously that's the case.

CI Another area where people look at the same evidence and draw quite different conclusions is the debate over contingency and convergence. In thinking about the emergence of brains, intelligence, and technology, do you think convergence arguments are persuasive?

NB I'm not sure. My intuition is that once a certain advanced level of evolution is reached, it's easier from there on. Everything happens more quickly. It's a matter of probability, but it certainly seems easier to get from ape to human than it is to get from prokaryote to eukaryote. I think the convergence argument gets more plausible the farther down the road you get.

CI Let me move to the Fermi question. "Where are they?" is a legitimate question, but should we be surprised by the lack of evidence for technological civilizations?

NB The most likely explanation is that they're so far away that they haven't had time to get here, and probably never will. It's the simplest explanation that's perfectly consistent with all we know.

CI Another aspect of the Fermi question is the timing argument. Being alone is one reasonable answer to Fermi's question, but if we're not alone in a technological-civilization sense, is it almost certain that the nearest civilizations are vastly more advanced than we are?

NB Yes.

CI What are the most serious selection effects that affect astrobiology?

NB The simplest ones are relatively obvious – for example, the fact that evolution produced intelligent life here gives us virtually no evidence about how likely that process is.

CI What about the Doomsday argument? We emerged as an intelligent civilization late in an eleven-billion-year galactic chronology. Does that imply anything about our longevity?

NB It's a controversial area. The Doomsday arguments fall quite far down that ladder of scientific robustness, because we have to make assumptions about what's included in our reference class. Our reference class consists of other intelligent observers that are like us in some relevant respect. For the Doomsday argument to work, we've got to assume that future observers, if they exist, are in our reference class. Future observers could be different from us, including the fact that they would live in the future and therefore be in different epistemic states – they might also have used technology to change themselves. We must assume a specific definition of the reference class to make the Doomsday argument work.

The situation is different with regard to other applications of anthropic reasoning, such as the one mentioned earlier: the fact that life evolving here doesn't give us evidence that it's common. We can make that kind of inference without making so many assumptions about the exact nature of the reference class, so it's much stronger.

CI   What's your favorite version of the Doomsday argument?

NB   The one developed by Brandon Carter and John Leslie. We can explain the core idea using an analogy with balls and urns. You don't know how many balls are in the urn, it's either ten or a million, numbered from one up to the maximum. You randomly select a ball and it's number seven. That gives you strong evidence that you are faced with a ten-ball urn. The idea then is that we should think of our own birth rank, our position in the time order of the human species, as analogous to a randomly selected ball from an urn. This idea is known as the "self-sampling assumption." There are independent arguments for accepting the self-sampling assumption, or something like it.

CI   So it's the difference between drawing the number seven from one million items or from ten items. Is it as simple as that?

NB   The conditional probability of drawing seven is different if you have ten balls in the urn than if you have a million. That's standard Bayesian probability theory.

CI   But doesn't the idea of ranking add a rule to the situation?

NB   No. Whether the human species goes extinct sooner or much later, there will still have been 6 billion people alive at one point. The two Doomsday hypotheses include one part that is the same, and then different extra amounts of stuff – like the urns, where both of them contain at least ten balls and one contains a lot more than that. Instead of numbers on the balls, they could have their own specific pattern or color. As long as you drew a ball that you know would be included among those in the ten-ball urn, then the fact that you drew that ball would be evidence for the ten-ball-urn hypothesis.

CI   Beyond birth order in a set of humans living and dying, the overlay of technology creates unstable and nonlinear conditions. That must feed into our expectations in some way.

NB   In the Carter–Leslie version of the Doomsday argument, that's what we might call the prior probability of the hypothesis. With the urn case, suppose you know that there's a two-thirds probability that you're faced with a million-ball urn. Maybe there are twice as many such urns as there are ten-ball urns, and one of the urns was randomly selected and put in front of you. That creates a prior probability favoring the million-ball-urn hypothesis.

Similarly, there are different Doomsday hypotheses about how long the human species will last, and your prior probability of those hypotheses includes your estimate of how likely it is for nuclear war to wipe us out, or a meteorite to strike us down, or all the other particular disaster scenarios you might imagine. If the Doomsday argument is valid, you would take all of that information and use that as your prior. Once you take into account your relatively low birth rank – your early place in the human species – you will revise that prior probability and arrive at the more pessimistic posterior.

CI   That makes sense. Our potential control over biology might exempt us from natural selection, or we might even transition to a post-biological future. That presumably changes the calculations.

NB   It changes the prior probability. If we become more careful, then the prior probability of Doom goes down. One possibility is that future people would be radically technologically modified, rather than facing extinction. The Doomsday argument is sensitively dependent on the reference class, and for that reason is not a compelling argument.

CI   Anthropic ideas seem like rich terrain for a philosopher, but it's surprising that so many physical scientists and cosmologists have swooned into the arms of anthropic thinking in the last decade or so. Is anthropic thinking relevant to the question of life in the universe and the likelihood that it's widespread?

NB   It's key to that question, as well as to questions about contemporary cosmology, but it needs to be interpreted in the right way. In my dissertation, I counted thirty different anthropic principles, and I'm sure there were a lot more. They all meant different things: some were nonsensical, others tautological, some contradictory, some expressed an empirical hypothesis, some seemed to express a wish …

CI   … so maybe we should let Darwinian selection operate on anthropic ideas, and see which ones survive.

NB   At the heart of all these different anthropic principles, there is a core of genuinely valuable methodology, which is observation-selection theory. In its simplest form, it's the injunction that we need to take observation-selection effects into account in our reasoning, wherever they might exist. This develops into a probabilistic framework.

CI   After studying the resonance states of carbon and oxygen to find out how heavy elements were made in stars, Fred Hoyle said that a super-intelligent being was monkeying with the universe, because these resonances were so well tuned. How should we view fine-tuning arguments in physics?

NB   People are tuning in to multiverse cosmologies, and inflationary cosmological models, where our universe is just one of many. It seems like all popular theories have that feature, but it's not the only conceivable way to explain fine-tuning. We might find a unifying theory later on where all these numbers will pop out from simple assumptions, simple symmetry configurations, and then everything else will be derived from that. Failing that, however, the multiverse explanation for fine-tuning looks promising. We could potentially explain why conditions are right for life in this universe without assuming that the totality of existence – the multiverse – is itself fine-tuned.

CI   Doesn't it go against a pillar of epistemology, the idea of parsimony and Occam's razor? Hypothesizing vast numbers of universes seems very extravagant.

NB   That's a misunderstanding of Occam's razor. It doesn't exclude postulating lots of entities. It's against introducing too many free parameters in your theory. It's against overcomplicating your theory, as opposed to thinking that the world is big. The multiverse theory potentially could be very simple. You hope for a simple theory that makes a small number of simple, elegant assumptions, from which we can derive predictions that match what we observe. It doesn't matter whether the multiverse is big; it's the *complexity* of the theory that is the measure to which Occam's razor should be applied.

CI   What about the criticism that the anthropic ideas and the related multiverse theories are overly anthropocentric? Perhaps complexity and self-organization as it defines life doesn't need carbon or long-lived stars.

NB   Here again there is a spectrum of cases. We're on firmer ground at one end of the spectrum and on much shakier ground at the other. Take the gravitational constant: if it turns out that with a slightly greater gravitational constant, the universe would have recollapsed shortly after the big bang, that would give us fairly good grounds for thinking that there wouldn't be many observers in that scenario. Or if the universe had just been increasingly diluted hydrogen gas as time went by – atoms sliding ever further apart because gravity was too weak – there, too, we would have fairly strong reasons to believe that there would be a low density of observers. As we move on to other parameters, the claim that observers could not have existed, or that there wouldn't be very many of them, becomes more speculative. The case for fine-tuning depends on exactly which parameter we're looking at.

CI   Are we limited in our ideas of what life might look like or how it might function?

NB   Theoretically. But I don't assign a high probability to finding life in the dust around planets, or in black holes or neutron stars. Everywhere we've looked, other than Earth, we have found no life at all. It seems difficult to generate life even under the best circumstances – a nice, suitable planet – so we shouldn't be too eager to think it would arise in places that seem less hospitable.

CI   But the universe is so magnificently large and complex that it seems hard to anticipate what might have evolved.

NB   Oh, yes. When I say that my preferred answer to the Fermi paradox is that they are very far away, I mean that if the universe is infinite, then there will be lots of aliens out there. In fact, all possible life forms will exist *somewhere* in an infinite universe; if the probability in each draw is nonzero and you have an infinite number of draws, you end up with a probability of one. But in my view, they would be so far away as to be outside our backward causal light-cone, and our future one as well.

CI   The Drake equation is focused on the Milky Way. If the longevity of civilizations is hundreds of millions or billions of years, there's plenty of time to send signals. The Milky Way is a *piece* of the landscape. How do we conceive of civilizations that might be millions or even billions of years more advanced than us?

NB   There are some things we can say about them with moderately high probability. From the outside, a civilization might simply look like a sphere centered on the planet where it originated, expanding at a significant fraction of the speed of light. They would send out colonization probes in all directions that reproduce themselves and send out more probes. What happens within that expanding sphere is harder to know. There are different scenarios, from mindless replicators making more replicators – and not doing anything with all these resources – to optimistic scenarios where they make new forms of life and new civilizations, perhaps by building vast computer systems, where most of the life would exist in virtual realities and in digital form. You can squeeze so much more into that than if you run it biologically. That would be my guess of how it would look.

CI You've alluded to the simulation hypothesis. What was your motivation for that work?

NB It seemed to me a significant discovery. The simulation argument says that one of three possibilities is true; it doesn't tell us which one. One is that almost all civilizations at our stage go extinct before they become technologically mature. The second is that there is a strong convergence among all technologically mature civilizations, such that they all lose interest in creating ancestor simulations. An ancestor simulation is a computer simulation of people like our historical ancestors, one that is sufficiently fine-grained that the simulated people are conscious. You can think of it as the sum of all the thought processes of a human, or nonhuman, civilization. The third possibility is that we are living in a computer simulation.

CI Could you verify or disprove any of those premises with current information?

NB I don't think so. The original paper – as opposed to some of the spin-off popular newspaper articles – the one that was published in *Philosophical Quarterly* titled, "Are We Living in a Computer Simulation?" argues only that at least one of those three is true.

CI Is the first possibility suspect on the grounds of a timing argument?

NB No, I don't think so.

CI So there's no a-priori reason to be inclined for or against any of the possibilities?

NB We don't know which of the three possibilities is correct, and so we should distribute our credence to cover all of them. When you first explain the simulation argument, a lot of people will accept it, but they'll say, "Obviously it's number one that's true, we're developing technology and weapons." But somebody else will say, "Well, it's obviously possibility two that's correct." And others will say, "It's clearly possibility three." Different people pick different options.

CI So it's an ink-blot test.

NB Maybe it says something deep about their personality. If I had to assign more probability to just one possibility, it would be the second one, the convergence hypothesis – but only slightly more.

CI Substrate independence, or life being independent of "wet" biology, seems like a premise of the argument. What's our basis for expecting that in advanced stages of evolution?

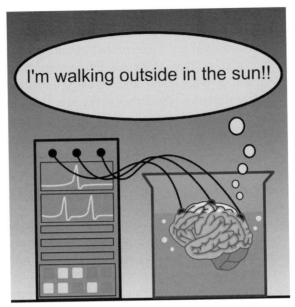

The simulation argument takes as a premise substrate independence: the idea that all conscious thought processes can be represented by computation without having a brain involved. Given the large number of likely intelligent civilizations, those that have advanced far beyond us will have the ability to create simulated entities, meaning that we should take seriously the possibility that we live in a simulation. The fact that a brain might function normally without being present in a body is the source of some major debates in philosophy (courtesy Wikipedia Foundation).

NB   Substrate independence is a philosophical position, not a clear-cut empirical prediction. It just says that if you had a suitably powerful computer, running a suitable program, sufficiently similar to the computational processes that take place in the human organic brain, then that would be conscious and have experiences just as we do. The part of the argument that is empirical is the claim that a technologically mature civilization would be able to create huge numbers of ancestor simulations. I argue for that by assessing various technologies we can already foresee are possible, and what they will be able to do computation-wise once they are developed, and comparing that to estimates of how much actual computing power it would take to create ancestor simulations.

CI   Interesting. It's a bizarre type of Copernican argument. Simulations are so trivial for advanced creatures to generate that we're unlikely to be in one of the few biological situations. Do you think we're heading towards a post-biological future?

NB   Assuming we don't go extinct in this century, or suffer some great collapse of our civilization, then by the end of this century, and possibly before that, we will have developed artificial intelligence, and uploading, and the other human modification and enhancement technologies that will take us into a "post-human" era.

34

Paul Davies

It's a rare career that includes debating science and religion with the Pope and the Dalai Lama, writing myriad articles and popular-science books, producing radio and TV shows, and being awarded the Templeton Prize by Prince Philip at Buckingham Palace. These are all features in the career of Paul Davies, selected as "one of Australia's ten most creative people" by *The Bulletin*. Davies earned his PhD from University College London and is a Visiting Professor of Physics at Imperial College London, an Adjunct Professor at the University of Queensland, and an Adjunct Professor at Macquarie University in Sydney. He has published over a hundred research papers and twenty-five books, and a science-fiction novel, *Fireball*. He was elected a Fellow of the Royal Society of Literature and a Fellow of the World Economic Forum. His awards include the Advance Australia Award and the UK Institute of Physics' Kelvin Medal, and he has been shortlisted for the Royal Society Science Book Prize. Asteroid 6870 is named in his honor. Davies currently leads the BEYOND Center for Fundamental Concepts in Science at Arizona State University.

CI   I'm gobsmacked that you've written more than two dozen books. How do you do it?

PD   People often ask how I manage to write so many books and do other things as well. I write about the things I'm thinking about anyway, in particular the topics of my research projects. Research is useful for writing a "popular" book – I like to think of it more as public outreach than popularization. It's a good way to really understand something. If I can explain it in simple sentences, I'm probably on top of the subject.

CI   That's similar to the reason some people teach outside of their discipline – you're explaining it to students, who ask obvious and direct questions.

PD   You're fast to get to the core issue, exactly. I also write quickly if it's a subject I know something about. I can do a three-thousand-word article on black holes in an afternoon. A short book on a topic I know well may take no more than a week or two. As anyone who has written a book knows, the easy part is the creative part, getting the core content down. Most of the labor goes into the many edits, tracking down references, getting names spelt correctly, artwork, the index, and so on.

CI   What about popularization in general? There's a mixed reaction among senior scientists as to whether it's a good or a bad thing to do. There's the famous story of Carl Sagan not being elected a member of the National Academy of Sciences because of his popularization.

PD   So I gather. When I started this in the seventies, a work colleague and friend said, "Stop writing these popular books; it'll seriously damage your career." There was a rule of thumb in theoretical physics that for every popular book you wrote, you subtracted ten papers from your publication list.

       Then a few things changed. Students turned away from physics in the seventies and eighties. Universities realized that the subject was going to die unless young people could be enthused. Suddenly, most university administrations became more supportive of communication with the wider public, particularly with young people, and of encouraging them to come into what were perceived as difficult subjects – physics, mathematics – by giving them some sex appeal.

       The other change was the Stephen Hawking phenomenon; *Brief History of Time* outsold almost every other book – it was on the London *Sunday Times* bestseller list for over four years. A lot of people felt, "If it's all right for Hawking, then it's all right for me: I can write a popular book." Everybody charged through the gap. Most distinguished scientists you meet today have written a popular book. Now it's okay to be a popularizer, but it was a long, hard struggle.

CI   Do you have an archetypal reader in mind?

PD   I have no particular person in mind. My readers need no specialist background knowledge in physics or cosmology, but I'd expect them to have a strong interest in science. It helps if they know what's going on.

There's always this problem of which words you can get away with. My wife is a great help because she's a radio science journalist, so she knows you can't use words like "isotopes." You can use "atom" and "black hole," and increasingly you can talk about DNA, even though most people wouldn't have a clue what the letters stand for. I regard these as the words or concepts that have currency, even though they're not understood; they're like little pegs upon which you can hang other things.

For example, time travel is a fun topic to imagine. How do we do it? It may not be possible, but one way is using a wormhole. The concept of a wormhole in space requires the general theory of relativity and knowledge of differential geometry and topology. These topics are far beyond what ordinary people are going to understand. But I can say, "You've all heard of black holes. A black hole is a one-way journey to nowhere, you fall into it and you can't get out again. Imagine something *like* a black hole, but with an exit as well as an entrance. It would be like a 'stargate' or a shortcut between two points in space a long way apart. Imagine going through it and coming out somewhere else suddenly."

CI   Astrobiology's tricky because, at least in the United States, the slate is not clean. People have been inculcated by the popular culture to believe not only that aliens exist, but that they've visited us. Getting back to a scientific reference point for astrobiology is even harder than in physics.

PD   That's quite true. It's even worse for SETI, which is the speculative end of astro-biology by anybody's standards, and yet the populist media is awash with alien abductions and UFO stories. The difficulty is that we like to play off that intrinsic fascination – as a teenager I was blown away by these stories, thinking that we're not alone, and we're surrounded by advanced alien beings. It's a thrilling con-cept, and it's unfortunate that when we look at the scientific evidence, all that stuff melts away. Yet there's still this sense that trying to find out whether we're alone in the universe, trying to locate a second genesis of life – even if it's only microbial life – is a wonderfully compelling goal. But how do we keep the public on our side without giving them the impression that we're looking for little green men? They're disappointed when we say, "No, it's just microbes." [Laughs]

CI   We have hundreds of extrasolar planets, but pictures for only a handful of them. We can't tether to the visual impact of the beautiful Hubble pictures. Astrobiology research, when it's represented for the public, is sometimes disappointing. The "little green men" expectation is a part of it, but there's also, "Show me! We can't see all these things you're talking about."

PD   It would take a very significant discovery to turn that around; perhaps if we saw chlorophyll in a spectrum from an extrasolar planet, but we're a long way from being able to do that. The search for Earth-like planets elsewhere in the Galaxy could be made the focus of an international movement. Searching for other places like home has uplifting appeal. It's not inconceivable that, decades ahead, we'll have instruments with the capability of imaging entire planets to a reso-lution where we could look at pictures and say, "Wow, there's another Earth out there." A lot of people could get behind the idea of exploring the cosmos beyond the now-familiar retinue of planets in the Solar System.

CI   You've had parallel interests for a long time. When did you get interested in astrobiology at a research level? What issues attracted you?

PD   I've often wondered how I got into this game. The interest goes back a long way, to my early teens and reading those UFO stories. When I was a student in the sixties, no scientist wanted to talk about life beyond Earth; it was regarded as absolutely inconceivable that there was any life out there. For me, a turning point was a Cambridge conference in 1983 – organized by Martin Rees, now President of the Royal Society – in which astronomers, cosmologists, and biologists were brought together. It was called "From Matter to Life." I realized there were whole subject areas that could be investigated. Freeman Dyson also traces his interest to that point; he wrote *Origins of Life* as a result of that meeting.

I had read Schrödinger's book, *What is Life?*, as a young postdoc at Cambridge. To a physicist, life looks like a miracle. [Laughs] I think physicists are much more intrigued by the fact of life than biologists. Biologists take it for granted – "Yes, of course it's living!" – because that's all they study. But it blows physicists away. I've always been fascinated by what it takes to make a living organism. The other half of the question is, when did stupid atoms start doing such clever things?

Then I met Duncan Steele in Australia, who is an expert on asteroid and comet impacts. He had done work with micrometeorites and was particularly interested in organics in meteorites. I learned a whole new dimension of astrobiology from him, that the Solar System is full of organic material, although the word "organic" has to be used carefully. It doesn't mean detritus of once-living things, it means it's a building block that we find in life. I put two and two together, as Jay Melosh did independently at University of Arizona, and realized that if rocks could be traded between planets, maybe microorganisms could be as well. In the early nineties, I wrote about and lectured on the possibilities of transport of life between Earth and Mars by this mechanism, but nobody would believe me.

CI   Now the paradigm has shifted to the point where everyone says, "Sure, there's a conveyer belt."

PD   Exactly. That all turned around. Why? Bill Clinton stood on the White House lawn in 1996 and proclaimed that NASA had evidence for life on Mars, based on the Allan Hills meteorite. Most of that evidence went away, but it brought to public attention the idea that a Mars rock could come here with microfossils, and so maybe it could come with live microorganisms.

I was asked to help set up the Australian Center for Astrobiology, which was founded around 2000. Before coming to Arizona State University, I spent five or six years getting to know people working in all aspects of astrobiology. I keep saying, "I'm just a physicist trying to make sense of this stuff – I blundered into this field from the outside."

CI   In the history of biology, there have been some interesting perspectives brought by physical scientists.

PD   You don't have to convince me that physics is the discipline that can illuminate everything!

CI   The Allan Hills meteorite reanimated the idea of panspermia. Is transfer between planetary systems possible if organisms could go into a hibernation state for long enough?

PD   Statistically, what's favorable is transfer between neighbor planets. It's easier to go from Mars to Earth, because of the lower surface gravity of Mars, but it can go the other way as well. Getting off Venus is hampered both by the higher gravity and the thick atmosphere. But big enough impacts could splatter rocks off any of these bodies. Some of those rocks will be ejected from the Solar System by Jupiter. It boils down to two things. Could microorganisms survive long enough to travel interstellar distances? The answer to that seems to be: maybe. But that's not the real issue. Much more significant is the question of the chances that a rock blasted off Earth would *ever* hit another Earth-like planet in another star system. The statistics for that are *incredibly* unfavorable; it's exceedingly unlikely. Rocky panspermia works well within a planetary system, but works very badly between planetary systems.

That's not to say it's never happened. It's possible with very favorable statistics, or at an early phase during which a lot of planetary systems formed close together. If perchance one of them had early life, it's not inconceivable that it could have spread to the others, and then those star systems moved apart. But generally speaking, life isn't going to spread across the Galaxy this way.

Another panspermia theory is quite different – they're often confused. It goes back to ancient Greece, but was popularized by Svante Arrhenius about a hundred years ago. It suggests that microorganisms could waft naked across the Galaxy, propelled by the pressure of sunlight and starlight. Microbe-sized, bacteria-sized particles can get across the Galaxy that way, but they're going to be dead on arrival because they're exposed to the harsh conditions of outer space, in particular ultraviolet radiation. UV radiation is absolutely, totally deadly. It's easy to screen out – a small rock will do it – but a truly naked microbe isn't going to make it.

CI   Where do you stand on the likelihood of abundant life in the cosmos? Most astrobiologists think of the hundreds of millions of habitable places in the Milky Way – more if you include moons of outer planets as well as terrestrial planets – and they say, "How could all those Petri dishes be dead?"

PD   Right. But the flaw in this argument is that you have to decide at the outset: did life form from random shuffling of the building blocks? We know it's easy to make building blocks out of amino acids – it's dead easy, you could make that in any high-school lab. It's the next step – putting them together in the exceedingly elaborate, highly specific, complex structures that we would recognize as

In the earliest form of the panspermia theory, dating to ancient Greece but popularized much later by Arrhenuis, naked microbes traverse the Galaxy propelled by radiation pressure. The propulsion mechanism is plausible, but UV radiation would render unprotected microbes dead after a short time in deep space (courtesy Edward Willett).

an autonomous, living thing – that gets tricky. We may be living in a universe that has 10^{20} potential Earth-like planets within the body of space that we can see. But 10^{20} is a *trifling* number compared to the odds against shuffling those molecules into the right formation. If it happened by chance shuffling, we're it.

CI   It makes me think of Fred Hoyle's comment on the implausibility of assembling a jumbo jet...

PD   ...by a whirlwind in a junkyard! It's a compelling image. We don't know if that's the way it happened. Maybe there is something like a Judeo-Christian deity, a cosmic imperative. That is, maybe life really is built into the laws of nature in some fundamental way. Maybe there are organizing tendencies that shortcut those odds enormously. In other words, maybe life is a natural outcome of a complexification of matter, in much the same way that the formation of a crystal is a natural outcome of the laws of physics. This type of biology, which is called predeterminism, is very popular. Many astrobiologists accept it as the default assumption: that life will out, given Earth-like conditions. But in the present state of ignorance it could be anywhere on the spectrum between happening only once in the entire universe, a stupendously improbable accident, up to being part of the natural workings of a fundamentally bio-friendly universe.

CI   Which is why going from the first example to the second example is so critical. Perhaps that second example will be the subsurface of Mars.

PD   The frustrating thing about Mars is that we might go there, find life, celebrate it, and find that it's good old Earth life. It got there from here or here from there: it's another branch on the known tree of life; it didn't start from scratch independent of life on Earth. To find a truly independent genesis, we may have to go beyond the Solar System.

There is one way of getting around all of this: if life forms readily in Earth-like conditions, shouldn't it have happened many times over, here on Earth? In other words, if there is a cosmic imperative or life principle, we can test it by looking for evidence of multiple geneses right here. My research in astrobiology is precisely directed to looking for evidence of a second genesis right under our noses on our home planet.

CI   The physical evidence more than 3.5 billion years old is pretty dicey, so this is difficult empirical work.

PD   We can't reconstruct those events. Many of the records have almost certainly been obliterated. I think it's inconceivable to identify the cradle of life – find the place where it happened, maybe a relic or a trace of what happened. I get into arguments with people about whether science can explain the origin of life. We may never know, because the origin of life is an exercise in chemistry, in physics, in earth sciences, in informational processing, computation, all sorts of things. It's also an exercise in history – there may have been certain sequences of events that were necessary for life to get going, all traces of which have been lost. That doesn't mean it was a miracle; it just means that, like a lot of history, the record peters out if you go back far enough.

I'm hopeful that we may find, right under our noses, extant organisms descended from an independent genesis. There could be microbes that are not *our* life, but life from some other origination event. Finding such microbes would immediately establish the cosmic imperative, the life principle. Then we *would* expect life all around the universe.

CI   What might that life look like? The alternative biologies that could result from similar starting points and building blocks are probably quite diverse.

PD   We're limited only by the powers of our imagination. Alternative forms of life here on Earth are likely to be microbial, because we'd notice if there were alien elephants wandering around. But you can't tell just by looking at microbes what they're made of – you have to explore their innards. The biochemical techniques used to study life as we know it are *customized* precisely to life as we know it, so there's circularity: we'll only discover life as we know it.

If there are alien microbes – not "alien" in the sense of having come from another planet, but alien in the sense of being an alternative form of life – life could differ in a large number of ways from life as we know it. It could be something as small as a different genetic code, a different sequence of amino acids or nucleotides. It could have opposite chirality, that is, it could be like a mirror form of known life. It could be based on a more extreme solvent than water. It could be nonchemical altogether.

CI   It might not need the cell as the unit of organization.

PD   Yes. We're obsessed with the idea that life is a little blob of something; there's no reason that has to be the case. I often wonder whether life, even life as we know it, could have started with something larger, a complicated chemical cycle that only later on became refined and microminiaturized and packaged into cells.

CI   I want to shift to the anthropic principle. There's a distinction between recognizing fine-tuning in the physical universe and jumping to the various levels of anthropic interpretation of those physical facts. Where do you stand on those arguments?

PD   It's a vast topic – on which I've written a book.

CI   [Laughs]

PD   It's hard to give a quick summary, but I'll make a few points. The first is that I do take life of mind seriously. Most physicists regard life as a bizarre aberration in the universe, not a phenomenon integral to its workings. I have always thought differently, largely for philosophical reasons, not scientific reasons. I think the emergence of life and the emergence of mind are no trivial accident or by-product; they are deeply embedded in the workings of things, which is why I'm so interested in finding evidence that life is widespread.

Giordano Bruno was burned at the stake for many heresies. One of them was preaching the idea of alternative inhabited worlds – the Catholic Church felt it was very threatening, and that a universe with a purposeful God would not have made other inhabited worlds. That's exactly wrong; a universe throughout which life flourishes, a universe that is in favor of life, is much more congenial to the notion of meaning and purpose than a universe in which life is a bizarre statistical fluke that happened on one planet and is of no significance in the great cosmic scheme of things. By temperament, I like the idea of a biological universe.

CI   I sense some optimism on the cosmic imperative towards biology. It's amenable towards scientific investigation, that's the key.

PD   That's right. I get tired of arguments about the deeper aspects of science and cosmology, when it's only words. So I ask, "Where is the science? How can we test this?" For example, in the subject of emergence people often say that there are certain thresholds of complexity above which we see phenomena that cannot be captured by the physics of the lower level. But what does it matter? Are these new laws of physics? We have a phenomenon of downward causation. It looks like higher levels of the system have causal efficacy over the lower levels; wholes can affect parts in a way that we can't understand by looking at the parts on their own. But does it matter to the humble foot soldier of physics – namely, the atom? Are there any forces being deployed that we could not understand without having to see emergence? If so, what do we look for? I want to go beyond philosophy and words and look at real science.

CI   Do physical scientists have good enough tools to address complex systems that they could apply them to biology and emergence?

PD   No, the field of complex systems is still in its infancy. One of the reasons we don't understand the origin of life is that we don't understand the principles that apply to the emergence of complex organization.

CI   It doesn't seem amenable to computation approaches, either.

PD   The problem with computation is that you often lose sight of the physics amid the welter of pictures and computer output. That's okay; it's like taxonomy at the time of Linnaeus. You collect plants, look at their shapes and classify them, but you don't understand the underlying principles. I'd like to know the principles shared by complex, organized systems. If we're to try to understand the secret of life – which we don't, we can describe life at the molecular level in great detail but the whole package still looks like a miracle – we probably need clues from nonliving systems that have organized complexity, in particular, emerging complexity.

35

Martin Rees

Martin Rees is the preeminent cosmologist of his generation and he has long been interested in the properties of the universe that permit and constrain the development of biology. After studying at the University of Cambridge, he held postdoctoral positions in the UK and the USA, before becoming a professor at Sussex University. He became a fellow of King's College and Plumian Professor of Astronomy and Experimental Philosophy at Cambridge, and served for ten years as director of Cambridge's Institute of Astronomy and for eleven years as a Royal Society Research Professor. He is Professor of Cosmology and Astrophysics and Master of Trinity College at the University of Cambridge, and holds the honorary title of Astronomer Royal. He is a foreign associate of the National Academy of Sciences, the American Philosophical Society, and the American Academy of Arts and Sciences, and is an honorary member of several other foreign academies. His awards include the Gold Medal of the Royal Astronomical Society, the Balzan International Prize, the Bruce Medal of the Astronomical Society of the Pacific, the Heineman Prize for Astrophysics, the Bower Award for Science of the Franklin Institute, the Cosmology Prize of the Peter Gruber Foundation, the Einstein Award of the World Cultural Council and the Crafoord Prize of the Royal Swedish Academy. He has been president of the British Association for the Advancement of Science and the Royal Astronomical Society. In 2005, he was appointed to the House of Lords and elected President of the Royal Society. Rees is the author or coauthor of over 500 research papers, seven books, and numerous magazine and newspaper articles.

CI You've been interested in the cosmic context for life for some time. Was that interested sparked by the conference in Cambridge that hooked many people?

MR Indeed. I had been interested for a long time, but I co-organized with Sidney Brenner a conference called "From Atoms to Life," held at Trinity, my college at Cambridge. We attracted a wonderful set of people for a day of discussion, including Freeman Dyson, Tony Gold, and Manfred Eigen.

CI When was this?

MR It must have been the mid eighties. Freeman Dyson subsequently did a series of lectures in Cambridge called the "Tarner Lectures," and published a book on the origin of life. In that book he acknowledges the initial impetus of this conference.

CI That was an influential book. Was it the first time a major physicist had taken on the concept of biology from a bottom-up approach?

MR Certainly not. There was the famous Schrödinger book, *What is Life*, in 1946.

CI What do physical scientists bring to the discussion of biology?

MR In the forties and fifties, physicists were providing some of the impetus to molecular biology. Many of the people who did the primary work in molecular biology, including Crick, started out as physicists. More recently, lots of people who were primarily interested in physics or astronomy have been fascinated by the origin of life and its relevance to the likelihood of life elsewhere. One thinks of Sagan, Shklovsky, and Morrison in particular.

CI What do you think of the perspective of physicists and astronomers, who look at the sheer numbers of habitable places and imagine that these events must have happened elsewhere?

MR I don't think physicists should do that. We have to accept that so long as we only know about life here in this one place, we can't say whether even simple life is likely or unlikely. Even if we think simple life is likely, we still wouldn't be able to say, in our present ignorance, how probable it is that simple life evolved into a biosphere and into intelligence as it did here on Earth. The only rational stance is to be agnostic about how likely it is for simple life to exist, and also how likely it is for simple life to become highly evolved. Within twenty years, we'll probably have a firmer estimate of the likelihood of simple life. We may find life elsewhere, or we may come to understand better how life began here. We'll have a feel for how much of a fluke it was. But even then, we won't know how likely it is that simple life evolves towards intelligence, rather than ending up with a planet covered with microbes or insects.

CI   In that distinction between microbial life and more advanced, more "interesting" kinds of life, SETI is somewhat controversial among scientists, in part because a null result is difficult to interpret. What's your perspective?

MR   I think it's fully worthwhile, because we have no idea what to look for. That being so, given how important and fascinating the question is, we should look in all possible ways. We should do the kind of SETI programs that are currently being pursued, but we should accept that, even if they fail, that doesn't mean there isn't life out there. It could mean that our imagination is limited and we haven't devised the optimum search procedures, or that the life out there is not of a kind we can recognize. Since we don't know exactly what to look for, the sensible strategy is to use all possible techniques.

CI   Compared to the cost of a new particle collider or a telescope in space, SETI is pretty inexpensive.

MR   It's cheap enough that it's funded by private donations, more or less. But given the importance and the public interest, even if you bet heavy odds against success, it's still worth pursuing. I'm glad they're doing it.

CI   Take a snapshot of the field of astrobiology right now, and the progress that's been made in the last five or ten years. What are the things that most excite you, or the areas where you expect advances in the near future?

MR   One key development for the benefit of astronomy in general is the detection of a huge number of planets around other stars, so we can start to understand planetary systems and their properties, and classify them. That provides a focus for trying to study what sort of planets exist, and what sort of atmospheres they might have. It's an important subject and it provides an impetus to astrobiology.

       At the same time, there are two developments on the biological side. One is that we are closer to understanding the origin of life and the transition from the nonliving to the living, because of a better understanding of microbiology and the idea that we can perhaps synthesize a simple organism. The other development is the discovery and study of extremophiles.

CI   That research implies that the traditional views of the habitable zone are probably wrong. Habitability must have a broad envelope.

MR   It could be that any complex or advanced life forms, or complex biospheres, can exist only in a narrow range of conditions, narrower than the extremophiles. On the other hand, we should be open-minded about whether even advanced life could exist in a wider range of conditions – Salpeter and Sagan, for instance, envisaged huge balloon-like creatures floating in the atmosphere of Jupiter – or creatures existing at much lower temperatures.

CI   Sometimes provocative questions lead to interesting ideas. The Fermi question has provoked thought for many years. Is that a well-posed question?

MR   It's not a well-posed question, but if you apply basic reasoning, then the fact that they're not already here – the aliens – does make us suspect that they are rather rare. There are ways around the argument. Stephen Webb wrote a book with fifty counterarguments, *If the Universe is Filled with Aliens, Where is Everybody?* But we should bear in mind that there are many stars around which life would have had a head start of at least 1–2 billion years, compared to life on Earth.

CI   Maybe we're limited in imagining the nature of advanced forms of life, especially if they move beyond the biological realm. You've recommended that people read first-rate science fiction over second-rate science. What books have you found enjoyable and stimulating?

MR   I was stimulated by some of Arthur C. Clarke's work. There's lots of other science fiction where the ideas are fascinating. What drove me to make that remark was the important point, which is often overlooked, that we should not think of humanity as being in any sense the culmination of evolution. It's taken 4 billion years for us to evolve from primordial life here on Earth, but life has billions of years ahead of it, on Earth and beyond. We should be open-minded about post-human life. The key questions are whether that life would remain on the Earth or go beyond, and whether it would remain organic, or whether some sort of silicon-based computers will take over. That brings the Fermi question into sharper focus, because life could have taken that route on some other planet, long before it could ever have happened here on Earth.

CI   When people parse the Drake equation and examine the timescale element, only a modest fraction of these intelligent civilizations need to become post-biological and essentially eternal for them to be quite readily available and communicable.

MR   Yes. In that perspective, one has to take the Fermi question seriously.

CI   How tuned or not is the universe for biology?

MR   We mustn't be too anthropocentric, obviously. Life doesn't have to be like the life here on Earth. But there are some generic features that make up anything complex enough to be called intelligence. That probably requires some kind of chemistry, in the sense of some element other than hydrogen. Of course, even that you could question – you could have Hoyle black clouds in space, where the structure is magnetic. Leaving that aside, you would need to have more than just hydrogen.

CI   If you are presuming that you need to go beyond the simplest element, which of the numbers that frame cosmology are most sensitive to allowing or disallowing chemistry?

MR   Nuclear physics is a competition between nuclear force and electromagnetic force. That determines which nuclei are stable. That's something where there's a certain degree of tuning. The most important and most general requirement is that there must be at least one very large number in the universe. Anything that is complex – especially if it's evolved to be complex – must be large, in the sense of consisting of huge numbers of the basic atomic entities. And it must have existed for a long time, compared to the basic timescale for a single reaction.

CI   Is that problematic for hypothetical speeded-up, small-scale universes?

MR   Yes. Supposing that you keep all the microphysics, the nuclear force, and the electric force the same, so that the chemistry is in principle the same, then you can still have stars, in the sense of gravitationally confined fusion reactions. But if gravity were much stronger, then the stars would be smaller because the number of particles that need to be crammed together in order for gravity to overwhelm other forces is smaller. Gravity wins in the end for sufficiently large assemblages, because everything has the same "sign" of gravitational charge. The number of particles you can pack together to make a star is roughly the three halves power of the ratio of the electrical forces to the gravitational forces between a proton. That number is 10^{38} or thereabouts. If you make gravity 10^{10} times stronger, so that the large number, instead of being 10^{38}, is 10^{28}, then stars would be 10^{15} times less massive. It's not quite straightforward to show this, but their lifetime would also be about 10^{10} times shorter.

In that hypothetical universe, two things work against the emergence of complex life. The first is that any organism much bigger than an insect would be crushed, because you have a much smaller planet, and much stronger gravity. Nothing could get as big as we are without being crushed by gravity. The second point is that stars in this small-scale, speeded-up universe would last a brief time. There would not be time for the huge number of basic chemical reactions that are essential in order to build up a single generation of organisms, and then have huge numbers of generations for them to evolve. In order to have large-scale universes and lots of time, there must be one large number, and that large number in our universe reflects the weakness of gravity on the microscopic scale. You have to pack together huge numbers of particles before gravity becomes competitive with the other forces.

The other general requirement is a certain amount of disequilibrium. Suppose our universe had recollapsed after the first million years, when it was still in the fireball state. You would never get complexity. You've got to have nonlinearities, and temperature contrasts between stars and cold, interstellar space.

CI   In the mode of Feynman's "There's Room at the Bottom," it seems like there's a potential for small, fast biologies in our universe. The catalytic rates of reactions can, in principle, be incredibly fast. We tend to think of evolution in the scale of time in which it happened on this planet, but perhaps hypothetical biologies, even conventional DNA-based ones, could run much faster.

MR   Possibly. There's a science-fiction story by Bob Forward about life on the surface of a neutron star. The argument is that complexity on the neutron scale, not the atomic scale, evolves much faster and overtakes ordinary life in its sophistication.

CI   You've also drawn attention to the Q parameter, the "graininess" of the universe. That's less familiar to most people, but it's relevant to the potential for life.

MR   Yes. If you imagine everything else staying the same in the universe – the same microphysics, and the same cosmological parameters – a characteristic feature of the universe is that it's not completely smooth. It has fluctuations, which are the seeds for structure formation. If those fluctuations weren't there at all, then after 10 billion years, the universe would still be uniform, cold hydrogen and helium, and nothing else. No galaxies and no stars. There must have been some fluctuations present in the early universe, and we know that in our universe these fluctuations had an amplitude, measured as a dimensionless number, which is about 10^{-5}. That's related to temperature fluctuations over the entire sky in the microwave background. That number determines the amplitude of fluctuations, and we know that amplitude is compatible with the emergence of the structures we see in the present-day universe, galaxies and clusters by gravitational instability. One of the triumphs of our understanding of structure formation is that the simulations, which feed in fluctuations at the recombination era with the amplitude we observe, all predict the scale of present-day clustering that we actually observe. That confirms we understand the essential features of structure formation in the universe.

What would happen if the early universe had been rougher or smoother than it was? If this parameter, Q, was zero, then we'd have a completely cold, uniform universe with nothing of interest in it. If the number was 10^{-6} or 10^{-7}, fluctuations would take much longer to condense out, they'd be much smaller, and the gravitational potential wells would be much shallower. If the number was 10^{-6}, you would have an anemic universe, where you'd get some small galaxies with rather shallow potential wells.

CI   It's not entirely sterile; is it just a much lower density?

The "graininess" of the universe, or the parameter Q, is measured by the fluctuation in the cosmic microwave background radiation, imaged by the WMAP mission. This radiation dates from 400 000 years after the big bang, when tiny inhomogeneities were beginning to form the first cosmic structures. If Q had been much higher, the universe would have evolved to be smooth, and biology would have been impossible. If it had been much lower, structure formation would have been anemic, and biology would probably be extremely rare (courtesy NASA/WMAP).

MR   A lot lower, yes. I would guess that the cutoff below which you get *nothing* of interest would be somewhere between 10^{-6} and 10^{-7}. If it's below 10^{-7}, you'd get no stars forming at all. If it was a bit more than 10^{-7}, you'd get some stars, but then you'd get no second-generation stars, because the ejecta from the first supernovae would blow out of the shallow potential well.

CI   So what would happen to all galaxies in that universe is what actually happened to the small, early galaxies in our universe.

MR   That's right. It would be interesting if the people who are doing these simulations could try some of these cases, but it's not in the cards at the moment.

Imagine a "rougher" universe where Q was higher than the actual 10^{-5}. If Q were a bit higher, you might get a rather more interesting universe, because you then get larger clusters, and huge disk galaxies, bigger than present-day clusters. But if the roughness got too excessive, two things would go wrong. First, galaxies, if they formed at all, would be compact and tightly bound. The stars would be close together and no planets could survive on undisturbed orbits.

CI   You'd have lots of black holes, presumably.

MR   Yes. If Q was larger by a factor of 10^{-3}, fluctuations would start to collapse when they were still relatively thick, and they'd trap their radiation and go to huge black holes and probably no stars. Most people think Q is fixed by quantum fluctuations and inflation, or something in the very early stages. Q is narrowly pinned down by observations, but it's got to lie within a certain range in order to allow stars and planets.

CI   Do we know enough about the inflation model to say whether a range of three orders of magnitude in Q for potential life amounts to fine-tuning or not?

MR   I don't think we do. One of the depressing things is that, despite twenty-five years of studying inflation, there's no agreed model that predicts why it should be 10^{-5}. You can take different models and fiddle with them to get whatever number you like. The generic idea is that the fluctuations are imprinted by quantum effects when the universe was of microscopic size. But we don't have any good theory to pin it down. We can't say that's fine-tuning. All we can say is that if you have an ensemble of universes where there are more possibilities, then within that broad range of fluctuation amplitudes, there's only a subset that allows astronomical systems of the kind we observe to develop.

The other issue is the cosmological constant, or dark energy. If that were much stronger, in other words if lambda was much higher, then the cosmic repulsion would overwhelm gravity at an earlier stage than it has done in our present universe. If the repulsion becomes dominant before galaxies form, then they will never form. There is an apparent fine-tuning in lambda, which can't be all that large, because otherwise it would have inhibited galaxy formation. Lambda could be larger without inhibiting galaxy formation if Q were larger, because galaxies then form earlier. If Q were 10^{-4} rather than 10^{-5}, the anthropic limit on lambda is a thousand times higher, because galaxies form at a redshift ten times higher.

CI   We're talking about hypothetical universes and their implications, counterfactual histories. Is that a useful mode of speculation in this subject?

MR It develops our intuition, like counterfactual history does. It would be good if the people who do simulations were to simulate some of these other models to see if they get the same answers, even when they can't compare the observations. It's a useful exercise even at that basic level. If one looks at the work that's been done on models of the early universe – unified theories, and string theory – then one encounters a genuine debate about whether our laws of nature are unique or not. We have to consider whether the idea of other big bangs governed by other laws is inevitable or irrelevant.

Eventually, people might come up with some unique theory that pins down all these numbers exactly; on the other hand, it may turn out, as some theorists suspect, that there will *never* be a theory that pins down all these numbers exactly – that there are many big bangs, and what we call the fundamental laws are just bylaws in our cosmic patch. That's an important question, and it features in a book edited by Bernard Carr, called *Universe or Multiverse?*, which contains papers from three conferences, two of which I hosted, on questions of whether the laws of nature are unique.

CI The two of you triggered a lot of this with your *Nature* paper in 1979. It led people to start accumulating the coincidences or general anthropic arguments. It's spawned a lot of speculation. What is the epistemological status of the multiverse idea?

The multiverse is a cosmology idea that accommodates the possibility of other big bangs and other space-times, unobservable by us. In the multiverse, the fundamental parameters that govern the properties of the universe take a wide range of values, only some of which are conducive to long-lived stars or the formation of heavy elements or the evolution of biology. The apparent fine-tuning of physical constants around values conducive to biology would be explained by us living in one universe of a vast ensemble of universes (courtesy *Nature*).

MR   We should be open-minded. It depends crucially on whether the laws of nature are such that physical constants are unique or whether they're little accidents. There's a decision tree that shows the options. We must first ask whether there's one big bang or many, and second, if there are many big bangs, are the laws in the aftermaths of all the big bangs the same or not? We don't know the answer to either of those questions. If it goes one way, anthropic reasoning's irrelevant. If it goes the other way, we have no better understanding of why the laws of nature are the way they are than we can derive from an anthropic argument.

CI   These underlying theories leading to a multiverse usually have string theory beneath them as a framework for the fundamental nature of matter and energy. Are they predictive at any level, in terms of the cohort of realities or universes that might emerge?

MR   At the moment, string theories are not predictive at all – except that in a generic sense they "predict" a force like gravity. The hope is that they will be predictive. We're never going to directly observe other universes. But that's not required for the multiverse concept to be regarded as physics, not metaphysics. In order for something to be scientific, it's got to be based on a theory which has credibility. If we had a theory which we believed in because it explains why we had neutrinos or some other as-yet unexplained features of our universe, and if that theory predicts that the inflationary potential has the form that Linde hypothesizes in his multiverse model, then we would take that model seriously. For example, Linde's "eternal inflation" is a well-defined model. You put in a certain assumption about the inflation potential, and you get that model out.

But we don't know whether the physics that has to go into that model is the right physics or not. We might test a theory in other ways, just as we believe in the physics of primordial nucleosynthesis, because it's based on nuclear physics we can test in the lab. We might have a theory we *can* test, which pins down what the inflation potential is, and tells us whether it leads to eternal inflation or not. If it does, then we should take the idea of multiple big bangs very seriously.

CI   One parameter of the universe that seems relevant to biology is the number of space dimensions. There's a particular kind of complexity that's required to code information in atoms and molecules. Is the dimensionality of space that survives the big bang an anthropic quantity?

MR   I think so. Two dimensions might not be enough. You've got to have three. There are other special mathematical features in three dimensions, compared to four – nondispersive waves, and the number of rotations equaling the dimensionality of space itself. There are lots of special things about three dimensions. I don't know whether we can rule out even more complicated universes in some extra dimensions. That's one of the many things we aren't clever enough to answer. In this speculation, we have to bear in mind that there may be scientific questions our brains just can't cope with.

CI   Speaking of brains, you made a provocative comment in the past about us only being able to detect a fraction of the brains in the universe. What did you mean by that?

MR   There could be brains that "package" and understand reality differently from ours. Also, there could be brains that can understand things that we can't conceive. There's no reason to think the human brain is in any sense the culmination of evolutionary intelligence. Just as we can understand things that a dog can't, there may be things that are as beyond our comprehension as modern physics is to an animal.

CI   Of course, if some fraction of the hypothetical biological experiments out in the universe goes through that transition to post-biology, then all bets are off.

MR   Indeed. You've got to rethink the Fermi question in a more sophisticated way, and realize that there could be types of intelligence we wouldn't recognize at all if they're vastly beyond ours.

CI   What about the bizarre argument that conventional biology may not be always necessary as a pathway to a brain – the idea of big enough set of possibilities in the universe that you can spontaneously generate a so-called Boltzmann brain? Is that an idea we should take seriously?

MR   There are models in which it's more likely to have a brain emerge spontaneously than in the kind of universe that leads to one by the path it happened here. I don't know how seriously that should be taken. Intelligence might not always evolve the way it did here on Earth.

CI   The epitome of these radical ideas is the one that says that the limitations of our brains are due to the fact that we are created entities by a superior civilization, the so-called "simulation hypothesis." I take it you don't believe in evolutionary convergence towards a particular human form of intelligence?

MR   It seems slightly unlikely to me, but I'm not an expert. It is an important question.

CI   You've joked that, as the Astronomer Royal, you don't have to cast the Queen's horoscope. But do you have to advise the Queen on protocols if there is an ET contact or visitation?

MR   I don't think so. Some have suggested that Commission 49 of the International Astronomical Union should be the welcoming committee.

CI   The astronomers have it covered, one way or the other.

MR   Hopefully.

CI   Will you be on the welcoming committee?

MR   Yes!

36

Ben Bova

    As the author of more than 115 books set in the future, Ben Bova has got many things right; examples include the space race, solar power, organic chemicals in space, cloning, ice on the Moon, and eBooks. Bova was editor of *Analog Science Fiction* and *Omni* magazine, and has written for many publications, including *The Wall Street Journal* and *Scientific American*. His series of novels includes "Exiles," "Sam Gunn," and the immense (and immensely popular) "Grand Tour." He is a President Emeritus of the National Space Society and a past president of Science Fiction Writers of America, and he belongs to the American Association for the Advancement of Science, the National Space Society, the Planetary Society, and he served on the NASA/Space Transportation Association steering committee on space tourism. Bova has taught science fiction at Harvard and at the Hayden Planetarium, and he is an avid fencer. His numerous literary awards include the Lifetime Achievement Award of the Arthur C. Clarke Foundation, six Hugos, the Isaac Asimov Memorial Award, a Balrog, an Inkpot, and the first annual Ben Bova Award for outstanding achievement in science-fiction literature in 2008.

CI   You've been writing for more than fifty years. How did you start?

BB   In tenth grade my English teacher told us to write an essay. I did. He liked it. He suggested in no uncertain terms that I write for the school newspaper – he was the faculty advisor, so it was more of a command. I loved it. I've been writing ever since.

CI   Did you have a regular job before you were a writer?

BB   Almost all writers have other jobs when they start. It's difficult to support yourself or a family just by writing. Fortunately I was able to get interesting jobs. I woke up early and wrote for a couple of hours at home before work.

CI   How did you get your job at *Analog*? That was a breakthrough.

BB   *Analog Science Fiction Magazine* was run by John W. Campbell from 1937 until he died suddenly in 1971. When he took over, it was just a pulp magazine called *Astounding Stories of Super Science*. John, who had trained in physics, wanted to make stories about the future that were based on real science. What we call science fiction today is what John Campbell envisioned. He gave up a successful career as a writer and spent the rest of his life running that magazine. He read every manuscript that came in, and there were hundreds each week. He spent a lot of time helping new writers learn how to write. He used to say, "The real job of an editor is to find a good writer in a bad story." Find someone with talent, with promise, even if the story you're reading can't be published. He'd give the writers advice, fill them with ideas, and eagerly await good stories from them. He was a man of strong opinions and a chain smoker. He didn't believe this nonsense from the Surgeon General about cigarette smoking causing heart trouble, and he died of a massive heart attack at the age of 61.

CI   Did you have to think twice about taking over at the magazine?

BB   I thought a long time about it. It was like being drafted to run for president. I was terribly afraid of mucking it up, but I couldn't say no. *Analog* was and still is the best science-fiction magazine in the Solar System. It was a great responsibility, but I took the chance. I had a career in the aerospace industry and moved to New York and edited the magazine; it worked out very well.

CI   Did you find, as he must have, that talent is scarce?

BB   No, the great problem is that talent is abundant. *Skill* is scarce. There are many talented writers, but they've never learned how to write a story, a commercial story that people will want to read. That was the real problem. It was an educational experience for me, seeing so much good writing that didn't work. It taught me what skills a writer needs and how a story should be constructed.

CI   That implies that writing workshops are not a bad idea: people need those skills.

BB   That's right. No one can teach talent. But you can be taught skill. Most beginning writers, whatever age they begin, need to learn the skills of constructing a good story.

CI   It must have been gratifying to see people with talent and not much skill move to the top.

BB   It was very rare. I'd see very talented writers. I'd give them the best advice I could. They'd send the story back, and it would be worse. You ask yourself, "Is it me, or did they misunderstand?" We write stories from deep emotional needs, possibly needs that the writer doesn't understand. Trying to change the story means knocking on the door of something deep in that writer's heart. It takes a special kind of person with very special talents to understand how to change a story and keep the message that he or she originally wanted to get across.

CI   It must have been a consuming job. How much time did it leave for *your* writing?

BB   Not much. I was still in my old habit of getting up early in the morning and writing for an hour before going to the office. I started writing for newspapers – first my high school newspaper, then commercial newspapers in the Philadelphia area. I learned very early that you get the job done. There is a deadline, with the accent on *dead*. Humphrey Bogart said, "A professional is a guy who gets the job done whether he feels like it or not." You may not always be doing your best work, but you get the work done.

CI   Does it get any easier? You've written more than a hundred books.

BB   The discipline stays with you. The projects get harder because you're constantly striving for something you haven't done before. You're always trying to stretch your muscles, and it is difficult. Mario Andretti said, "If everything is under control, you're not going fast enough." If it's easy, you're probably not doing it right.

CI: I love that quote. What was the community of science-fiction writers like when you were editing *Analog*, before science fiction permeated popular culture? If you told someone on the street what you did, would they even understand?

BB   It depends on the community. In Houston they'd understand, or Palo Alto, where there's a university or a big research center. Those people read science fiction. But in Naples, Florida, where I live, most people don't read science fiction, and they don't get it.

CI   Back in the sixties and seventies it must have been a small community.

BB   It was, and most of the writers knew each other. It was a tight-knit family. We constantly complained that we were trapped inside a ghetto. We were trying to break down the walls, and we did eventually.

CI   Writing is so solitary. Having that community must have helped.

BB   Although we were close emotionally, we were geographically far apart. Gordon Dickson said that this was a good thing. He said, "Writers should live far enough from each other so that they can visit only with the greatest of difficulty," because otherwise it's too easy to go visit and have some fun with your pals – instead of writing.

CI   How has the landscape of science fiction changed over the years?

BB   The problem is that most people have never read any science fiction, but they have seen science-fiction films or what Hollywood calls "sci-fi flicks." These are usually old Bela Lugosi scripts where, instead of a vampire in a castle, you've got an alien in a spacecraft. It's the same crap.

CI   It's identikit. It's doesn't have an essence of science at its core.

BB   Worse than that, in most Hollywood sci-fi flicks it turns out that "science is bad for you," and science is apt to be evil, full of bumbling oafs or megalomaniacs who want to rule the world. I've known a lot of scientists and they're too smart to want to rule the world. Take a successful film like *Star Wars*. What is the message of that film? At the ultimate moment, at the crisis point, Alec Guinness whispers in Luke's ear, "Use the force, Luke." So he pushes aside the computer and feels his way. That's baloney. That's a philosophy for slaves.

CI   It's either that or *Dr. Strangelove*.

BB   *Dr. Strangelove* was at least funny.

CI: What are the films dealing with science that you admire, that get it right?

BB   Stanley Kubrick's *2001*, because he had the brains to go to a real science-fiction writer for the story: Arthur Clarke. Arthur always has a mystical side to him, and that comes across in the film beautifully. Another one I enjoyed was *Galaxy Quest*. It's a comedy, a spoof of the *Star Trek* phenomenon. It's beautifully done, and funny.

CI   In this variegated landscape, how do you define "hard science fiction?"

BB   Science fiction in which the stories are based on real, known science. You can extrapolate. I'm perfectly free to invent anything I like, as long as nobody can prove it's wrong. If I say Mars is blue, people are going to object, and they'd be right to object. But I will say Mars was once inhabited by an intelligent race, and nobody can say that's wrong, not yet.

CI   That's an interesting way to frame it. Our exploration of space is just beginning, so that leaves a lot of latitude for speculation.

BB   That's why I've written a few historical novels in areas where we know something of what happened but not enough to straitjacket us. The novelist wants room to develop the characters and the situation. Real science forms a background. In my series of novels, the so-called "Grand Tour of the Solar System," I'm using the different planets and asteroids as we know them, but I feel free to project what it would be like there and to show you things we don't know.

CI   Do scientists lobby you for the topic of your next book?

BB   The chief scientist for the Pluto Express mission once asked me, long before his mission took off, if I planned to do a novel about Pluto. I didn't know that it was his life, and I said, "I don't think so. Pluto's kind of dull." He was very hurt.

CI   The Solar System is still rich ground. There may be a half-dozen potentially habitable places right here. There's a lot going on.

BB   A century ago, writers did adventure novels set in Africa or in the Amazon jungle. You can't do that anymore. If Jack London were alive today, he wouldn't be writing about the Klondike. He'd be writing about Mars. That's where you can write stories about human beings facing hostile nature and struggling to survive.

I don't think of my novels as science fiction. I think of them as historical novels that haven't happened yet, and if we wait a while, perhaps they will. I'm

Return to Mars is part of Ben Bova's "Grand Tour" series of novels, a fictional treatment of human colonization of the Solar System in the late twenty-first century. Seventeen books are involved, although chronology overlaps and the books do not have to be read in a specific order. The series present a realistic and also an optimistic view of our future in space (courtesy Ben Bova).

writing for people who've never read science fiction. I don't intend to bowl the reader over with my erudition. I'm writing romances and adventure novels set in exotic places, and I hope these novels show people what it would be like to be there and deal with the problems they would face. As in any decent novel, the major problems are people's interactions with each other. Whether you're in a research lab orbiting Jupiter or walking across the boiling surface of Venus, people are still going to be people, with the same relationship problems we have here.

CI   Characters in your books face problems or difficult situations beyond Earth, but there's a sense of optimism. This is our future. We can go there, and we can live and work there. It's our destiny.

BB   Look back on the history of the human race and see the problems we've solved. We've lived through ice ages. We've lived through enormous plagues. We've lived through wars, and we have learned and solved problems. We create new problems for ourselves, but I think we do have an optimistic future. Einstein once said that the eternal mystery of the universe is that it's understandable. We can understand the universe. We're in the midst of a great era of exploration and understanding. My motto is, "The best is yet to come."

CI   Some commentators look at the short history of the space program and they are disappointed. In the sixties, we went to the Moon by harnessing the talents of the best engineers and the best scientists. Now we're struggling to get back to the Moon with crude chemical rockets, the Shuttle is dangerous and on its last legs, and nobody wants the Space Station. How do you view the current state of space exploration?

BB   Our progress in space has been governed by government. So far it has been 99 percent government operations, and government operations are always guided by politics. It was politically expedient for the United States to put Americans on the Moon before the Russians. It was politically expedient for the Soviet Union to pretend, after the race was over, that they were never racing at all. As long as the government is running the show, political considerations will decide what can and cannot be done. If we had continued at the pace of the Apollo program in the sixties, and NASA had follow-on programs in the planning stage, we would be having this conversation on Mars today. There's nothing magical about it. You have to spend the money and build the technology.

CI   Is it realistic to expect the public to support space exploration as an ideal, that visionary challenge of releasing our bonds from the Earth?

BB   Almost everyone I've ever talked to is excited about exploring space, but it's not their prime priority. They're much more worried about garbage collection, crime, taxes, the quality of their schools. But nobody's really *against* space. Politicians have played a cruel game of either–or. Every dollar we spend on space is a dollar "taken away" from school lunches. That's baloney. Every dollar we spend on space has brought back ten, twenty, a hundred dollars to the economy. And the government doesn't pay any bills. You and I pay the bills.

CI   It's been a limited pool of players. The privatization of space is exciting. Imagine what Burt Rutan or Richard Branson would do with NASA's 16 billion dollars.

BB   Unfortunately, they'd probably get just as bureaucratic as NASA. Don't give them that kind of money. Let them go lean and hungry and invent some new things; there is a motivation that is ignored, but could be helpful. We can build solar-power satellites, huge satellites in high orbits where they're always in sunlight and convert that sunlight into electrical power and beam it back down to the

Earth in the form of microwaves. One such solar-power satellite could close down all the power plants in the southwestern United States. We wouldn't need them, and we'd be getting this energy cleanly, not burning any fuel, not messing up the atmosphere. Our biggest power plant is the Sun, 93 million miles away, and it's chugging along. Why not tap more sunlight and bring that energy to the Earth – clean, renewable, inexhaustible energy – and stop burning fossil fuels?

CI   I have a feeling you have the answer to that question.

BB   There are two major barriers. One is that it's an expensive project. We're talking about billions of dollars. Second, nobody's certain they'll get their money back. Nobody's done this before. If Thomas Edison came to you while he was still a telegraph clerk and said, "I want to build electrical power generators and replace all the gas lighting with electric lighting," would you have invested your hard-earned money? Probably not. Alexander Graham Bell begged Mark Twain to invest in the telephone. He wouldn't do it. Twain invested in some other dubious schemes and lost most of his money and hated the telephone as a result. You've got to invest a large amount of money, on the order of the kind of money that oil companies invest every year in exploration, which results mostly in dry holes. It's the kind of money corporations *do* invest. It's not beyond the scope of investment, but there's always the question of how long it will be before there's a profit.

CI   Can you imagine a time when the cost of launch drops to the point where we have the analogy of a dot-com boom, an entrepreneurial flowering into space?

BB   Yes. The technology that needs to be developed is the space elevator. We go into space now by burning chemical fuels in rockets; it's inefficient. Arthur C. Clarke pointed out, back in the heyday of the Apollo program, that the amount of energy needed to land an astronaut on the Moon and bring them back to Earth would cost a few thousand dollars at your local electric utility, based on unit price per watt-hour. The fact that it actually costs billions shows how inefficient rocket engines really are. Now a space elevator, an elevator extending up to high-Earth orbit, would be able to lift you for the price of electricity. We're finally coming to the point where we will have materials to build such an elevator. Buckyball cables have the necessary strength-to-weight ratio.

CI   Carbon nanotubes. So new materials are going to make this possible?

BB   We need a high strength-to-weight ratio, but we have these new materials now in the laboratory. Making them practical and making them sixty thousand miles long is, as von Braun used to say, "an engineering detail." It can be done. Then we get into space for the price of an elevator ride.

CI   That would transform the prospects. One of the technologies in your books, long before it existed, was virtual reality. Should space exploration be something we do physically, or should we accept assistive and projective technologies and telepresence to explore that way?

BB   I once testified before a Congressional committee about building a space station in Earth orbit. A friend of mine, a professor from MIT, had just testified that we should

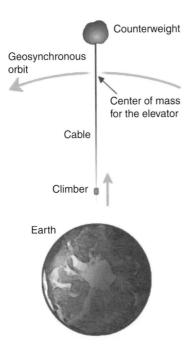

Counterweight

Geosynchronous
orbit

Center of mass
for the elevator

Cable

Climber

Earth

The space elevator is a concept that could transform the economics of space exploration by allowing material to move to Earth orbit for a tiny fraction of the present cost. A cable tethered to the Earth has a center of gravity at the altitude of a geosynchronous orbit and high above a counterweight. The concept dates to 1895 and a design of Konstantin Tsiolkovsky. Material is moved by climbing up the elevator, which in the current technological landscape might be made of something like carbon nanotubes. NASA has held workshops on space elevators and development may be spurred by competitions like the Ansari X Prize (courtesy Wikipedia).

do it robotically, that putting people in space is too expensive and dangerous. I said, "Professor, when you can roboticize your campus laboratory so that you don't need graduate students, *then* I'll believe you can build a space station robotically." The truth is, we have never had a completely robotic mission. They're all under human control, and they need humans to guide and adjust them. We don't know how to build robots that are that good. On the other hand, there have been no purely human missions. We go surrounded by technology, a lot of wonderful machinery.

Think about exploring Mars: commands from Earth take five or ten minutes to get to Mars. If a Mars rover is heading for the edge of the big Grand Canyon and asks, "What should I do?," by the time the answer comes back, it is tumbling down the cliff. As we start exploring Jupiter and farther into the Solar System, the time lag will be too long. We're going to need people on the scene. It's a false argument, man versus machine. They're partners.

CI   It's also important in a psychological and emotional sense for us, the Earth-bound people, to stay engaged. It's important to have our emissaries out there.

BB   Humans are much better at facing the unknown than preprogrammed machines. I had this argument for years with Carl Sagan. He said, "Send robots to Mars. We don't need people," and I said, "Carl, when those robots start sending back data, you're going to want to go. You're going to want to sit in the Martian sand with your pail and shovel and see what there is to be seen." I'm not denigrating the robots. They're important and valuable, but for emotional reasons, and because human beings are smarter and more flexible, we'll send people. It's a pain in the neck to support them, but we could start to use ideas like those developed by Robert Zubrin, of developing the resources *in situ*. Using Martian resources to produce life-support materials cuts the cost of doing this in half.

CI   Why is it so difficult for people to think of the future?

BB   Human societies are built to maintain the status quo. Every institution we have developed is backward-looking. Law, government, custom, religion – they're all meant to keep things as they were yesterday, no matter how lousy yesterday might have been. Along comes science. By nature, science is forward-looking. Science always tries to discover new stuff, and scientists are always in trouble with their governments because the government doesn't want new stuff. The government wants things under control. Scientists are saying, "Okay, it's under control, but look: I can make you invisible."

It's like history. There's always more of it, more of the same. Science, on the other hand, offers enormous capabilities. It's like the old story of the wizard, or Dr. Faust. You sell your soul for these powers. We've gotten the powers. We haven't sold our souls. We are far better off now than we were before science came into our lives. The problems we face today are largely problems of our own success. We have come to the point where we can support more than 6 billion people on this planet. We're not running out of Earth or resources, but we are starting to strain the social fabric that decides how we share those resources. The problems we have today are problems over who gets what part of the Earth resources, and at what price. If we use the resources of space – energy and the raw materials that are out there – and enlarge the human race's supply of resources, we make the pie bigger and everybody benefits.

CI   That's an optimistic scenario: that we move into space not because we've trashed our planet, but because it's part of our natural growth and development.

BB   We can take a biological look and say that organisms tend to expand to fill every ecological niche they can reach. We now can fill an ecological niche off the planet. We have the technology to do that. We don't adapt to the conditions of space, but we bring our technology along so that we can go into space with an Earth-like environment. Due to the biological drive alone, we'll push into space.

CI   What will space exploration look like in fifty or a hundred years?

BB   We will have a private space industry. We will have factories in orbit. We will be in the process of bringing our Earthly industries into space where we can use free energy from the Sun. Our pollution problems will be practically non-existent, and we will begin the conversion of the Earth into, in real-estate terms, a class-A residential zone. If we can make this planet clean and green again, we can support six, ten, God knows how many billions of people. We will be building habitats in space. Only a few people will leave the planet, but what those few do will transform the whole world, just as surely as the few people who left Europe for the New World transformed Europe, and transformed the whole world. We don't know what we're going to find out there. No European expected to find the potato in South America, but the potato changed the world's eating habits.

CI   You don't think cultural instability, or tribalism, or other social forces will stop this happening?

BB   That may delay it. That may warp it in one way or another. But I think it's inevitable.

CI   Will colonizing, or going off-Earth in general, be a democratizing force, or just an opportunity for dynasties and despots who have a larger landscape to deal in?

BB   I was going to say we'll see. But even at the speed of light, it takes ten or fifteen minutes to talk to somebody in the Asteroid Belt. The distances and the difficulty of communications make it difficult to build an empire. Independent societies will grow. Carl Sagan made a beautiful point, years ago, that building space societies will give us a chance to experiment – that different groups of people will invent different forms of governance. For the first time since modern democracy, we'll see the development of new forms.

CI   We're at an interesting point in human history, being close to going off-Earth and exploring our larger environment. Do you think this has happened elsewhere in the Galaxy and beyond?

BB   The universe is so large that it's difficult to say we're the only intelligent creatures. On the other hand, the evidence so far points to the probability that intelligent life is rare – unless it's so intelligent that it doesn't want to talk to us. If *Star Trek* has the idea that we shouldn't interfere with lesser civilizations, perhaps there are vastly smarter aliens that have picked up our radio signals. They may be saying, like Jane Goodall, "Leave them alone. Let's see how they work."

CI   As a writer, you're aware of the anthropocentric nature of a lot of science fiction in the popular realm, in TV and movies especially. Do science-fiction writers have enough imagination to think of what exobiology might have led to out there?

BB   No, I think we're terribly limited by the only example of life we have. Some people will be disappointed when we find life on other worlds, and it's bacteria. As Gould said, "Bacteria are the most populous form of life on Earth. It's what we should expect to find elsewhere." Are there worlds with multicelled creatures? Perhaps. Are there worlds where we'll find intelligence? That might be very rare. We have an innate assumption that if life begins on a planet, given enough time, an intelligent species will arise. Intelligence may be an offshoot, like the horns on a stegosaurus – an adaptation that is not general.

CI   If it turns out that intelligent, sentient creatures are rare in the universe, in what sense does that make us special?

BB   A lot of real-estate developers would welcome that kind of news, but I think it gives us an enormous responsibility. If there are no other intelligent creatures in the universe, *we* will populate the universe with our progeny.

CI   Are we up to that responsibility?

BB   We'll find out. We have made something of a mess of this planet, and we are in one of the greatest periods of annihilation the world has ever seen. Species are being wiped out by the thousands every year. But we're learning, and we're realizing there is a problem – and that's the first part of the solution.

37

Jennifer Michael Hecht

Jennifer Michael Hecht is a historian and philosopher who made her name writing nonfiction but who self-identifies as a poet. She earned her PhD in the History of Science and European Cultural History at Columbia University and she teaches at the New School University. Her nonfiction books are *Doubt: A History*, *The Happiness Myth*, and *The End of the Soul*, the latter winning the Phi Beta Kappa Society's Ralph Waldo Emerson Award. Hecht's volumes of poetry are *The Next Ancient World*, which won the Poetry Society of America's Norma Farber First Book Award, and *Funny*, which won the University of Wisconsin's Felix Pollak Poetry Prize. Her book reviews have appeared in *The New York Times* and the *Washington Post*, and she has served as a Fellow of the New York Institute of Humanities. She was inspired by her tenure as a Templeton Research Fellow at the University of Arizona to think deeply about the implications of contact with intelligent forms of life in the universe, both for us and for them.

CI   Are the different aspects of your writing like children, in that you feel differently towards them, but love them all equally?

JMH   I love all of it. At this point I don't write anything I don't love, though I took longer to warm up to some areas. But poetry is the one. When I allow myself time for poetry, I feel happy. Poetry writing looks like doing nothing most of the time, and it took me a long time to feel comfortable with that. I have to remind myself that I'm working even when I'm just sitting there. When I'm not working, even for a couple of hours, I feel jumpy. Luckily I've managed to think of raising my children as partially an art project. I don't need to do that anymore, but at first I did, to give them a tremendous amount of time without feeling too jumpy.

CI   Is that a Calvinist thread in you, or is that the pressure of academic culture?

JMH   That's me dealing with the sense of time passing, and my determination to use time well. I've gotten better at valuing things that look like they're not work, like the time it takes to think, even if I'm just sitting there and screwing around for a while. But to do nothing whatsoever, to waste time, is really hard for me. Poetry is so precious that when I write it, and it's good, I feel good about all the time that went into creating it. I know it's good because I want to read it again. That's the only test.

CI   Right away, you want to read it again.

JMH   I want to read it so many times, you'd never believe. I listen to it as many times as teenagers listen to their favorite new song; over and over and over. If I don't love it, then I don't want to read it over and over, and even if I thought it was brilliant at first, it loses status. That keeps me honest. If everyone else loves it, but I don't want to reread it, it doesn't stay. I throw it out. Pleasure is huge with poetry. I love the way the words come up. I feel like I'm guiding the stream.

CI   That channeling aspect is something I've heard from a lot of poets. If writing nonfiction is staring at something and analyzing it, and fiction is averted vision, what's poetry?

JMH   Poetry is definitely more like a chant. Because there isn't much money to get out of it, there's a profound way in which you're writing for yourself. You're the important audience. No one else in my mind gets to judge the quality of the poetry, which isn't true for nonfiction. Nonfiction is also creative, though. I never know exactly what I'm going to do before I do it. If I did, then I'd be too bored to start. It's always good for me to write about subjects I don't know too well, so that I'm still excited about explaining it to myself.

CI   You had a lot of physics and science around you. What hooked you the other way into the darker arts?

JMH   All three of us kids were set in a competitive situation, where it was made clear we couldn't win. My father was a physicist and didn't want us to compete with him in his field. But he did have a value for poetry, so there was room for me there. Straight psychological drama is what comes to mind when asked what got me started. But then I read a lot of romantic poetry. From the age of twelve, I could recite *Kublai*

Khan and "The Raven." I loved that stuff – rhymey, childish in some ways. When I learned what modern poetry was about, I was shocked – first that anyone was talking about such personal things, and second that they had abandoned almost all of the best stuff about poetry. I decided I was going to write rhymey pieces that gave me that same kind of pop-music pleasure, but that I was going to attempt to talk about things that mattered to me.

CI   That selfish element is similar to comedy. A great comedian laughs at their own jokes.

JMH   How else can you judge the importance of your work in those areas? It's different from quantifiable fields, where a physicist has discovered some definite thing that other people hadn't found before. Then they dress goofily, almost to show people, "My status is so high that I don't have to match my socks for you." In the softer sciences, they dress up in tweed jackets and do everything they can to signal to everybody that they have prestige, intellect, and gravitas. But in these deeply creative fields, it's a judgment call. You have to believe that you're not just wasting your time with poetry. Frankly, I believe most poets are *not* in love with their own work.

CI   You did history of science. It must be hard to live in a culture that is becoming more amnesiac as the years go by. How do you deal with that?

JMH   I find it absolutely remarkable. If you're pouring milk in your coffee and you don't know what pasteurization is, and you don't know when Pasteur lived, and you don't know where the milk comes from, and you don't know when we first started using caffeine, then you are missing the incredible depth of the world around you. When you study all this, especially history of science, there's a tremendous depth in everything you look at – and the people around you are dealing with it all as if it's always been this way.

CI: In the big picture of our history, technology is on such a cusp that it drives people to understand our cultural path. Is there a sensible way to talk about our cultural evolution over large spans of time?

JMH   I think about the future in terms of the word "reset." Instead of slogging through our grandparents' mistakes, is there a way to step back and start again? For instance, can we rip up the roads? They had the fortitude and strength of character to put them down; we can have the fortitude and strength of character to rip them all up, and be done with cars. When you list the problems with cars, it's shocking: wars for oil, pollution, global warming, and long commutes are a major cause of depression. They also kill us; between ages two and thirty-three it's the main source of death in the USA, and after that it only trails heart disease and cancer. Meanwhile, here we are doing all these studies to figure out whether you should eat a kiwi or a lemon to get the most vitamin C.

To think about cultural evolution, part of the imaginative leap has to be not only new things, but things that would be new for us. What if we had all the technology and knowledge we have now, and we could go back to 1880? If we have steel-reinforced concrete and the ability to make panes of glass, would we

	ever build the suburbs? What about a law that you have to live within walking distance to where you work? We could make it possible to do that.
CI	We don't like to admit as a species, as a global culture, that we made a mistake.
JMH	It's impossible to even think of it.
CI	The Drake equation hinges on our longevity as a technological civilization. Frank Drake's license plate is "N EQLS L." It's as simple as that: we're either sparsely alone or we have tons of pen pals, depending on that number. Do we have it in us to live a million years, or will we be lucky to last a thousand? What bet would you place on our longevity as a technological civilization?
JMH	My bet is that we last a long time. Having our numbers bottleneck down to a very small amount doesn't seem to do us terrible damage. In some cases it can do us good. The genetic difference in South Africa is still much richer than elsewhere, because that's where we started, and it was only a few guys who left. Race is going to change profoundly the longer we last. We're going to all brown up and not have to worry about it anymore. I'm hoping that sex will cure racism, and that we'll look back on today's divisions as being crazy.
CI	If we were to get the message in a bottle – a few of which we've sent out ourselves – it would probably be an incomprehensible sliver of another culture. How many elements of life and intelligent life might be universal?
JMH	The relationship I've had with physicists and the physics world has made me keenly aware that life elsewhere is likely to be very different. My guess is that the ways we feel are necessarily linked to the ways we think. If a creature gets to the point where it can analyze things and teach another creature of the same type, that ability probably comes with a sense of fairness and right and wrong. With intelligence comes a sense of morality.
	That doesn't mean people respect that morality, but most humans, except for the sociopaths, know what's generous and what's obnoxious, what's right and what's wrong. When we go against it, it's because we don't care, or we've decided against it, or we've put up a meta-analysis: "I was done wrong on such a profound level that I could never do enough harm back to this world." I think we on Earth are in an intolerable situation, because we have morality, and we have a system wherein we have to eat each other all the time. We're all in this unbelievable schism where we have a feeling of our own deservedness and we're able to extrapolate that to others, and yet we're hungry. The only reason it shouldn't make us leap off a cliff is that there's an equal cacophony of pleasure and cooperation; there's as much birth as there is death, and we have to ride it.
CI	The science-fiction trope says that if another civilization were more advanced than us, which statistically they're likely to be, they'll have overcome the limitation of resources. Then they can relax, and the problems that go with our competition for resources will ebb away.
JMH	There are fewer border wars in the United States of America, since everybody has enough bananas. The state is the monopoly of force. Peace is a combination of having enough resources so that you don't need competition, and having one

central, designated monopoly of force so the rest of us have to put down our arms. There are many ways for that to get more sophisticated than it is now.

But there's always this heart. The heart wants to be special, to be loved, to be different, to be the same but needed in your specificity. If we passed out of that, we would so far pass out of humanity that it's almost not worth talking about. Our experience is so filled with longing, so filled with the desire to be special, which is the other mocking schism: we aren't special. There are 6 billion of us. Individuals don't count much at all, but we all want to. That tension will always be there, and it will always make for both pain and rejoicing. It's not just resources, it's also affection. Affection is an endless resource, but you have to earn it. I could make ten different people feel like they were the greatest person in the world, and I could add an eleventh without taxing myself much. But they would have to do the work of being worthy of that, and it's not easy.

CI What about communication with the other? We're alone within our organism amongst the 6 billion, and that carries with it a lot of baggage and problems and obstacles. And we're unable to communicate with creatures with whom we share 99 percent of our DNA. Astronomers posit the question of being alone in the universe onto Earth clones and humanoids, onto beings with which we might be able to communicate, with whom we might share a bond. How should we think about being alone?

JMH We have to direct our attention back to the planet. We have to allow this subject to direct our attention back to ourselves and to the other creatures. If we had no whales or elephants here on Earth, and we went to a planet and found them, they would be so astounding, compelling, and wonderful; we would be so careful not to step on their toes, to respect and hear them. Their longevity, the length of their gestational period, we would take as majestic. It's not that we don't do that here, but it's like growing up with something strange. It takes a while to see it for what it is. It even takes looking at the universe to see what we have here. This is where the creatures are. It's crazy to look for an octopus and an ant and an anteater in outer space when they're here. We have an embarrassment of riches. It's hard to appreciate because it's overwhelming.

But the other side of it: "alone." So long as we are the only creatures who make Marx brothers movies, the only creatures doing this level of material cultural production, it's going to feel weird. In our daily lives as individuals, we are never the top dog. Even when you're the top dog, you've got to listen to your wife, or your dog. When you're top dog in everything, you're toppled, so you were only top dog for a moment of your life. Most of us never get anywhere *near* being the top dog. Being human is first and foremost a lesson in humility. And that is so at odds with our position on the planet that it's breathtaking. How could it be that *we're* the smartest? I'm never the best at anything; how could I be of the species that's the absolute best? And yet it seems I am.

Whenever I hear the odds scientists like to play with, that it would be rare to be at the beginning or the end, therefore we must be in the middle, I think it's

Elephants have a brain larger than humans and a complex social life, with much time spent raising their young and teaching them social skills. If we ever discovered creatures as interesting and complex and intelligent as elephants elsewhere in the universe, it's likely we would treat them better and with more curiosity and respect than we do the elephants on our own planet (courtesy Alexander Klink and Wikipedia).

bullshit. It would be rare to be the highest creature on a planet of hundreds of millions of different species. But we are having this conversation. We think of ourselves as the only important being, but it's still shocking. That's why people are comforted by the idea of God. It's a rational act of humility to realize it's unlikely that we are the top, that there must be an oversight committee; there always is down here. If we were to be confronted with people from another part of the universe who knew things we didn't, we would feel so much more existentially comfortable.

CI   There seem to be two desires vis-à-vis "out there." There's the desire for kinship, however that gets expressed. That's mitigated by the fact that we haven't tried very hard for kinship with these amazing creatures on our planet. Then there's the quasi-religious projection, which is easy to understand: we can be saved, so we make them the repository of all our needs and fears.

JMH   The desire for kinship is similar. It's the idea that a friend from this other place will be a true friend in a way we haven't been able to find down here. But if it happens, it's likely to be ordinary, in some way we haven't imagined. It'll be amazing in some way, but then it will get ordinary. We'll either use them badly, or they'll use us badly, or we'll make friends, but it'll be the kind of friends who hide the last chocolate for themselves. But it would put my mind at ease if I thought there was somebody else out there, somebody I could communicate with, who was more sophisticated and who I could rely on to make judgment calls. That sounds soothing, to throw up our hands and be less ultimately responsible.

CI   Some people would probably be angry if the companionship were mundane.

JMH   Mundane is real, it's good. Once you go into fantasy, there's something wrong. That's why it's fantasy – there's something broken about it. If you're bored, then you're in the right place. You've got to find a way to get excited, but it's a sign of reality in a way.

Sometimes we correct things too well and it takes a long time to stop. Scientific thinking in the eighteenth and nineteenth centuries was damaged by anthropomorphizing, so we set up a diligent police against it. That's one of the reasons we don't have respect for other creatures, because we are *so* careful not to say that the lion wants to kiss its friend. But the lion flirts and walks over and kisses the friend. I love shutting off the sound on the nature shows and doing my own dialog.

CI   Woody Allen style.

JMH   It's so clear, especially the male–female interactions. Some anthropomorphizing becomes so obviously political and culturally determined that it's dangerous. Nevertheless, we've gone too far. We say, "I don't want to anthropomorphize by saying that this is singing. It just sounds likes singing, looks like singing, and feels like singing, but it's not singing."

CI   Do you think "meat that thinks" is unexceptional in the universe, that all meat that thinks is going to have similar issues?

JMH   I do. Meat that thinks is weird, but it's the situation.

CI   We're meat that thinks and we're not out of nature. It seems these mechanisms are not mystical, not magical. It happened here more than a few times.

JMH   And in similar ways. Eyes have developed in many different ways, but they have a lot of similarities. I have not met anything that thinks without a gray, smushy, curlicued thing that looks like a brain. Do I expect such things to develop elsewhere? Yes, given fantastic amounts of time.

CI   There is a recognition problem. Communication depends on having overlapping periods of time when two intelligent species might both be alive and able to communicate. Even if there's overlap, if the life spans and speeds are very different, there will be a gulf. One of my favorite *Star Trek* episodes is one with creatures that are like mosquitoes, and they know something's going on with us, but their lives as individual entities are so short compared to ours that we seem static.

JMH   I wondered for a while whether it was reasonable that longer-lived creatures might be the more sophisticated beings. But we have turtles and trees that live hundreds of years, and they don't seem to have too much going on. A creature that lives a long time and goes very slowly could live happily with a whole population of creatures that move very fast and have short lives, because the fast ones would go around the other ones.

If we went to a place like that, with whom would we feel most comfortable, or the most recognition? We're so freaked out by mortality that the idea of an intelligent creature that lives for a much shorter time than we do upsets us. We don't want to think they're smart, because it would break our hearts. A fruit fly is born, wants to see the world, and it's dead twenty minutes later – it's tragic. But with these longer-lived creatures, we get the idea of wisdom. And what if a whole civilization existed and is gone now? Would you feel kinship? Why would I feel more kinship with them than with the people of ancient Rome?

CI   Time and space constraints set the minimum longevity of civilizations for real communication. Less than 3000 years and you're temporally isolated, you're extinct, and it's only dead messages and runes. But if you have a long tail of extremely long-lived civilizations, then you throw the whole Drake equation out, because the durable civilizations can travel between galaxies. If they are to us what we are to bacteria, what would that even look like? Did they go through a thinking meat phase?

JMH   All we really care about is how it all resounds on us, how it feels. We twist things until they feel okay. It's so abstract. If you go into a library and take a hundred books off the shelf without looking at them, and open them and read a few pages, one after another, it's going to be boring and weird, especially if they're all written two hundred years ago. But every once in a while, you open one and it's a friend. Every once in a while, it's Cicero. It's somebody who seems to be sitting right next to you and you don't understand how they could speak to you, and these other ones try so hard, but they just aren't interesting anymore. Read any of the other people who were writing like Dickens at the time Dickens was writing. It's unreadable. What makes the popular ones popular at the time is often the same thing that makes them last. Not always; it's inexplicable that you can find a friend in one out of a thousand books from a period.

   If we get word from something out there in any way, it will either touch us or it won't. If it touches us, it doesn't have to be much, and it doesn't have to be at the same time. If it has that magical thing, you open it up and you feel they're right there with you, and they have the same concerns, the same way of talking or looking. We don't know how that works, why there are so many millions of volumes of poetry in the world, and most of it could be burned without loss. Really, a tremendous amount of it's just trash, and the guy in the next cubicle is writing stuff that's going to last for a couple of million years. It's like love, it only has to work once. And my guess is it will work.

CI   With the fecundity of the universe, it's almost a given.

JMH   That's my feeling. I don't give any credence to this Rare Earth business at all. Different things had to come together to make this particular situation, and it was clearly an unusual event. But it seems to me that life is maybe less unusual than we give it credit for.

CI   There are a lot of good poker hands and they're not all the same.

JMH   Exactly. Life is more likely than not-life. Something is more likely than nothing. Something is more stable than nothing. Nothing is rare and unlikely, and there's no time in nothing. So even if there were some way to conceive of there being a great deal of time between nothings, there isn't *any* time between nothings. If there's nothing in the universe, there is no time, because time is a rate of change. A completely full universe also has no time, because nothing changes. Between those two extremes, things burble and bubble. In that context I expect patterns. Patterns are more stable than non-patterns. Entropy is a misleading idea, because things fall apart, but they fall into other things. What fall apart are arrangements.

CI   Your best laid plans fall apart.

JMH   That's just your best laid plans. Things that fall into patterns tend to stay in patterns because they're much more stable than a mishmash. A mishmash changes all the time. There's every reason to think there's thought out there. But there's no reason to think that it's come here yet. I know what genuine surprise looks like, and what I've seen so far is what it looks like when people are lying.

CI   Would you put a bet on humor in the universe?

JMH   Absolutely.

CI   The whole range – pratfalls to irony – in different guises?

JMH   I like its chances, right next to love. Both are ridiculously bizarre, but I like the chances. It's not even complicated. I have come to understand humor primarily through the notion of misunderstandings, which are a reminder that we're working in an invented code, that all of our knowings are an invented code, and that we slip between codes that don't make sense together with these little maneuvers. Humor is when they overlap and you can see both parts.

CI   It's a high form of self-awareness. There cannot be humor without that.

JMH   You can't have coherent systems. If you have any systems, they are partly incoherent, so there will have to be moments where we go, "Huh!" I don't know whether you're always going to be exhaling or inhaling to make that noise, and I don't know how much other creatures are going to enjoy it. We enjoy that. We can't remember jokes because they are a crack in the phenomenal world where the phenomenal is leaking out, and we've got to patch it back up. We need to forget the jokes. Jokes are profoundly frightening because of that, because they showed us to have been mistaken.

CI   We imagine it's a shell game, it's been nested. Higher and higher jokes are going to be revealed, and we don't really want to go there.

JMH   That's why I take the next step and say that if you want to remember a joke, you should philosophize about it. You'll make it not funny, but you'll also be able to remember it forever.

Glossary

Accretion	In the history of the Solar System, the early process by which gas and dust steadily clumped by gravity and grew into moons and planets.
Allen Array	When complete, a set of 350 antennas designed to conduct a sensitive search of radio signals from distant civilizations.
Amino acids	The building blocks of proteins, as specified by the genetic code. Life on Earth uses twenty out of a much larger possible set.
Anthropic principle	The idea that certain characteristics of the physical universe are carefully tuned to allow the existence of carbon-based life forms.
Archaea	The most ancient of the three major branches of life on the Earth; the others are Eukarya and Bacteria.
Arecibo	A 305-meter-diameter radio dish in Puerto Rico, sometimes used to send SETI signals.
Astrobiology	The study of life in the universe, including the history and limits of life on Earth.
Bacteria	The smallest type of living organisms.
Biochemistry	The chemistry of life.
Biomarker	Indirect tracer of extraterrestrial life, usually anticipated to be the spectral signature of gases in a planet atmosphere that indicate metabolic processes.
Burgess Shale	Located in Canada, one of the rare places with well-preserved fossils from the Cambrian.
Cambrian	A major period of geological time, when life proliferated in the oceans of the Earth, running from 542 to 488 million years ago.
Carbon cycle	The cycling of carbon dioxide between the atmosphere and the Earth's crust.

Carbon-based life	Life that requires carbon for its critical functions by using it in long, information-storing molecules.
Cassini	Highly successful mission to Saturn and its moons, including the Huygens lander sent to the surface of Titan.
Catalysis	Speeding up a chemical reaction by introducing an agent that is not changed by the reaction.
Cell	The smallest unit of life processes; highly organized chemical factories.
Cephalopods	Invertebrates that can have well-developed senses and large brains, such as the octopus.
Cetaceans	Mammals well suited to aquatic life, including species with complex and sophisticated behaviors, like whales and dolphins.
Chemical reactions	Processes where elements and compounds combine and separate. Chemical reactions affect electrons but not atomic nuclei.
Chemistry	The study of the composition, structure, properties, and reactions of atoms and molecules.
Comet	Small Solar System body made of rock and ices, occupying a spherical cloud far beyond the orbit of Neptune.
Complexity	An important, but varied, concept in biology. It can refer to sophistication of genes, metabolic pathways, brain architecture, or functions of the organism.
Convergent evolution	Similarities of organisms that arise when they evolve in similar environments.
Copernican revolution	Profound change in thought in the sixteenth century, when the Earth was understood not to be the center of the universe.
Core accretion	The standard theory of planet formation, where a rocky core steadily accretes gases in the cool outer part of a solar system.
Cryovolcanism	The eruption of water or other volatiles onto the surface of a planet or moon due to internal heat.
Cyanobacteria	The photosynthetic bacteria that produced the oxygen in the Earth's atmosphere.
Dark energy	Enigmatic component of the universe causing cosmic acceleration.
Dark matter	Enigmatic component of the universe that permeates galaxies and the space between them. Dark matter outweighs normal matter by a factor of five or six.

Design	The idea that features of the universe and living organisms have an intelligent cause, rather than being the result of undirected processes.
Domains	The major classifications of life: Eukarya, Bacteria, and Archaea.
Doomsday argument	A statistical argument that we are a substantial way through the entire span of the human species.
Drake equation	Astrobiology pioneer Frank Drake formulated this way of calculating the number of intelligent civilizations in the Milky Way.
Eccentricity	The amount by which the orbit of a planet or moon deviates from a circle.
Enceladus	A small moon of Saturn, only 500 kilometers in diameter, with subsurface water that occasionally erupts as geysers.
Entropy	The measure of disorder in a physical system, which tends to increase with time. Entropy is also related to the number of possible states of a system.
Enzyme	A protein that catalyzes, or accelerates, a chemical reaction.
Eukarya	The domain of life that includes all cells with a nucleus containing chromosomes and organelles found by membranes. The first eukaryotes arose about 1.7 billion years ago.
Eukaryotic cell	A kind of cell with a nucleus separated by a membrane from the rest of the cell, the most complex cell type.
Europa	Sizeable moon of Jupiter that is covered with a fractured icy crust overlying a water ocean, perhaps the most likely place to find life beyond Earth.
Evolution	In biology, the change in inherited traits of a population from generation to generation.
Exoplanet	Also called an extrasolar planet, any planet orbiting a star beyond the Solar System.
Extremophile	Organisms that like to live in what are considered hostile environments to humans, such as extremely saline or hot environments.
Fermi paradox	Enrico Fermi's idea that if extraterrestrial intelligence did exist, we should know of its existence. Also called Fermi's question: "Where are they?"
Fine-tuning	The fact that many of the constants of nature have values within a narrow range suited to the existence of carbon-based life.

Fossil	The remains of a living organism that have been turned to stone over a long period of time.
Fossil record	The story of the Earth's geological history as told through fossils.
Gaia hypothesis	The idea that living and nonliving parts of the Earth operate as a complex interacting system, though not actually as a single organism.
Gas-giant planets	In the Solar System: Jupiter, Saturn, Uranus, and Neptune. In general, a planet made primarily of gaseous hydrogen and helium.
Gene	The minimum amount of genetic material that expresses a characteristic of a living organism. A gene is a sequence of several thousand bases along the DNA molecule.
Genetic code	The way DNA base pairs are interpreted to provide instructions for the organism's genes.
Genome	The full sequence of DNA base pairs in an organism.
Geology	The study of the history, origin, and structure of the Earth.
Gravitational instability	An alternative view of planet formation, where top-down collapse quickly leads to a planet.
Habitability	A broad concept that includes locations that span the full range of biological organisms, which may be even broader beyond Earth.
Habitable zone	The area surrounding any star in which an orbiting planet or the moon of that planet could have liquid water on its surface.
Hubble Space Telescope	The premier observing facility in astronomy. HST is NASA's flagship mission, and an important contributor to the study of exoplanets.
Huygens probe	The lander from the Cassini mission to Saturn and its moons, which transmitted data from the surface of Titan.
Hydrothermal vents	Places where volcanoes emerge from the deep seafloor and support complex ecosystems living in total darkness.
Impacts	Randomly occurring collisions of the Earth with space debris.
Information	A quantity that can be measured and transmitted. In biology, information is stored in the genetic code.
Intelligence	The capacity for abstract thought, coupled with the mastery of tools or technology. Intelligence is only

	generally found in animals with large and complex brains.
Interferometry	Combining radio or optical telescopes to achieve the angular resolution equivalent to a single huge telescope.
James Webb Space Telescope (JWST)	The successor to the Hubble Space Telescope; will have the potential for imaging and spectroscopy of exoplanets.
Kepler mission	Launched by NASA in 2009, it is designed to detect Earth-like planets if they exist using the eclipse method.
Kingdom	The second largest classification of living organisms.
Lander	A spacecraft designed to land on planetary surfaces.
Last common ancestor	The root point of a diverged set of species. The last universal ancestor is the hypothetical single-celled organism that gave rise to all life on Earth.
Life	Challenging to define in any way that has meaning beyond the Earth, but probably requiring the localized use of energy and the storage of information in molecular forms, evolution, and adaptation to the environment.
Many worlds	The idea that the Earth is just one among many worlds in space, including potential applications of geology and chemistry and biology beyond the Earth.
Mars Exploration Rovers	The twin rovers Spirit and Opportunity have explored Mars since 2003, providing much of the evidence that Mars has hosted water.
Mars Global Surveyor	This mission from mid nineties provided the first strong evidence that Mars had hosted surface water in the past.
Mars Science Laboratory (MSL)	Scheduled to reach the red planet in 2012; MSL is a large rover with sophisticated life-detection capabilities.
Martian meteorite	The three dozen or so Martian meteorites demonstrate that material can move around the Solar System. There was a claim in 1995 that the Allan Hill 840001 meteorite contained fossilized life forms.
Mass extinction	At least five times in the history of life on Earth a sizeable percentage of species have been extinguished in a geologically short time, potentially by an impact.

Metabolism	The chemical reactions that govern the functioning of living things.
Metallicity	The proportion of all the elements heavier than hydrogen and helium in an astronomical object.
Meteorite	A stony or sometimes metallic object landing on the Earth from space that represents relatively pristine material from the formation of the Solar System.
Microbe, microorganism	A living creature too small to be seen by the naked eye.
Microlensing	Temporary brightening of a star's light when an unseen planet passes in front of it, focusing its light.
Milky Way	The large system of several hundred billion stars, including the Sun.
Miller–Urey experiment	The classic "life in a bottle" experiment, conducted by Stanley Miller and Harold Urey in the early fifties. No life was created, but the results included many of life's ingredients, such as amino acids.
Molecular fossil	Highly indirect evidence of life, in the forms of isotopic imbalances in rock that indicate the presence of an ancient metabolism at work.
Multiverse	A speculative theory suggesting that conditions at the time of the big bang led to a suite of universes, each with different physical properties.
NASA	National Aeronautics and Space Administration, the US governmental agency that launches most planetary missions and telescopes in space.
NASA Astrobiology Institute (NAI)	A virtual institute, consisting of dispersed labs and university groups working on NASA-funded astrobiology projects.
Natural selection	A mechanism of Darwin's theory of evolution, where individuals well adapted to the environment reproduce more than those less well adapted.
Organic chemistry	The chemistry of molecules containing carbon.
Origin of life	A historical event on Earth, probably occurring about 4 billion years ago. The origin of life is subject to investigation but the details may never be known.
Paleomagnetics	The study of magnetic minerals and tracers in rock as a way of revealing the geological history of the Earth.
Panspermia	The hypothesis that life on Earth and elsewhere may have been seeded by material that travels between habitable bodies.

Periodic table	A way of organizing all the elements according to the number of outer electrons. Elements in a particular column share many chemical properties.
Periods	The third largest division of Earth's geological history.
Phoenix Mars Lander	This NASA mission to the north polar region of Mars found evidence for subsurface ice and a complex hydrological cycle.
Photosynthesis	Probably the most important biochemical pathway of life on Earth, the use of sunlight to produce sugar, and then ATP, the fuel for all living things.
Phylogeny	The study of the history of life on Earth based on the systematic evolution and divergence of the base-pair sequence of DNA or RNA.
Physical universe	Space containing all matter and energy; may be substantially larger than the observable universe.
Physics	The study of the forces of nature and the laws that govern the way matter and radiation interact.
Plate tectonics	A theory of geology where continental-size plates of the crust and upper mantle move over the semi-liquid rock layer below.
Project OZMA	The first attempt, in 1960, to detect radio transmissions from an extraterrestrial civilization.
Racemic mixture	Equal amounts of the left-handed and right-handed versions of a chiral molecule. Life on Earth uses only left-handed amino acids.
Radioactive decay	The random process where an atomic nucleus spontaneously decays.
Rare Earth	The idea that the conditions that led to complex life on Earth are so rare that we may be the only planet in the Galaxy with complex life.
RNA World	A hypothesis that on the early Earth a phase of RNA-based life preceded the current DNA-based life.
Simulation hypothesis	An argument, based on logic and a few assumptions, that we might possibly be the simulated computational creations of an advanced civilization.
SETI	The search for extraterrestrial intelligence, in either radio or optical signals.
Singularity	A term referring to a hypothetical time in the future when humans will attain a post-biological state.
Snowball Earth	The term referring to long ice ages that engulfed our planet several times between 750 and 580 million years ago, and probably also 2.2 billion years ago.

Space Interferometry Mission (SIM)	This recently descoped NASA mission aims to use extremely high precision measurement of star position to detect Earth-like planets around other stars.
Species	A basic unit of biological classification; for multicelled organisms a population whose individuals can breed and produce fertile offspring.
Spectroscopy	The technique of dispersing light into an array of wavelengths in order to see the narrow atomic or molecular features that are indicative of chemical composition.
Spontaneous generation	The early idea that living organisms emerge fully formed and do not develop over time from simple to more complex.
Stromatolites	Fossilized bacterial colonies that first developed about 3 billion years ago.
Substrate independence	The idea that life could exist computationally or in machine form, without reference to "wet" biology.
Super-Earth	An exoplanet about 3–5 times the mass of the Earth, marking the current limit of exoplanet detection.
Super-Jupiter	An exoplanet more massive than Jupiter, ranging up to the mass of a brown dwarf. Many of the first exoplanets discovered were super-Jupiters.
Supernova	The violent death of a massive star produces many of the heavy elements on which life depends, and can affect life on a planet sufficiently nearby.
Terrestrial Planet Finder (TPF)	NASA's ambitious future mission to discover and characterize Earth-like exoplanets.
Terrestrial planets	In the Solar System: Mercury, Venus, Earth, or Mars. In general, a small planet made primarily of rocky material.
Tidal heating	Heating caused when a small body is in a tight elliptical orbit around a larger body.
Titan	Large moon of Saturn with a thick nitrogen atmosphere and shallow seas made of liquid ethane and methane.
Trace fossil	Indirect evidence of a living organism rather than the fossilized organism.

Transit	Situation where an exoplanet periodically passes in front of its parent star, dimming it very slightly for a short time in a partial eclipse.
Tree of life	In biology, a metaphor for the steady diversification of life from a common ancestor.
UFOs	Unidentified flying objects, purported to be visitations of aliens, but there is no compelling evidence to support this assertion.
Virtual Planetary Lab	A team of NASA-funded scientists building computer-simulated Earth-sized planets to learn the range of signatures of such planets around other stars.
Volatiles	Compounds with low boiling points, often found in comets.
Von Neumann machines	Hypothetical self-replicating space probes that could be constructed by a civilization not much more advanced than ours, and then could explore the Galaxy.
Voyager	Twin spacecraft that explored and then exited the Solar System carrying gold records with sounds and images of Earth.
Zircon	Very stable crystal, a sample of which is the oldest rock found on Earth.

Reading list

Part I Introduction

Bennett, J. and Shostak, S. (2007). *Life in the Universe*, 2nd Edition. New York: Addison Wesley.

Chyba, C. F. and Hand, K. (2005). Astrobiology: The Study of the Living Universe. *Ann. Rev. Astron. Astrophys.*, 43, 31–74.

Dick, S. J. (1996). *The Biological Universe: The Twentieth-Century Extraterrestrial Life Debate and the Limits of Science*. Cambridge: Cambridge University Press.

Ferris, T. (2000). *Life Beyond Earth*. New York: Simon and Schuster.

Gilmour, I. and Sephton, M. A., eds. (2003). *An Introduction to Astrobiology*. Cambridge: Cambridge University Press.

Grinspoon, D. (2003). *Lonely Planets: The Natural Philosophy of Alien Life*. New York: HarperCollins.

Lunine, J. I. (2005). *Astrobiology: A Multi-Disciplinary Approach*. San Francisco: Pearson/Addison Wesley.

NASA Astrobiology Institute. http://astrobiology.nasa.gov/

NASA Astrobiology Roadmap (2008). http://astrobiology.nasa.gov/roadmap/

Sullivan, W. T. III and Baross, J. A. (2007). *Planets and Life: The Emerging Science of Astrobiology*. Cambridge: Cambridge University Press.

Part II Earth

Conway Morris, S. (1998). *Crucible of Creation: The Burgess Shale and the Rise of Animals*. Oxford: Oxford University Press.

Fry, I. (2000). *The Emergence of Life on Earth: A Historical and Scientific Overview*. Piscataway: Rutgers University Press.

Hanlon, R. and Messenger, J. B. (2000). *Cephalopod Behaviour*. Cambridge: Cambridge University Press.

Hazen, R. (2007). *Genesis: The Scientific Quest for Life's Origins*. Washington: Joseph Henry Press.

Knoll, A. H. (2003). *Life on a Young Planet: The First Three Billion Years of Evolution on Earth*. Princeton: Princeton University Press.

Lunine, J. I. and Lunine, C. J. (1999). *Earth: Evolution of a Habitable World*. Cambridge: Cambridge University Press.

Reznikova, Z. (2007). *Animal Intelligence: From Individual to Social Cognition*. Cambridge: Cambridge University Press.

Schopf, J. W. (2002). *Life's Origin: The Beginnings of Biological Evolution*. Berkeley: University of California Press.

Wharton, D. A. (2002). *Life at the Limits: Organisms in Extreme Environments*. Cambridge: Cambridge University Press.

Part III Solar System

Bova, B. (2001). *Jupiter: A Novel*. New York: Tor.

Bova, B. (2008). *Mars Life (Grand Tour)*. New York: Tor.

Carr, M. H. (2006). *The Surface of Mars*. Cambridge: Cambridge University Press.

Clancy, P., Brack, A., and Horneck, G. (2005). *Looking for Life, Searching the Solar System*. Cambridge: Cambridge University Press.

Greenburg, R. (2008). *Unmasking Europa: The Search for Life on Jupiter's Ocean Moon*. Berlin: Springer.

Hester, J., Blumenthal, G., Smith, B., Burstein, D., and Voss, H. G. (2007). *21st Century Astronomy: The Solar System*, 2nd Edition. New York: Norton.

Jones, B. W. (2004). *Life in the Solar System and Beyond*. Berlin: Springer.

Lang, K. R. (2003). *The Cambridge Guide to the Solar System*. Cambridge: Cambridge University Press.

Lorentz, R. and Mitton, J. (2008). *Titan Unveiled: Saturn's Mysterious Moon Explained*. Princeton: Princeton University Press.

Morton, O. (2002). *Mapping Mars: Science, Imagination, and the Birth of a World*. New York: Picador.

Part IV Exoplanets

Boss, A. (2009). *The Crowded Universe: The Search for Living Planets*. Philadelphia: Perseus.

Casoli, F. and Encrenaz, T. (2005). *The New Worlds: Extrasolar Planets*. Berlin: Springer-Praxis.

Extrasolar Planets Encyclopedia. http://exoplanet.eu/

Halpern, P. (2003). *The Quest for Alien Planets: Exploring Worlds Outside the Solar System*. New York: Basic Books.

Mason, J. W. (2009). *Exoplanets: Detection, Formation, Properties, Habitability*. Berlin: Springer-Praxis.

Mayor, M., Frei, P.-V., and Roukema, B. (2003). *New Worlds in the Cosmos: The Discovery of Exoplanets*. Cambridge: Cambridge University Press.

NASA Kepler Mission. http://kepler.nasa.gov/

PlanetQuest Exoplanet Exploration. http://planetquest.jpl.nasa.gov/

Scharf, C. A. (2009). *Extrasolar Planets and Astrobiology*. Herndon: University Science Books.

Ward, P. D. and Brownlee, D. (2000). *Rare Earth: Why Complex Life is Uncommon in the Universe*. New York: Copernicus.

Part V Frontiers

Davies, P. (1999). *The Fifth Miracle: The Search for the Origin and Meaning of Life*. New York: Touchstone.

Davies, P. (2006). *Cosmic Jackpot: Why Our Universe is Just Right for Life*. London: Penguin.

Hecht, J. M. (2001). *The Next Ancient World*. North Adams: Tupelo.

Kurzweil, R. (2000). *The Age of Spiritual Machines: When Computers Exceed Human Intelligence*. New York: Penguin.

Kurzweil, R. (2005). *The Singularity is Near: When Humans Transcend Biology*. New York: Penguin.

Rees, M. J. (1997). *Before the Beginning: Our Universe and Others*. New York: Perseus.

Rees, M. J. (1999). *Just Six Numbers: The Deep Forces that Shape our Universe*. New York: Perseus.

Sagan, C. (1978). *Murmurs of Earth: The Voyager Interstellar Record*. New York: Ballantine Books.

Sagan, C. (1985). *Contact*. New York: Simon and Schuster.

Shostak, S. (2009). *Confessions of an Alien-Hunter: A Scientist's Search for Extraterrestrial Intelligence*. Washington: National Geographic.

Webb, S. (2002). *If the Universe is Teeming with Aliens…Where is Everyone? Fifty Solutions to Fermi's Paradox and the Problem of Extraterrestrial Life*. New York: Copernicus.

Index